Large Covariance and Autocovariance Matrices

MONOGRAPHS ON STATISTICS AND APPLIED PROBABILITY

Editors: F. Bunea, P. Fryzlewicz, R. Henderson, N. Keiding, T. Louis, R. Smith, and W. Wong

For more information about this series please visit:
https://www.crcpress.com/Chapman--HallCRC-Monographs-on-Statistics--Applied-Probability/book-series/CHMONSTAAPP

Large Covariance and Autocovariance Matrices

Arup Bose

Monika Bhattacharjee

CRC Press
Taylor & Francis Group
Boca Raton London New York

CRC Press is an imprint of the
Taylor & Francis Group, an **informa** business

A CHAPMAN & HALL BOOK

CRC Press
Taylor & Francis Group
6000 Broken Sound Parkway NW, Suite 300
Boca Raton, FL 33487-2742

First issued in paperback 2020

© 2019 by Taylor & Francis Group, LLC
CRC Press is an imprint of Taylor & Francis Group, an Informa business

No claim to original U.S. Government works

Version Date: 20180608

ISBN 13: 978-0-367-73410-7 (pbk)
ISBN 13: 978-1-138-30386-7 (hbk)

Visit the Taylor & Francis Web site at
http://www.taylorandfrancis.com

and the CRC Press Web site at
http://www.crcpress.com

TO THE MEMORY OF J.K. GHOSH

...AB

TO MY PARENTS

...MB

Contents

Preface

Many areas of science now routinely generate data where both the dimension and the sample size are large. Theoretical and practical study of such type of data has attracted recent attention of researchers since most of the methods in finite dimensional setting do not work in these cases, even asymptotically. This monograph is a collection of ideas and results in high-dimensional models with emphasis on the sample covariance and the sample autocovariance matrices. It is laid out in three parts.

Part I (Chapters 1, 2, and 3) is on estimation of large covariance and autocovariance matrices using banding, tapering, and thresholding. Part II (Chapters 4, 5, 6, and 7) covers the essentials of random matrix theory, non-commutative probability and then discusses at length limiting spectral distribution of generalized covariance matrices. Part III (Chapters 8, 9, 10, and 11) provides results on limit spectra and asymptotic normality of traces of symmetric polynomial functions of sample autocovariance matrices in high-dimensional linear time series models. It also uses these results to develop graphical and significance tests for hypotheses on parameter matrices of one or more such independent high-dimensional time series.

The material is based on recent advances in the theory and application of large covariance and autocovariance matrices. It is gleaned from recent books and from articles in leading statistics, probability, and econometrics journals. It should be of interest to people in econometrics and statistics (large covariance and autocovariance matrices and high-dimensional time series), mathematics (random matrices, free probability), and computer science (wireless communication). Post-graduate courses in high-dimensional statistical inference and in random matrix results in high-dimensional covariance matrices and auto-covariance matrices can be developed based on the material covered. Researchers involved in developing statistical methods in high-dimensional time series models will also find it useful.

Some knowledge of the following topics is helpful to follow the material. Multivariate analysis (p-variate normal distribution, sample covariance matrix, eigenvalues, as in Anderson [2003]); time series analysis (stationary time series, vector moving average and autoregressive processes, autocovariance matrices, as in Brockwell and Davis [2009] and Hannan [1970]); different modes of stochastic convergence, central limit theorem, weak law of large numbers, as in Billingsley [1995], Ash [2000], and Serfling [2002]; random matrix theory, including ideas of limiting spectral distribution, moment method and method of Stieltjes transform, as in Bai and Silverstein [2009], however, the essentials needed are covered in Chapter 4; non-commutative probability, including ideas of moments, free cumulants, non-commutative convergence, free independence, as in Nica and Speicher [2006], however the essentials are covered in Chapter 5.

ARUP BOSE, *Cincinnati* and MONIKA BHATTACHARJEE, *Gainesville*

March 16, 2018

Acknowledgments

I am immeasurably thankful to students who have helped me to learn the subject by attending and actively participating in my lectures on random matrices and free probability in recent years: Debapratim Banerjee, Monika Bhattacharjee (co-author), Ayan Bhattacharya, Sohom Bhattacharya, Biltu Dan, Samir Mondal, and Sukrit Chakraborty.

Discussions with, and advice from, Octavio Arizmendi, Arijit Chakrabarty, Rajat Subhra Hazra, Iain Johnstone, Carlos Vargas Obieta, Debashis Paul, Jack Silverstein, Roland Speicher, and Jeff Yao are gratefully remembered.

Generous and understated hospitality of 5875 Taylor Ridge Drive as well as the continued interactions over academic matters every year for over more than a decade have been a major catalyst in my academic endeavors.

We are grateful to our editor John Kimmel for giving us a long leash so that we could take our time and do our best. We are thankful to Shashi Kumar for his help in sorting out all manners of formatting issues.

...AB

My co-author has been my PhD supervisor. He has offered me guidance, valuable advice, and has inspired me to join in this collaboration. I am thankful to Ayan Bhattacharya for his continuous encouragement and push to complete the work. I convey my thanks and gratefulness to Moulinath Banerjee, George Michailidis, Rajat Subhra Hazra, and Parthanil Roy for their inspiration and suggestions. Special thanks to Octavio Arizmendi, Carlos Vargas Obieta, and Debashis Paul for valuable discussions.

...MB

Introduction

Chapter 1. Large covariance matrix I (i.i.d. observations). Suppose the observations $\{X_i : 1 \leq i \leq n\}$ are identically distributed with mean 0 and the $p \times p$ covariance matrix Σ_p. In the high-dimensional setting, the dimension $p = p(n) \to \infty$ as the sample size $n \to \infty$. Estimation of the covariance matrix Σ_p is crucial in statistical analysis. This estimation, when $\{X_i\}$ are independently and identically distributed (i.i.d.) random vectors, is discussed in Chapter 1. An estimator $\hat{A}_{p,n}$, based on a sample of size n, is called *consistent* (in operator norm) for A_p if

$$||\Sigma_p - \hat{A}_{p,n}||_2 \xrightarrow{P} 0, \quad \text{as} \quad n \to \infty, \tag{1}$$

where $|| \cdot ||_2$ is the *operator norm* of a matrix. In the finite-dimensional case i.e., when p is fixed, the sample covariance matrix $\hat{\Sigma}_p$ is a consistent estimator of Σ_p under some modest assumptions.

However, in the high-dimensional setting, as many researchers have shown, $\hat{\Sigma}_p$ fails to estimate Σ_p consistently, even for i.i.d. $\{X_i\}$. This inconsistency is due to the increase in the number of unknown parameters along with the sample size.

A possible remedy is to have some restrictions on the parameter space and modify the basic estimator $\hat{\Sigma}_p$. This is usually called *covariance regularization*. Several covariance regularization techniques such as parameter shrinkage, penalized likelihood maximization and thresholding are available. The first two techniques have appeared while studying different aspects such as regression, classification or principal component analysis. Two excellent references where they are discussed extensively are [Bühlmann and van de Geer, 2011] and [Pourahmadi, 2013]. The focus of the present monograph is on thresholding. In particular, the ideas and results on banding and tapering ([Bickel and Levina, 2008a]), hard thresholding ([Bickel and Levina, 2008b]) and generalized thresholding ([Rothman et al., 2008]) are presented in details. These regularization on $\hat{\Sigma}_p$ provide consistency when Σ_p is bandable, Toeplitz or sparse. Rates of convergence results for these sample covariance matrix estimators are presented with proofs.

In addition to methodological advances in estimation of covariance matrices, there have been significant studies of the minimax risk for different structured parameter spaces. We shall touch on this aspect also.

Chapter 2. Large covariance matrix II (dependent observations). The independence assumption on $\{X_i\}$ that is presented in Chapter 1, is often inappropriate. The literature is replete with such examples. [Efron, 2009] proposed the matrix-variate normal distribution as a model for dependent $\{X_i\}$. It appears that there are very few articles that focus on the dependent situations and estimate Σ_p. After a brief discussion on the covariance regularization idea of [Allen and Tibshirani, 2010], estimation of Σ_p is discussed when $\{X_i\}$ obey certain weak dependence structures.

Three general dependent models (called *weak models*) which allow cross-covariance dependence are introduced. In the first, the dependence is via restriction on the growth of the powers of the trace of certain matrices derived from the cross-covariance structure. In the second, the dependence among any two columns weakens as the lag between them increases and in the third, weak dependence is imposed among the high-indexed columns. In each of these models, explicit upper bounds for the rate of convergence of the banded and tapered sample covariance matrix are presented. It turns out that in the first case, the convergence rate of the banded estimator is the same as in the i.i.d. case under a *trace condition*. Easy sufficient conditions for this condition to hold are provided. In particular, these estimators continue to remain consistent in operator norm. The permissible growth rate of p and the obtained convergence rates for the tapered and the banded estimators are in general slower than those in the i.i.d. case. There is a trade-off between these rates and the extent of dependency.

Estimation of sparse Σ_p in the presence of dependence is also discussed in Chapter 2. A natural *measure of stability*, based on the spectral density of the process captures the dependence in these cases. Under appropriate assumptions on this measure, thresholded sample covariance matrix achieve consistency.

Chapter 3. Large autocovariance matrices. High-dimensional data with dependent observations are often time series in nature. The most common assumptions made in modelling such data is stationarity. Let $\{X_{t,p} : t = 0, \pm1, \pm2, \ldots\}$ be p-dimensional random vectors with $E(X_{t,p}) = 0$ for all t. It is called weak or covariance stationary if and only if, for all $u \geq 0$, the $p \times p$ matrix

$$\Gamma_{u,p} = E(X_{t,p} X^*_{t+u,p}) \tag{2}$$

does not depend on t and is a function of only u. $\Gamma_{u,p}$ is called the (*population*) *autocovariance matrix* of order u. Note that $\Gamma_{0,p}$ is the (population) covariance matrix of $X_{t,p}$ for every t. Many well-known weak stationary high-dimensional time series models are special cases of the infinite-dimensional moving average process of order ∞, MA(∞), which is given by

$$X_t = \sum_{j=0}^{\infty} \psi_j \varepsilon_{t-j}, \quad \forall t, \tag{3}$$

where X_t and ε_t are p-dimensional vectors, $\{\varepsilon_t\}$ are i.i.d. with mean 0 and $p \times p$ covariance matrix Σ_p, and $\{\psi_j\}$ are $p \times p$ *parameter matrices*. Moreover, as in Chapters 1 and 2, the dimension $p \to \infty$ as the sample size $n \to \infty$. Note that the dimension of the above models are not infinite but tend to ∞ with the sample size. However, it has become customary to call them "infinite-dimensional". In this chapter, the interest is to estimate $\{\Gamma_u\}$ for the process (3), based on a sample $\{X_{t,p} : t = 1, 2, \ldots, n\}$ from this model. Both p and n grow so that it is a high-dimensional setting.

If $\psi_j = 0$, $\forall j > q$, then it is an infinite-dimensional MA(q) process. The i.i.d. process is nothing but the MA(0) process. The IVAR(r) process is given by

$$X_t = \sum_{i=1}^{r} A_i X_{t-i} + \varepsilon_t, \quad \forall t, \tag{4}$$

where X_t and ε_t are p-dimensional vectors, $\{\varepsilon_t\}$ are i.i.d. with mean 0 and $p \times p$ covariance matrix Σ_p, and $\{A_i\}$ are $p \times p$ *parameter matrices*. As already indicated, under appropriate conditions on $\{A_i\}$, (4) can also be expressed in the form (3).

A method of moment estimator of $\Gamma_{u,p}$ is given by the $p \times p$ matrix

$$\hat{\Gamma}_{u,p,n} = \frac{1}{n} \sum_{t=1}^{n-u} X_{t,p} X_{t+u,p}^*. \tag{5}$$

$\hat{\Gamma}_{u,p,n}$ is called the *sample autocovariance matrix*. For simplicity this matrix is often written as $\hat{\Gamma}_u$. Since the size of the population autocovariance matrices $\{\Gamma_u\}$ increases as $p \to \infty$, the number of unknown parameters (entries in $\{\Gamma_u\}$) increases. Consequently, just like the sample covariance matrix in Chapters 1 and 2, $\hat{\Gamma}_u$ fails to consistently estimate Γ_u.

From the experience of Chapters 1 and 2, to extract consistent estimators, two things are needed—suitable restrictions on $\{\psi_j\}$ and Σ_p and, appropriate modifications such as banding or tapering on $\hat{\Gamma}_u$. Taking a cue from Chapter 2, it is assumed that $n^{-1} \log p \to 0$ and suitable restrictions are imposed on the parameters $\{\psi_j\}$ and Σ_p. These restrictions are directly borrowed from the development of Chapter 2. However, they are cumbersome and may be difficult to check.

At first it is shown that the banded and tapered $\{\hat{\Gamma}_u\}$ are consistent for $\{\Gamma_u\}$ under appropriate parameter spaces, and some rate of convergence results are also established. Then the focus is on providing simpler restrictions on the parameter spaces for the particular cases of MA(r) and IVAR(r). Consistency of the banded and tapered $\hat{\Gamma}_u$ are established with these parameter spaces. These results are then used to obtain consistent estimator for $\{A_i\}$ in the IVAR(r) process. These estimates are used in Chapter 10 for statistical inference. The above theoretical results are supported by simulations. Indeed, it appears from these simulations that the rate of convergence is also quite sharp.

Even though $\{\hat{\Gamma}_u\}$ are not consistent, their asymptotic properties are interesting in their own right. Moreover, they can also be used for statistical analysis. Before this can be explored, some tools from random matrix theory (RMT) and non-commutative probability (NCP) theory is needed. The essential concepts are available in the two books Bai and Silverstein [2009] and Nica and Speicher [2006]. However, in Chapters 4 and 5 a concise and a unified description of the tools and the results that are needed along with additional information and links between the different results, are provided.

Chapter 4. Spectral distribution. This chapter collects the basic concepts and results that are needed from RMT. These are crucially used in Chapters 6–10. Note that $\{\hat{\Gamma}_u\}$ are random matrices. The empirical spectral distribution (ESD) of a $p \times p$ (random) matrix R_p is the (random) probability distribution with mass p^{-1} at each of its eigenvalues. The so-called spectral statistics, useful in statistical applications, are functions of this spectral distribution. It is also a very useful object in wireless communication. Many classical test procedures in multivariate statistics use traces of polynomials of the sample covariance matrix, which also fall in this category.

If the ESD converges weakly (almost surely) to a (non-degenerate) probability distribution, then the latter is called the limiting spectral distribution (LSD) of R_p. LSD results for various random matrices have a central position in the RMT literature. They are of primary interest to those who are working in large random matrices. Incidentally, the study of the limit spectrum of non-hermitian matrices is extremely difficult and very few results are known for general random non-hermitian sequences. In this monograph, LSD of only real symmetric matrices are dealt with. The two methods that are used to establish LSD are the method of Stieltjes transformation and the moment method.

Two specific matrices play a central role in RMT—the Wigner matrix and the independent matrix. A *Wigner matrix* W_p, of order p, is a square symmetric random matrix with independent mean 0 variance 1 entries on and above the diagonal. An *independent matrix* Z, of order $p \times n$, is a rectangular matrix with all independent mean 0 and variance 1 entries. Let A_p be a $p \times p$ symmetric non-negative definite matrix, whose LSD exists. LSD results for the following matrices are discussed in this chapter. They require appropriate assumptions that are skipped in this discussion.

(a) $p^{-1/2}W_p$.

(b) $p^{-1/2}A_p^{1/2}W_pA_p^{1/2}$.

The classical RMT model for Z assumes $p, n = n(p) \to \infty$ and

$$\frac{p}{n} \to y \in [0, \infty). \tag{6}$$

For the $y > 0$ case, the LSD of the following matrices are known

(c) $n^{-1}ZZ^*$.

(d) $n^{-1}A_p^{1/2}ZZ^*A_p^{1/2}$.

The LSD results for the case $y = 0$ are quite different from the case $y > 0$. Let $\{B_n\}$ be $n \times n$ square symmetric norm bounded non-negative matrices with $\lim_n n^{-1}\mathrm{Tr}(B_n^k) < \infty$, $k = 1, 2$. The LSD of the following matrices are known:

(e) $\sqrt{np^{-1}}(n^{-1}ZZ^* - I)$.

(f) $\sqrt{np^{-1}}(n^{-1}A_p^{1/2}ZZ^*A_p^{1/2} - A)$.

(g) $\sqrt{np^{-1}}(n^{-1}A_p^{1/2}ZB_nZ^*Z_P^{1/2} - n^{-1}\mathrm{Tr}(B_n)A_p)$.

Moreover, LSD of (e) and (f) are respectively identical with the LSD of (a) and (b) above.

Chapter 5. Non-commutative probability. This chapter serves as a brief introduction to *non-commutative probability* and related notions that will be used in later chapters, specially in Chapters 6–10. In the previous chapter the convergence of a single sequence of matrices was discussed in terms of the convergence of their ESD. The method of Stieltjes transform can deal with one sequence of matrices at a time. What happens if, instead of one matrix at a time, one wishes to study the joint convergence of several sequences of matrices together? Two most natural ways to study the joint convergence of a collection of random matrices are through the following interrelated ideas.

(1) Limiting spectral distribution (LSD) of (symmetric) polynomials in these matrices and,

(2) convergence in the non-commutative sense of these symmetric (or often non-symmetric) polynomials.

The convergence in (2) above needs explanation. As is known, commutative (classical) random variables are attached to a probability space $(\mathcal{S}, E)$, which consists of a σ-field $\mathcal{S}$ and an expectation operator E. However, matrices are non-commutative objects, so appropriate non-commutative notions are needed. A *non-commutative $*$-probability space* (NCP) $(\mathcal{A}, \varphi)$ consists of a unital $*$-algebra $\mathcal{A}$ and a linear functional $\varphi : \mathcal{A} \to \mathbb{C}$ (called *state* of $\mathcal{A}$) with $\varphi(1_{\mathcal{A}}) = 1$. Elements of $\mathcal{A}$ are called non-commutative variables. *Free variables* are non-commutative variables and are analogous to independent variables in classical commutative probability. LSD in complicated situations are often described in terms of the simpler free variables.

Convergence of non-commutative variables is defined as follows. As $N \to \infty$, a collection of non-commutative variables $\{a_i^{(N)}, a_i^{*(N)} : i \geq 1\}$ in $(\mathcal{A}_N, \varphi_N)$

converges jointly to $\{a_i, a_i^* : i \geq 1\}$ if for all polynomials Π,

$$\lim_{N \to \infty} \varphi_N\big(\Pi(a_i^{(N)}, a_i^{*(N)} : 1 \leq i \leq t)\big) \;=\; \varphi\big(\Pi(a_i, a_i^* : 1 \leq i \leq t)\big). \qquad (7)$$

If a set of $p \times p$ random matrices is given, then the class of all polynomials generated by these matrices with the state defined as $\phi_p(A) = \frac{1}{p}E\mathrm{Tr}(A)$ turns into a non-commutative $*$-probability space (NCP). Therefore, by joint convergence (in the non-commutative sense) of a collection of $p \times p$ random matrices $\{A_i : i \geq 1\}$, is meant the convergence of $p^{-1}E\mathrm{Tr}(\Pi(A_i : i \geq 1))$ as $p \to \infty$ and for all polynomials Π.

The convergences in (1) and (2) above are related as follows. To establish the LSD of a $p \times p$ symmetric random matrix A_p by the moment method, a crucial step is to show $\lim p^{-1}E\mathrm{Tr}(A_p^k) < \infty$, $\forall k \geq 1$. Therefore, the convergence of A_p in the non-commutative sense with some additional effort, yields the LSD of A_p. Similarly, the NCP convergence of $\{A_i, A_i^* : i \geq 1\}$ with some additional effort, yields the LSD of any symmetric polynomial in $\{A_i, A_i^* : i \geq 1\}$.

The idea of freeness can be extended to *asymptotic freeness*. The important result of asymptotic freeness between Wigner and deterministic matrices is also discussed in this chapter.

Chapters 6 and 7. Generalized covariance matrices I and II. Chapters 4 and 5 briefly discuss a few results on the LSD and NCP convergence for some specific type of random matrices. Now the scope can be broadened significantly and LSD for much more general matrices, which are called *generalised sample covariance matrices*, can be established, both for the cases $p/n \to y \in (0, \infty)$ and $p/n \to 0$. The results of this chapter should be of interest to researchers in random matrix theory, non-commutative probability and wireless communication. This generality is needed not just for theoretical elegance or completeness. This is the cornerstone on which development of the asymptotic properties of the sample autocovariance matrices and their usage in statistical applications rest in Part III.

Suppose there are matrices $Z_u = ((\varepsilon_{u,t,i}))_{p \times n}$, $1 \leq u \leq U$, where $\{\varepsilon_{u,t,i} : u, i, j \geq 0\}$ are independent with mean 0 and variance 1. Note that each Z_u is an independent matrix and moreover, they are independent among themselves. Also suppose $\{B_{2i-1} : 1 \leq i \leq K\}$ and $\{B_{2i} : 1 \leq i \leq L\}$ are constant matrices of order $p \times p$ and $n \times n$ respectively. Without loss of generality, assume that these collections are closed under the $*$ operation.

Consider all $p \times p$ matrices

$$\mathbb{P}_{l,(u_{l,1}, u_{l,2}, \ldots, u_{l,k_l})} = \prod_{i=1}^{k_l} \big(n^{-1} A_{l,2i-1} Z_{u_{l,i}} A_{l,2i} Z_{u_{l,i}}^*\big) A_{l,2k_l+1}, \qquad (8)$$

where $\{A_{l,2i-1}\}$, $\{A_{l,2i}\}$ and $\{Z_{u_{l,i}}\}$ are matrices from the collections $\{B_{2i-1}, B_{2i-1}^* : 1 \leq i \leq K\}$, $\{B_{2i}, B_{2i}^* : 1 \leq i \leq L\}$ and $\{Z_i : 1 \leq i \leq U\}$

respectively. As the sample covariance matrix $n^{-1}ZZ^*$ (without centering) is a special case of the above matrices, these matrices are called *generalized covariance matrices*.

Consider the sequence of NCP $(\mathcal{U}_p, p^{-1}E\mathrm{Tr})$, where

$$\mathcal{U}_p = \mathrm{Span}\big(\mathbb{P}_{l,(u_{l,1},\dots,u_{l,k_l})} : \ l, k_l \geq 1\big). \tag{9}$$

Note that $\mathcal{U}_p$ forms a $*$-algebra. All the matrices discussed in the previous chapter that involved Z and Z^* belong to this algebra. With the use of the machinery developed in Chapter 5, it is shown that the NCP $(\mathcal{U}_p, p^{-1}E\mathrm{Tr})$ converges. At the same time, recall that NCP convergence with some additional effort guarantees existence of the LSD.

First consider the case $p, n(p) \to \infty$, $p/n \to y > 0$. In Chapter 5, asymptotic freeness of Wigner and deterministic matrices was mentioned. Using this result and appropriate embedding, one can show that the LSD of any symmetric polynomial in $\{\mathbb{P}_{l,(u_{l,1},u_{l,2},\dots,u_{l,k_l})}\}$ exists and the limit can be expressed in terms of some freely independent variables. The Stieltjes transform of the LSD is also derived for a large class of polynomials. Finally a list of LSD results for specific matrices is also provided. The results (a)–(d) mentioned earlier follow as special cases.

Now consider the case where $p, n(p) \to \infty$ but $p/n \to 0$. In this case, the embedding technique does not work since the growth of p and n are not comparable. Therefore, a different approach is needed. At the same time, keeping in mind (e), (f) and (g) quoted earlier, it can be concluded that very different centering and scaling on $\{\mathbb{P}_{l,(u_{l,1},u_{l,2},\dots,u_{l,k_l})}\}$ are needed to get non-degenerate limits. Taking a cue from those results, define the centered and scaled matrices

$$\mathcal{R}_{l,(u_{l,1},\dots,u_{l,k_l})} = (n/p)^{1/2}(\mathbb{P}_{l,(u_{l,1},u_{l,2},\dots,u_{l,k_l})} - \mathbb{G}_{l,k_l}), \text{ where} \tag{10}$$

$$\mathbb{G}_{l,k_l} = \Big(\prod_{i=1}^{k_l} n^{-1}\mathrm{Tr}\,(A_{l,2i})\Big)\prod_{i=0}^{k_l} A_{l,2i+1} \tag{11}$$

are the *centering matrices*. Now consider the convergence of the sequence of NCP $(\mathcal{V}_p, p^{-1}E\mathrm{Tr})$ where

$$\mathcal{V}_p = \mathrm{Span}\{\mathcal{R}_{l,(u_{l,1},\dots,u_{l,k_l})} : l, k_l \geq 1\}. \tag{12}$$

Note that $\mathcal{V}_p$ forms a $*$-algebra. The main result states that the above sequence of NCP converges. The limiting NCP can be expressed in terms of some free variables. Further, the LSD of any symmetric polynomial in $\{\mathcal{R}_{l,(u_{l,1},\dots,u_{l,k_l})}\}$ exists and can be expressed in terms of free variables. Several applications of these results to specific models are then provided. As before, the results (e)–(g) mentioned earlier follow as special cases.

One of the major uses of the above results is that the LSD of any symmetric polynomial in the sample autocovariance matrices $\{\hat{\Gamma}_u\}$, along with their joint convergence can be tackled by the above results. This is done in details in the later chapters.

High-dimensional time series is a very new, active and important area where researchers are working desperately to develop statistical methods and techniques. Chapters 8–11 are based on recent research in this area and constitute Part III of the monograph. By using the ideas, methods and results developed in Part II, in Chapters 8 and 9 first the asymptotic properties of (symmetrised) sample autocovariance matrices are derived respectively for one, and more than one, independent infinite-dimensional MA(q) processes. These results are interesting in their own right. Then in Chapter 10, it is shown how these results can be used for statistical inference in one and two sample problems involving high-dimensional time series. In Chapter 11, asymptotic normality of polynomials in sample autocovariance matrices are presented with their potential application to construct significance test of hypotheses in one and two independent high-dimensional time series.

Chapters 8 and 9. Spectra of autocovariance matrices I and II. As discussed earlier, sample autocovariance matrices are important in high-dimensional linear time series. In addition to just $\hat{\Gamma}_u$ and $\hat{\Gamma}_u^*$, one may also be interested in functions of these. For example, if one wishes to study the *singular values* of $\hat{\Gamma}_u$, the matrix $\hat{\Gamma}_u\hat{\Gamma}_u^*$ needs to be considered. Likewise, as may be recalled, in the one-dimensional case, all tests for white noise are based on quadratic functions of autocovariances. The analogous objects in our model are quadratic polynomials in autocovariances. Thus, one is naturally led to the consideration of matrix polynomials of autocovariances. Chapters 8 and 9 focus on the LSD of polynomials in sample autocovariance matrices $\{\hat{\Gamma}_u\}$ respectively for one and more than one independent infinite-dimensional moving average processes. However, due to the reasons discussed in the previous chapters, only symmetric polynomials are considered.

General results developed in Chapters 6 and 7 now come quite handy. Let us briefly indicate how. Recall the $p \times n$ independent matrix Z and the sequence of parameter matrices $\{\psi_j\}$ from (3). Let $\{P_j : j = 0, \pm1, \pm2, \ldots\}$ be a sequence of $n \times n$ matrices where P_j has entries equal to one on the j-th upper diagonal and 0 otherwise. Note that $P_0 = I_n$ where I_n is the $n \times n$ identity matrix, and $P_j = P_{-j}^*$, $\forall j$. Define

$$\Delta_u = \frac{1}{n} \sum_{j,j'=0}^{q} \psi_j Z P_{j-j'+u} Z^* \psi_{j'}^*, \quad \forall u = 0, 1, 2, \ldots. \tag{13}$$

It turns out that $\{\Delta_u\}$ approximates $\{\hat{\Gamma}_u\}$ as far as the LSD and joint convergence are concerned. Note that the matrices $\{\Delta_u\}$ fall under the general setting of Chapters 6 and 7 and hence the results of those chapters are fruitfully applied to deal with the sample autocovariance matrices. In Chapter 8, it is shown that under most reasonable conditions on $\{\psi_j\}$, the LSD of any symmetric polynomial in $\{\hat{\Gamma}_u\}$ exists for both the cases $p/n \to y \in (0, \infty)$

and $p/n \to 0$. In the latter case, appropriate scaling constants and centering matrices are used. The limits are described in terms of some free variables and Stieltjes transform for some polynomials are derived. Similar results for more than one independent infinite-dimensional moving average processes are presented in Chapter 9.

General LSD results are not known for the non-symmetric polynomials. Simulations in a few non-symmetric cases are provided to convince the reader that LSD results in these cases remain to be discovered.

Chapter 10. Graphical inference. In this chapter it is demonstrated how the LSD results obtained in Chapters 8 and 9 can be used in statistical graphical inference of high-dimensional time series. This includes estimation of unknown order of high-dimensional MA and AR processes.

In the univariate setting, a plot of the sample autocovariances provides a method to identify the order of an MA process. If the sample autocovariances are close to zero for order $u > \hat{q}$, then $\hat{q}$ is taken to be an estimator of the unknown order. Based on LSD results in Chapter 8, an analogous graphical method in high-dimensional setup to determine the order of a moving average process is described. A similar idea is used on the residual process to determine the unknown order of an IVAR process.

Chapter 11. Testing with trace. Statistical inference in high-dimensional models is often based on linear spectral statistics of the form $\sum f(\lambda_i)$ where f is a suitable function and $\{\lambda_i\}$ are eigenvalues of a matrix. There is a large literature on the spectral statistics of high-dimensional covariance matrices and its application in statistical inference. However, there does not appear to be any results known for linear spectral statistics of general autocovariance matrices. This can be an important topic of future research.

While such general spectral statistics are not considered in this book, asymptotic normality results for the trace of any polynomial (which corresponds to $f(x) = x$) in sample autocovariance matrices for one or several independent MA processes are first developed in both cases $p, n = n(p) \to \infty$, $p/n \to y \in (0, \infty)$ and $p, n = n(p) \to \infty$, $p/n \to 0$.

These results are used in significance tests for different hypotheses on coefficient matrices involving one or more such processes. The inference methods for the two cases $p/n \to y \in (0, \infty)$ and $p/n \to 0$ differ significantly in the details.

Appendix. Supplementary proofs. Some of the longer proofs and full details of proofs in the main chapters have been relegated to an Appendix.

Part I

Part I

Chapter 1

LARGE COVARIANCE MATRIX I

High-dimensional data is often represented in the form of a *data matrix*

$$X_{p \times n} = \begin{bmatrix} x_{11} & x_{12} & x_{13} & \cdots & x_{1n} \\ x_{21} & x_{22} & x_{23} & \cdots & x_{2n} \\ \vdots & \vdots & \vdots & \vdots & \vdots \\ x_{p1} & x_{p2} & x_{p3} & \cdots & x_{pn} \end{bmatrix} \tag{1.1}$$

where the dimension p is assumed to be increasing with the sample size n i.e., $p = p(n) \to \infty$ as $n \to \infty$. Let

$$C_{ip} = (x_{1i}, x_{2i}, \ldots, x_{pi})^*, \ 1 \leq i \leq n, \tag{1.2}$$

be the columns of X. We assume that they are i.i.d. (independently and identically distributed) random vectors with mean 0 and covariance matrix Σ_p. In the next chapter we shall allow dependence between the columns.

Many statistical procedures such as classification, principal component analysis, discriminant analysis are based on an estimate of Σ_p. When p is fixed, the *sample covariance matrix*

$$\tilde{\Sigma}_{p,n} = \frac{1}{n} \sum_{i=1}^{n} C_{ip} C_{ip}^* - \left(\frac{1}{n} \sum_{i=1}^{n} C_{ip} \right) \left(\frac{1}{n} \sum_{i=1}^{n} C_{ip} \right)^* \tag{1.3}$$

is a consistent estimator of Σ_p. Here $*$ denotes the transpose of a vector or a matrix. However, when p increases with n, this is not necessarily the case. It is known that $\tilde{\Sigma}_{p,n}$ fails to estimate Σ_p consistently where consistency is defined in some natural way.

The remedy is twofold. First, restrictions are imposed on the parameter space. Second, modifications, called *covariance regularization* are done on the estimator $\tilde{\Sigma}_{p,n}$. Restricted parameter spaces include a variety of families of structured covariance matrices which often find their motivation in problems arising in bio-science, finance and other areas. At the same time, regularization methods have been developed to exploit these structural assumptions. These include banding and tapering, thresholding, penalized likelihood, and even regularization of principal components.

In this chapter we will discuss well conditioned, approximately bandable, Toeplitz and sparse covariance matrices and their estimators using covariance regularizations like banding, tapering and thresholding. Good references for other regularization methods such as spiked covariance matrices and penalized likelihood are Bühlmann and van de Geer [2011] and Pourahmadi [2013]. In addition to methodological advances in estimation of covariance matrices, there have been significant studies of the (asymptotic) minimax risk for different structured parameter spaces. We shall touch upon this aspect also.

1.1   Consistency

For any matrix M let

$$M(i,j) = m_{ij} = \text{ the } (i,j)\text{-th element of } M. \tag{1.4}$$

We often write

$$M = ((m_{i,j}))_{k\times l} \text{ or } M = ((M(i,j)))_{k\times l}. \tag{1.5}$$

For any real symmetric matrix A_p, let

$$\lambda_{\max}(A_p) = \text{ the largest eigenvalue of } A_p, \tag{1.6}$$
$$\lambda_{\min}(A_p) = \text{ smallest eigenvalue of } A_p. \tag{1.7}$$

The L_2 *norm* or the *operator norm* of A_p is defined as

$$||A_p||_2 = \sqrt{\lambda_{\max}(A_p^* A_p)}. \tag{1.8}$$

For a given sequence $\{a_i\}$, let $\text{diag}(a_1, a_2, \ldots, a_k)_{k\times k}$ be the diagonal matrix of order k with diagonal elements $\{a_1, a_2, \ldots, a_k\}$. Let

$$I_k = \text{diag}(1, 1, \ldots, 1)_{k\times k}, \tag{1.9}$$
$$J_k = ((1))_{k\times k}. \tag{1.10}$$

Example 1.1.1. We choose three different Σ_p, namely

(a) $\Sigma_{1,p} = \text{diag}(1, 2, \ldots, p)$,

(b) $\Sigma_{2,p} = \text{diag}(1, 2^{-1}, \ldots, p^{-1})$,

(c) $\Sigma_{3,p} = 0.5I_p + 0.5J_p$.

Note that these are three different kinds of covariance matrices, where

$$\sup_p \lambda_{\max}(\Sigma_{1,p}) = \infty, \quad \inf_p \lambda_{\min}(\Sigma_{2,p}) = 0 \tag{1.11}$$

and dependence among the i-th and the j-th variables in $\Sigma_{3,p}$ does not decrease with increase in $|i - j|$.

Let $n = 10, 15, 20, 25$ and $p = [e^{n^{0.2}}]$. For each n, consider independent random samples $\{X_{ijk} : i = 1, 2, 3, \ 1 \leq j \leq n, \ 1 \leq k \leq 1000\}$, where

$$X_{ijk} \sim \mathcal{N}_p(0, \Sigma_{i,p}). \tag{1.12}$$

For each model, consider the estimator $\tilde{\Sigma}_{p,n}$ defined in (1.3). Let,

$$R_{in} = \frac{1}{1000} \sum_{k=1}^{1000} ||\frac{1}{n} \sum_{j=1}^{n} X_{ijk} X_{ijk}^* - \frac{1}{n} \sum_{j=1}^{n} X_{ijk} \left(\frac{1}{n} \sum_{j=1}^{n} X_{ijk}^*\right) - \Sigma_{i,p}||_2. \tag{1.13}$$

We compute R_{in} by simulations for the three models and for the above choices of n and p. The results are reported in the following table.

Table 1: Value of $\{R_{in}\}$

n	10	15	20	25
p	8	21	55	149
R_{1n}	13.82	48.23	260.67	1862.19
R_{2n}	9.78	15.88	59.81	89.69
R_{3n}	15.04	41.37	85.08	205.94

Observe that R_{in} increases with n and hence $\tilde{\Sigma}_{p,n}$ fails to estimate Σ_p accurately in the high-dimensional setting.

We now introduce the notion of consistency that we will work with. Let $\xrightarrow{P}$ denote *convergence in probability*.

Definition 1.1.1. *An estimator $\hat{A}_{p,n}$, based on a sample of size n, is called consistent in operator norm for A_p if*

$$||A_p - \hat{A}_{p,n}||_2 \xrightarrow{P} 0, \quad as \ n \to \infty. \tag{1.14}$$

Throughout this book, by a consistent estimator we always mean consistency in the above sense unless otherwise mentioned.

1.2 Covariance classes and regularization

To overcome the inconsistency which is primarily due to the high-dimensionality, structural assumptions are imposed on the covariance matrix. Various families of structured covariance matrices have been introduced in recent years. These give rise to different covariance classes.

1.2.1 Covariance classes

Three kinds of covariance matrices will be considered.

(a) *Bandable covariance matrices.* In areas such as time series, climatology and spectroscopy there is a natural order or distance between indices of the variables. See, for example, Friston et al. [1994] and Visser and Molenaar [1995]. Often the dependence between the variables decreases as the distance becomes larger. We consider settings where σ_{ij}, the (i, j)-th entry of Σ_p, becomes smaller as $|i-j|$ becomes larger. In other words, for each k, the variables X_{ik} and X_{jk} are nearly uncorrelated when the distance $|i - j|$ between them is large. Such matrices are said to be bandable. Appropriate classes of these matrices and their estimation are discussed in Section 1.3.

(b) *Toeplitz covariance matrices.* A covariance matrix Σ_p is a Toeplitz matrix if it is of the form $((\sigma_{|i-j|}))$. Toeplitz covariance matrices arise naturally in the analysis of stationary stochastic processes. If $\{X_i\}$ is a real-valued stationary process then the covariance matrix of $X_1, \ldots, X_k$ (the autocovariance matrix) is a Toeplitz matrix. As a consequence, Toeplitz covariance matrices have a wide range of applications in many fields, including engineering, economics and biology. See, for instance, Franaszczuk et al. [1985], Fuhrmann [1991], and Quah [2000] for specific applications. Toeplitz matrices are bandable when $\sigma_k \to 0$ as $k \to \infty$. Section 1.4 focuses on estimation of large Toeplitz covariance matrices.

(c) *Sparse covariance matrices.* In many applications such as genomics, distance between indices of the variables have no natural meaning. On the other hand, the covariance between most pairs of variables are insignificant. When we assume that most of the entries in each row and each column of the covariance matrix are zero or negligible, we get a sparse covariance matrix. Clearly this class is very different from the bandable matrices and Toeplitz matrices. Section 1.5 deals with estimation of such covariance matrices.

1.2.2   Covariance regularization

In the high-dimensional setting, the estimate $\tilde{\Sigma}_{p,n}$, defined in (1.3), is not necessarily consistent for Σ_p even when we impose enough structural assumptions like (a)–(c) above on the parameter space. To achieve consistency, we additionally need appropriate regularization of $\tilde{\Sigma}_{p,n}$ which can control its corners. This control is achieved via two very familiar notions of *banding* and *tapering*. Similarly, when Σ_p is sparse, $\tilde{\Sigma}_{p,n}$ may not be so. In this case, consistency is achieved through appropriate *thresholding* of $\tilde{\Sigma}_{p,n}$. We now describe these three ideas in brief.

(a) *Banding.* For any matrix $M = ((m_{ij}))$, its *k-banded version* is defined as

$$B_k(M) = ((m_{ij}I(|i - j| \leq k))). \qquad (1.15)$$

When a matrix is banded, the extent of control over the corners is determined by the choice of the *banding parameter* k. Clearly a banded covariance matrix need not remain positive semi-definite.

(b) *Tapering.* Positive semi-definiteness is a desirable property for any co-variance matrix estimate. Recall that the Schur product or component wise product of two positive semi-definite matrices is positive semi-definite. This fact is at the heart of a tapered estimate. The banded matrix estimate is a Schur product of $\tilde{\Sigma}_p$ with another matrix whose entries are 0 and 1. However this 0-1 matrix is not positive semi-definite. This drawback is removed by modifying this 0-1 matrix and filling it up with decaying entries, in such a way that the positive semi-definiteness is preserved.

Definition 1.2.1. *A function $g : \mathbb{R}^+ \cup \{0\} \to \mathbb{R}$ is called a positive definite function if for all $x_1, x_2, \ldots, x_k \in \mathbb{R}$ and $k \geq 1$, the $k \times k$ matrix $((g(|x_i - x_j|)))$ is positive semi-definite. If in addition the range of g is $\mathbb{R}^+ \cup \{0\}$ then we call it a kernel.*

Let g be a kernel which is continuous, non-increasing, such that

$$g(0) = 1, \quad \lim_{x \to \infty} g(x) = 0. \tag{1.16}$$

Suppose $\{\tau_p\}$ is a sequence of positive numbers called the *bandwidth*. The positive definiteness means that for every p,

$$R_p = ((g(\frac{|i - j|}{\tau_p})))_{p \times p}$$

is positive semi-definite. One such choice is $g(x) = e^{-|x|}$.
Now, for a $p \times p$ matrix A, we define the *tapered* version of A as

$$
\begin{aligned}
R_{g,\tau_p}(A) &= A * R_p \quad \text{where} & (1.17) \\
* &= \text{Schur or component wise product of matrices.} & (1.18)
\end{aligned}
$$

Note that if A is positive semi-definite and g is a positive definite function, then $R_{g,\tau_p}(A)$ is also positive semi-definite.

(c) *Thresholding.* For a $p \times p$ matrix $A = ((a_{ij}))$, its t_p-thresholded version is given by

$$T_{t_p}(A) = ((a_{ij}I(|a_{ij}| > t_p))). \tag{1.19}$$

The number t_p is called the *thresholding parameter*. Note that if A is positive semi-definite, then $T_{t_p}(A)$ need not be so.

1.3 Bandable Σ_p

This section presents appropriate classes of bandable covariance matrices and provides explicit convergence rates of banded and tapered sample covariance matrices when Σ_p is in any of these classes.

1.3.1 Parameter space

For all $k \geq 1$, let $\{A_k\}$ be a nested sequence of square matrices where A_k is of order k. Define

$$A_\infty \quad = \quad \text{the } \infty \times \infty \text{ extension of } \{A_k\}. \tag{1.20}$$

Note that the relation between A_∞ and $\{A_k\}$ is a bijection.

Well conditioned covariance matrices. A covariance matrix Σ_∞ is called *well conditioned* if its eigenvalues are bounded away from both 0 and ∞. For any $\epsilon > 0$, the set of all ϵ-*well conditioned* covariance matrices is given by:

$$\mathcal{W}(\epsilon) = \{\Sigma_\infty : 0 < \epsilon < \inf_p \lambda_{\max}(\Sigma_p) \leq \sup_p \lambda_{\max}(\Sigma_p) < \epsilon^{-1} < \infty\}. \tag{1.21}$$

Hence, a covariance matrix Σ_∞ is *well conditioned* if

$$\Sigma_\infty \in \mathcal{W} := \bigcup_{\epsilon > 0} \mathcal{W}(\epsilon). \tag{1.22}$$

Clearly, this class avoids matrices like $\Sigma_{1,p}$ and $\Sigma_{2,p}$ in Example 1.1.1.

Covariance matrices with polynomially decaying corner. We consider the settings where σ_{ij} ((i,j)-th entry of Σ_p) is close to zero when $|i - j|$ is large. In other words, the variables X_{ik} and X_{jk} are nearly uncorrelated when the distance $|i - j|$ between them is large. This kind of covariance matrices were proposed in Bickel and Levina [2008a] and Cai et al. [2010].

Definition 1.3.1. *The k-corner measure of $A = ((a_{ij}))$ is defined as*

$$T(A, k) = \sup_j \sum_{i:|i-j|>k} |a_{ij}|. \tag{1.23}$$

So, $T(A, k)$ is the maximum column sum of the thresholded matrix $((a_{ij}I(|i - j| > k))$. For weak dependence among i-th and j-th variable as $|i-j|$ grows, the corner measure $T(\Sigma_\infty, k)$ should decay as k grows. A covariance matrix Σ_∞ is said to have a *polynomially decaying corner*, if

$$\Sigma_\infty \quad \in \quad \mathcal{X} := \cup_{\alpha, C > 0} \mathcal{X}(\alpha, C), \text{ where} \tag{1.24}$$

$$\mathcal{X}(\alpha, C) \quad = \quad \{A : T(A, k) \leq Ck^{-\alpha}, \forall k \geq 1\}, \ \alpha, C > 0. \tag{1.25}$$

Clearly, the class $\mathcal{X}$ avoids matrices like $\Sigma_{3,p}$ in Example 1.1.1.

The k-banded version $B_k(A)$ of a matrix A (defined in (1.15)), is connected to the corner measure $T(A, k)$ via the $(1, 1)$ *norm*. The latter is defined as

$$||A||_{(1,1)} = \sup_j \sum_{i \geq 1} |a_{ij}|. \tag{1.26}$$

This is nothing but the maximum column sum of the matrix A. Then the following lemma is immediate.

Lemma 1.3.1. *If $A \in \mathcal{X}(\alpha, C)$, then*

$$||B_k(A) - A||_{(1,1)} = T(A, k) \leq Ck^{-\alpha}.$$

As a consequence, any $A \in \mathcal{X}(\alpha, C)$ can be approximated by $B_k(A)$. This is why any $A \in \mathcal{X}(\alpha, C)$ is called *approximately bandable*.

Let

$$\mathcal{U}(\epsilon, \alpha, C) = \mathcal{W}(\epsilon) \cap \mathcal{X}(\alpha, C), \quad \epsilon, \alpha, C > 0 \quad \text{and} \quad (1.27)$$

$$\mathcal{U} = \bigcup_{\epsilon, \alpha, C > 0} \mathcal{U}(\epsilon, \alpha, C). \quad (1.28)$$

For a function f on any domain A, we define

$$||f||_\infty = \sup_{x \in A} |f(x)|. \quad (1.29)$$

Let $f^{(m)}$ denote the m-th order derivative of f.

Example 1.3.1. Consider the symmetric Toeplitz matrix $T = ((t_{ij} = t_{|i-j|}))$, where $\sum_{u=-\infty}^{\infty} |t_u| < \infty$. Its spectral density is given by

$$f_T(x) = \frac{1}{2\pi} \sum_{u=-\infty}^{\infty} t_u e^{iux}, \quad \forall \, 0 \leq x < 2\pi.$$

For any $\epsilon, C > 0$ and $m \geq 1$, consider the following class of Toeplitz covariance matrices

$$\mathcal{L}(\epsilon, m, C) = \{\Sigma_\infty = ((\sigma_{ij})) : \sigma_{ij} = \sigma_{|i-j|} \text{ with spectral density } f_{\Sigma_\infty},$$
$$0 < \epsilon < ||f_{\Sigma_\infty}||_\infty < \epsilon^{-1}, \, ||f_{\Sigma_\infty}^{(m)}||_\infty \leq C\}.$$

Bickel and Levina [2008a] proved that

$$\mathcal{L}(\epsilon, m, C) \subset \mathcal{U}(\epsilon, m-1, C). \quad (1.30)$$

Example 1.3.2. For any $\epsilon, m, C, C_1, C_2 > 0$, $m_1 \geq 1$ and $m_2 > 2$, consider the following class of covariance matrices

$$\mathcal{K}(m, C) = \{\Sigma : \sigma_{ii} \leq Ci^{-m}, \, \forall i\},$$
$$\mathcal{T}(\epsilon, m_1, m_2, C_1, C_2) = \{\Sigma_\infty = A + B : A \in \mathcal{L}(\epsilon, m_1, C_1), B \in \mathcal{K}(m_2, C_2)\}.$$

Bickel and Levina [2008a] proved that

$$\mathcal{T}(\epsilon, m_1, m_2, C_1, C_2) \subset \mathcal{U}(\epsilon', \alpha, C_3), \text{ where} \quad (1.31)$$
$$\epsilon' \leq \epsilon^{-1} + C_2$$
$$\alpha \leq \min\{m_1 - 1, \frac{m_2 - 2}{2}\}$$
$$C_3 \leq \frac{C_1}{m_1 - 1} + \frac{2C_2}{m_2 - 2}.$$

1.3.2 Estimation in $\mathcal{U}$

In this section we present two types of results. First we discuss some rate of convergence results for banded and tapered estimators which in particular yield their consistency. Then we present some results on minimax rates of these estimators. In the next sections we shall discuss similar results under the additional Toeplitz structure and the sparse structure of the covariance matrices.

Banded estimator.

We first state and prove a result on the consistency of $B_k(\tilde{\Sigma}_{p,n})$. We shall see extensions of this result for dependent models in the next chapter. Hence, it is worthwhile to see a detailed proof of this result.

Let $\{a_n\}$ and $\{b_n\}$ be two positive sequences. Then by $a_n \asymp b_n$, we mean

$$-\infty < K_1 < \liminf \frac{a_n}{b_n} \le \limsup \frac{a_n}{b_n} \le K_2 < \infty.$$

Theorem 1.3.1. *(Bickel and Levina [2008a]) Suppose $\{C_{ip}\}$ are i.i.d. $\mathcal{N}(0, \Sigma_p)$ and $\Sigma_\infty \in \mathcal{U}(\epsilon, \alpha, C)$. Then for $k_{n,\alpha} \asymp (n^{-1} \log p)^{-\frac{1}{2(1+\alpha)}}$ and $\log p = o(n)$,*

$$||B_{k_{n,\alpha}}(\tilde{\Sigma}_{p,n}) - \Sigma_p||_2 = O_P(k_{n,\alpha}^{-\alpha}). \tag{1.32}$$

To prove Theorem 1.3.1, we need the following two lemmas. For any square symmetric matrix $M = ((m_{ij}))$, let

$$c_j(M) = \text{ the number of non-zero entries in the } j\text{-th column of } M. \tag{1.33}$$

Define the $|| \cdot ||_\infty$ norm of M as

$$||M||_\infty = \max_{i,j} |m_{ij}|. \tag{1.34}$$

Lemma 1.3.2. *(Golub and van Loan [1996])*

$$||M||_2 \le ||M||_{(1,1)} \le (\sup_j c_j(M))||M||_\infty.$$

The following lemma is easy to prove. Later we will see a more general version in Lemma 2.1.2.

Lemma 1.3.3. *For a chi-square variable χ_n^2 with n degrees of freedom,*

$$P(|\chi_n^2 - n| \ge x) \le e^{-\frac{x^2}{4(2n+x)}}, \ \forall x > 0. \tag{1.35}$$

Now we are prepared to present the proof of Theorem 1.3.1.

Proof of Theorem 1.3.1. Define the sample covariance matrix without centering as

$$\hat{\Sigma}_{p,n} = \frac{1}{n}\sum_{i=1}^{n} C_{ip}C_{ip}^*. \tag{1.36}$$

Let

$$\hat{G}_{p,n} = \left(\frac{1}{n}\sum_{i=1}^{n} C_{ip}\right)\left(\frac{1}{n}\sum_{i=1}^{n} C_{ip}\right)^*. \tag{1.37}$$

Therefore, by (1.3)

$$\tilde{\Sigma}_{p,n} = \hat{\Sigma}_{p,n} - \hat{G}_{p,n}. \tag{1.38}$$

Since $||\cdot||_2$ is a norm, by triangle inequality, we have

$$\begin{aligned} ||B_{k_{n,\alpha}}(\tilde{\Sigma}_{p,n}) - \Sigma_p||_2 &\leq ||B_{k_{n,\alpha}}(\hat{\Sigma}_{p,n}) - B_{k_{n,\alpha}}(\Sigma_p)||_2 \\ &+ ||B_{k_{n,\alpha}}(\Sigma_p) - \Sigma_p||_2 + ||B_{k_{n,\alpha}}(\hat{G}_{p,n})||_2. \end{aligned} \tag{1.39}$$

By Lemma 1.3.1,

$$||B_{k_n}(\Sigma_p) - \Sigma_p||_2 = T(\Sigma_p, k_{n,\alpha}) = O(k_{n,\alpha}^{-\alpha}). \tag{1.40}$$

Recall $c_j(M)$ in (1.33). By Lemma 1.3.2, as

$$\sup_j c_j(B_{k_{n,\alpha}}(\hat{\Sigma}_{p,n}) - B_{k_{n,\alpha}}(\Sigma_p)), \quad \sup_j c_j(B_{k_{n,\alpha}}(\hat{G}_{p,n})) \leq k_{n,\alpha},$$

we have

$$\begin{aligned} ||B_{k_{n,\alpha}}(\hat{\Sigma}_{p,n}) - B_{k_{n,\alpha}}(\Sigma_p)||_2 &\leq k_{n,\alpha}||B_{k_{n,\alpha}}(\hat{\Sigma}_{p,n}) - B_{k_{n,\alpha}}(\Sigma_p)||_\infty \\ &\leq k_{n,\alpha}||\hat{\Sigma}_{p,n} - \Sigma_p||_\infty, \end{aligned} \tag{1.41}$$

$$||B_{k_{n,\alpha}}(\hat{G}_{p,n})||_2 \leq k_{n,\alpha}||\hat{G}_{p,n}||_\infty. \tag{1.42}$$

Now we prove

$$||\hat{\Sigma}_{p,n} - \Sigma_p||_\infty = O_P(\sqrt{n^{-1}\log p}). \tag{1.43}$$

To show (1.43), we first prove that for some $C_1, C_2 > 0$,

$$P(||\hat{\Sigma}_{p,n} - \Sigma_p||_\infty \geq t_n) \leq C_1 p^2 e^{-C_2 n t_n^2}, \text{ if } \{t_n\} \text{ is bounded.} \tag{1.44}$$

To prove (1.44), note that by (1.34)

$$\begin{aligned} P(||\hat{\Sigma}_{p,n} - \Sigma_p||_\infty \geq t_n) &\leq P(\max_{j,k}|\frac{1}{n}\sum_{i=1}^{n} X_{ij}X_{ik} - \sigma_{jk}| \geq t_n) \tag{1.45} \\ &= P(\bigcup_{j,k}\{|\frac{1}{n}\sum_{i=1}^{n} X_{ij}X_{ik} - \sigma_{jk}| \geq t_n\}) \\ &\leq \sum_{j,k} P(|\frac{1}{n}\sum_{i=1}^{n} X_{ij}X_{ik} - \sigma_{jk}| \geq t_n). \tag{1.46} \end{aligned}$$

Let us define, for all $1 \leq i \leq n$ and $1 \leq j, k \leq p$,

$$Z_{ij} = \frac{X_{ij}}{\sqrt{\sigma_{jj}}}, \quad \rho_{jk} = \frac{\sigma_{jk}}{\sqrt{\sigma_{jj}\sigma_{kk}}}, \tag{1.47}$$

$$U^i_{jk} = \left(\frac{Z_{ij} + Z_{ik}}{\sqrt{2 + 2\rho_{jk}}}\right), \quad V^i_{jk} = \left(\frac{Z_{ij} - Z_{ik}}{\sqrt{2 - 2\rho_{jk}}}\right), \tag{1.48}$$

$$U_{jk} = (U^1_{jk}, U^2_{jk}, \ldots, U^n_{jk})^*, \quad V_{jk} = (V^1_{jk}, V^2_{jk}, \ldots, V^n_{jk})^*. \tag{1.49}$$

Now for some $C_1, C_2, C_3 > 0$, we have

$$P\left(\left|\frac{1}{n}\sum_{i=1}^{n} X_{ij}X_{ik} - \sigma_{jk}\right| \geq t_n\right)$$

$$= P\left(\left|\sum_{i=1}^{n} 4Z_{ij}Z_{ik} - 4n\rho_{jk}\right| \geq \frac{4nt_n}{\sqrt{\sigma_{jj}\sigma_{kk}}}\right)$$

$$\leq P\left(\left|\sum_{i=1}^{n}(Z_{ij} + Z_{ik})^2 - n(2 + 2\rho_{jk})\right| \geq \frac{2nt_n}{\sqrt{\sigma_{jj}\sigma_{kk}}}\right)$$

$$+ P\left(\left|\sum_{i=1}^{n}(Z_{ij} - Z_{ik})^2 - n(2 - 2\rho_{jk})\right| \geq \frac{2nt_n}{\sqrt{\sigma_{jj}\sigma_{kk}}}\right)$$

$$= P\left(|U'_{jk}U_{jk} - n| \geq \frac{nt_n}{\sqrt{\sigma_{jj}\sigma_{kk}} - \sigma_{jk}}\right)$$

$$+ P\left(|V'_{jk}V_{jk} - n| \geq \frac{nt_n}{\sqrt{\sigma_{jj}\sigma_{kk}} + \sigma_{jk}}\right) \tag{1.50}$$

$$= P\left(|U'_{jk}U_{jk} - n| \geq C_1 nt_n\right) + P\left(|V'_{jk}V_{jk} - n| \geq C_1 nt_n\right) \tag{1.51}$$

$$= 2P(|\chi^2_n - n| \geq C_1 nt_n) \tag{1.52}$$

$$\leq C_2 e^{-C_3 nt_n^2}, \quad \text{if } \{t_n\} \text{ is bounded.} \tag{1.53}$$

Here is a justification for the last few steps above.
(1.51) holds true as $\Sigma_\infty = ((\sigma_{ij})) \in \mathcal{W}(\epsilon)$ implies

$$\sigma_{jj} \leq \sup_p \lambda_{\max}(\Sigma_p) < \epsilon^{-1}, \quad \forall j, \quad \text{and} \tag{1.54}$$

$$2(\sqrt{\sigma_{jj}\sigma_{kk}} \pm \sigma_{jk}) \leq (\sigma_{jj} + \sigma_{kk} \pm 2\sigma_{jk})$$

$$\leq \sup_p \lambda_{\max}(\Sigma_p) < \epsilon^{-1}. \tag{1.55}$$

(1.52) holds because

$$U_{jk}, V_{jk} \sim \mathcal{N}_n(0, I_n), \; \forall j, k \implies U'_{jk}U_{jk}, V'_{jk}V_{jk} \sim \chi^2_n, \; \forall j, k. \tag{1.56}$$

(1.53) then follows from Lemma 1.3.3.

Hence, by (1.46) and (1.53), (1.44) is proved.

Now by taking $t_n = C_4\sqrt{n^{-1}\log p}$ for sufficiently large C_4, we have

$$C_2 p^2 e^{-C_3 n t_n^2} \to 0. \qquad (1.57)$$

Hence, the proof of (1.43) is complete.

We now show

$$||\hat{G}_{p,n}||_\infty = O_P(n^{-1}\log p). \qquad (1.58)$$

This proof is similar to the proof of (1.43). Note that by Lemma 1.3.3, for some $C_1, C_2 > 0$, we have

$$P(||\hat{G}_{p,n}||_\infty \geq t_n) \leq 2p^2 P(n|n^{-1}\sum_{i=1}^n X_{ik}|^2 > nt_n) \leq C_1 p^2 e^{-C_2 n t_n} \qquad (1.59)$$

Therefore, by taking $t_n = C_3 n^{-1}\log p$ for sufficiently large $C_3 > 0$, we have

$$P(||\hat{G}_{p,n}||_\infty \geq C_3 n^{-1}\log p) \to 0 \text{ as } n \to \infty. \qquad (1.60)$$

This completes the proof of (1.58).

By (1.41), (1.42), (1.43), and (1.58), we have

$$||B_{k_{n,\alpha}}(\hat{\Sigma}_{p,n}) - B_{k_{n,\alpha}}(\Sigma_p)||_2 + ||B_{k_{n,\alpha}}(\hat{G}_{p,n})||_2 = O_P(k_{n,\alpha}\sqrt{\frac{\log p}{n}}). \qquad (1.61)$$

Now to choose $k_{n,\alpha}$ appropriately, by (1.39), (1.55), and (1.61), we have

$$k_{n,\alpha}^{-\alpha} = k_{n,\alpha}\sqrt{n^{-1}\log p} \text{ which implies } k_{n,\alpha} = (n^{-1}\log p)^{-\frac{1}{2(1+\alpha)}}. \qquad (1.62)$$

This completes the proof of Theorem 1.3.1. $\qquad\square$

Remark 1.3.1. *The independence assumption on* $\{C_{ip}\}$*, first used in (1.52), may not hold in dependent cases. What happens in dependent cases? Later in Theorem 2.1.1, we shall see that under suitable "weak" dependence among* $\{C_{ip}\}$*, (1.53) directly follows from (1.51) and hence the same convergence rate as in Theorem 1.3.1 also holds.*

Let us go back to the proof of Theorem 1.3.1. Now suppose that $k_{n,\alpha}$ is some banding sequence, not necessarily the "optimal" one given in (1.62). If we follow the proof, it is easy to see that

$$||B_{k_{n,\alpha}}(\hat{\Sigma}_{p,n}) - \Sigma_p||_2 = O_P(k_{n,\alpha}n^{-1}\log p + k_{n,\alpha}^{-\alpha}). \qquad (1.63)$$

An improvement to this upper bound is given in the following result.

Theorem 1.3.2. *(Cai et al. [2010]) Suppose* $\{C_{ip}\}$ *are i.i.d.* $\mathcal{N}(0,\Sigma_p)$*,* $\Sigma_\infty \in \mathcal{U}(\epsilon,\alpha,C)$*. Then for any* $\tilde{k}_{n,\alpha} = o(n)$*,* $n = o(p)$*,* $\log p = o(n)$*, there exists* $C_1, C_2 > 0$*, so that*

$$\sup_{\Sigma_\infty \in \mathcal{U}(\epsilon,\alpha,C)} E||B_{\tilde{k}_{n,\alpha}}(\hat{\Sigma}_{p,n}) - \Sigma_p||_2^2 \leq C_1 \frac{\tilde{k}_{n,\alpha} + \log p}{n} + C_2 \tilde{k}_{n,\alpha}^{-2\alpha} \qquad (1.64)$$

In particular, when $\tilde{k}_{n,\alpha} = n^{\frac{1}{2\alpha+1}}$, we have

$$\sup_{\Sigma_\infty \in \mathcal{U}(\epsilon,\alpha,C)} E\|B_{\tilde{k}_{n,\alpha}}(\tilde{\Sigma}_{p,n}) - \Sigma_p\|_2^2 < C_1 \frac{\log p}{n} + C_2 n^{-\frac{2\alpha}{2\alpha+1}}. \qquad (1.65)$$

It may be noted that the convergence rate of $B_{\tilde{k}_{n,\alpha}}(\tilde{\Sigma}_{p,n})$ is sharper for $\tilde{k}_{n,\alpha} = n^{\frac{1}{2\alpha+1}}$ compared to $k_{n,\alpha} = (n^{-1}\log p)^{\frac{-1}{2\alpha+1}}$.

Proof of Theorem 1.3.2. Relations (1.39) and (1.40) in the proof of Theorem 1.3.1 continue to hold in this case. Therefore, it is enough to establish

$$E\|B_{k_{n,\alpha}}(\hat{\Sigma}_{p,n} - \Sigma_p)\|_2^2 \;\leq\; C_1 \frac{k_{n,\alpha} + \log p}{n} \quad \text{and} \qquad (1.66)$$

$$E\|B_{k_{n,\alpha}}(\hat{G}_{p,n})\|_2^2 \;\leq\; C_1 \Big(\frac{k_{n,\alpha} + \log p}{n}\Big)^2. \qquad (1.67)$$

Here we shall only show (1.66). Similar argument works for (1.67).

Proof of (1.66). Let $M = B_{k_{n,\alpha}}(\hat{\Sigma}_{p,n} - \Sigma_p)$ and $k = k_{n,\alpha}$. Write t for t_n for convenience. As $\Sigma_\infty \in \mathcal{W}(\epsilon)$, it is easy to see

$$E\|M\|_2^2 \leq Cp^2 \text{ for some } C > 0. \qquad (1.68)$$

Therefore, for any $t^{-1} = O(p)$ and for some $C_1, C_2 > 0$, we have

$$\begin{aligned}
t^{-1}E\|B_k(M)\|_2^2 &= t^{-1}E(\|B_k(M)\|_2^2 I(\|B_k(M)\|_2^2 \leq t)) \\
&\quad + t^{-1}E(\|B_k(M)\|_2^2 I(\|B_k(M)\|_2^2 > t)) \\
&\leq C_1 P(\|B_k(M)\|_2^2 \leq t) + C_2 p^3 P(\|B_k(M)\|_2^2 > t). \quad (1.69)
\end{aligned}$$

Hence, the proof of (1.66) will be complete if

$$p^3 P(\|B_k(M)\|_2^2 > t) \to 0 \qquad (1.70)$$

for $t = C_3 \frac{k+\log p}{n}$ and sufficiently large $C_3 > 0$.
To prove (1.70), note that $B_k(M)$ can be written as

$$\sum_{l=1}^{p-k} S_{l,k}(M) \quad \text{where } S_{l,k}(M) = ((m_{ij}I(l \leq i, j \leq l+k))_{p\times p}. \qquad (1.71)$$

Therefore, for some $C > 0$, it is easy to see

$$\|B_k(M)\|_2^2 \leq C \max_{1\leq l\leq p-k} \|S_{l,k}(M)\|_2^2. \qquad (1.72)$$

Recall the notation in (1.4). Let $S_l(M) = ((S_{l,k}(M)(l+i-1, l+j-1)))_{k\times k}$. Then

$$\|B_k(M)\|_2^2 \leq C \max_{1\leq l\leq p-k} \|S_l(M)\|_2^2. \qquad (1.73)$$

Therefore, to prove (1.70), it is enough to show that for $t = C_3 \frac{k + \log p}{n}$ and sufficiently large $C_3 > 0$,

$$p^3 P(\max_{1 \leq l \leq p-k} ||S_l(M)||_2^2 > t) \to 0. \tag{1.74}$$

Now we state a lemma which is necessary to prove (1.74). Let $(\mathcal{M}, d)$ be a metric space and $U \subset V \subset \mathcal{M}$. Then U is called an ϵ-net of V if for all $v \in V$ there is $u \in U$ such that $d(u, v) < \epsilon$.

Lemma 1.3.4. *Let $\mathcal{S}^{(k-1)}$ be a 0.5-net of the unit sphere in $\mathbb{R}^k$. Then*

$$|\mathcal{S}^{(k-1)}| \leq 5^k.$$

Now to continue with the proof of (1.74). By Lemma 1.3.4, we have

$$
\begin{aligned}
p^3 P(\max_{1 \leq l \leq p-k} ||S_l(M)||_2^2 > t) &\leq C_1 p^4 P(||S_l(M)||_2^2 > t) \\
&\leq C_1 p^4 5^k P(|v^* S_l(M) v| > \sqrt{t}) \\
&\qquad \qquad \text{for some } v \in \mathcal{S}^{(k-1)} \\
&\leq C_2 p^4 5^k P(|\chi_n^2 - n| > n\sqrt{t}) \\
&\leq C_4 p^4 5^k e^{-C_5 nt} \to 0
\end{aligned}
$$

for $t = C_3 \frac{k + \log p}{n}$ and sufficiently large $C_3 > 0$. This completes the proof of (1.74) and consequently that of Theorem 1.3.2. $\qquad \square$

Tapered estimator

Recall tapering of a matrix from Section 1.2.2. For any matrix M and a sequence of positive numbers $\{\tau_{n,\alpha}\}$,

$$R_{g,\tau_{n,\alpha}}(M) = ((m_{ij} g(|i - j|/\tau_{n,\alpha}))) \tag{1.75}$$

where the tapering function g satisfies the following assumptions.

(T1) The function $g : \mathbb{R}^+ \cup \{0\} \to \mathbb{R}^+ \cup \{0\}$ is positive definite, continuous, non-increasing such that

$$g(0) = 1 \text{ and } \lim_{x \to \infty} g(x) = 0.$$

Let

$$\Delta_{g,\tau_{n,\alpha}} = \sum_{j=0}^{n-1} g\left(\frac{j}{\tau_{n,\alpha}}\right). \tag{1.76}$$

We have the following Theorem for tapered estimators. The authors did not provide a detailed proof of the theorem. A proof can be constructed along the lines of the proof of Theorem 1.3.1. This is left to the reader.

Theorem 1.3.3. *(Bickel and Levina [2008a]) Under the assumptions of Theorem 1.3.1 and (T1), if $\Delta_{g,\tau_{n,\alpha}} \asymp (n^{-1}\log p)^{-\frac{1}{2(1+\alpha)}}$, then*

$$||R_{g,\tau_{n,\alpha}}(\tilde{\Sigma}_{p,n}) - \Sigma_p||_2 = O_P((n^{-1}\log p)^{\frac{\alpha}{2(1+\alpha)}}). \qquad (1.77)$$

Consider the particular tapered estimator $R_{g_{\tilde{k}_{n,\alpha}},1}(\tilde{\Sigma}_{p,n})$, where

$$g_k(x) = \begin{cases} 1, & \text{when } 0 \leq x \leq k/2 \\ 2 - \frac{2x}{k}, & \text{when } k/2 < x \leq k \\ 0, & \text{otherwise.} \end{cases} \qquad (1.78)$$

We then have the following result. Clearly, the convergence rate in this result is sharper than that in Theorem 1.3.3. Its proof is similar to the proof of Theorem 1.3.2. We leave it as an exercise.

Theorem 1.3.4. *(Cai et al. [2010]) Suppose $\{C_{ip}\}$ are i.i.d. $\mathcal{N}(0, \Sigma_p)$ and $\Sigma_\infty \in \mathcal{U}(\epsilon, \alpha, C)$. Then for $\tilde{k}_{n,\alpha} = o(n)$, $n = o(p)$, $\log p = o(n)$ there exist $C_1, C_2 > 0$, so that*

$$\sup_{\Sigma_\infty \in \mathcal{U}(\epsilon,\alpha,C)} E||R_{g_{\tilde{k}_{n,\alpha}},1}(\tilde{\Sigma}_{p,n}) - \Sigma_p||_2^2 \leq C_1 \frac{\tilde{k}_{n,\alpha} + \log p}{n} + C_2 \tilde{k}_{n,\alpha}^{-2\alpha}. \quad (1.79)$$

In particular, when $\tilde{k}_{n,\alpha} = n^{\frac{1}{2\alpha+1}}$, we have

$$\sup_{\Sigma_\infty \in \mathcal{U}(\epsilon,\alpha,C)} E||R_{g_{\tilde{k}_{n,\alpha}},1}(\tilde{\Sigma}_{p,n}) - \Sigma_p||_2^2 \leq C_1 \frac{\log p}{n} + C_2 n^{-\frac{2\alpha}{2\alpha+1}}. \qquad (1.80)$$

1.3.3   Minimaxity

In the previous sections we have seen convergence rates for different banded and tapered estimators. Now the question is whether these estimators are optimal in the following sense.

The maximum risk of an estimator T for estimating $\Sigma_p \in \mathcal{U}(\epsilon, \alpha, C)$ is given by

$$\sup_{\Sigma_\infty \in \mathcal{U}(\epsilon,\alpha,C)} E||T - \Sigma_p||_2^2.$$

The *minimax risk* is defined as

$$\inf_T \sup_{\Sigma_\infty \in \mathcal{U}(\epsilon,\alpha,C)} E||T - \Sigma_p||_2^2 \qquad (1.81)$$

where the infimum is taken over all $p \times p$ symmetric matrix estimators. An estimator T will be called (asymptotically) *optimal/minimax* for $\Sigma_p \in \mathcal{U}(\epsilon, \alpha, C)$ if it attains the infimum risk in (1.81) asymptotically.

The objective of this section is twofold: (1) obtain an asymptotic lower bound for (1.81) and then (2) search for an estimator of Σ_p which attains this lower bound.

The following theorem provides a lower bound for the risk (1.81).

Theorem 1.3.5. *(Cai et al. [2010]) Suppose $\{C_{ip}\}$ are i.i.d. $\mathcal{N}(0, \Sigma_p)$. Then for $p \leq e^{\gamma n}$ and $\gamma > 0$, we have $C_1 > 0$ such that*

$$\inf_T \sup_{\Sigma_\infty \in \mathcal{U}(\epsilon, \alpha, C)} E\|T - \Sigma_p\|_2^2 \geq C_1 \frac{\log p}{n} + C_1 n^{-\frac{2\alpha}{2\alpha+1}}. \tag{1.82}$$

Therefore, $B_{\tilde{k}_{n,\alpha}}(\tilde{\Sigma}_{p,n})$ and $R_{g_{\tilde{k}_{n,\alpha}},1}(\tilde{\Sigma}_{p,n})$ with $\tilde{k}_{n,\alpha} = n^{\frac{1}{2\alpha+1}}$ and the tapered function g_k in (1.78), are optimal for $\Sigma_p \in \mathcal{U}(\epsilon, \alpha, C)$.

Proof of Theorem 1.3.5. The main idea is to carefully construct two finite collections of multivariate normal distributions $\mathcal{F}_1$ and $\mathcal{F}_2$ so that for some generic constant $C > 0$,

$$\inf_T \sup_{\Sigma_\infty \in \mathcal{F}_1} E\|T - \Sigma_p\|_2^2 \geq Cn^{-\frac{2\alpha}{2\alpha+1}} \text{ and}$$

$$\inf_T \sup_{\Sigma_\infty \in \mathcal{F}_2} E\|T - \Sigma_p\|_2^2 \geq C\frac{\log p}{n}.$$

Construction of $\mathcal{F}_1$. For given integers k and m with $1 \leq m \leq k \leq p/2$, define the $p \times p$ matrix $B(m, k) = ((b_{ij}))_{p \times p}$ where

$$b_{ij} = I(i = m, m+1 \leq j \leq 2k \text{ or } j = m, m+1 \leq i \leq 2k).$$

Set $k = n^{\frac{1}{1+2\alpha}}$ and $a = k^{-(\alpha+1)}$. We now define the collection of multivariate normal distributions whose covariance matrices belong to the following finite collection:

$$\mathcal{F}_1 = \left\{ \Sigma_\infty(\theta) : \Sigma_p(\theta) = I_p + \tau a \sum_{m=1}^k \theta_m B(m, k), \ \theta = \{\theta_m\} \in \{0, 1\}^k \right\} \tag{1.83}$$

where I_p is as in (1.9) and $0 < \tau < 2^{-\alpha-1}C$. Without loss of generality we assume that $\epsilon < 1 < \alpha$. Otherwise we replace I_p in $\mathcal{F}_1$ by βI_p where $0 < \beta < \min(\epsilon^{-1}, \alpha)$. For $0 < \tau < 2^{-\alpha-1}C$ and large n and p, it is easy to see that

$$\mathcal{F}_1 \subset \mathcal{U}(\epsilon, \alpha, C). \tag{1.84}$$

We first show that for some $C_1 > 0$

$$\inf_T \sup_{\Sigma_\infty \in \mathcal{F}_1} E\|T - \Sigma_p\|_2^2 \geq C_1 n^{-\frac{2\alpha}{2\alpha+1}}. \tag{1.85}$$

We need the following lemma which gives a lower bound for the maximum risk over the parameter set $\Theta = \{0, 1\}^k$ to the problem of estimating an arbitrary quantity $\Psi(\theta)$, belonging to a metric space with metric d. We need the following two notions.

Let H be the *Hamming distance* between $\theta, \theta' \in \{0,1\}^k$, which counts the number of positions at which they differ:

$$H(\theta, \theta') = \sum_{i=1}^{k} |\theta_i - \theta_i'|.$$

Suppose that P and Q are two probability measures with a dominating measure μ. Let p and q be the respective densities. The *total variation affinity* between P and Q is defined as

$$||P \wedge Q|| = \int \min(p,q) d\mu = 1 - 0.5 \int |p - q| d\mu.$$

Note that

$$\left(\int |p - q| d\mu \right)^2 = \left(\int \frac{|p-q|}{q} q \, d\mu \right)^2 \leq \left(\int \frac{|p-q|^2}{q^2} q \, d\mu \right) = \int \frac{p^2}{q} d\mu - 1.$$

Therefore,

$$||P \wedge Q|| \geq 1 - 0.5 \left(\int \frac{p^2}{q} d\mu - 1 \right)^{1/2}. \tag{1.86}$$

We shall need the above inequality later.

Lemma 1.3.5. *(Assouad [1983]) Let $\Theta = \{0,1\}^k$ and let T be an estimator based on observations from a distribution in the collection $\{P_\theta : \theta \in \Theta\}$. Then for all $s > 0$*

$$\max_{\theta \in \Theta} 2^s E_\theta d^s(T, \Psi(\theta)) \geq \min_{H(\theta,\theta') \geq 1} \frac{d^s(\Psi(\theta), \Psi(\Theta'))}{H(\theta, \theta')}$$

$$\times \frac{k}{2} \min_{H(\theta,\theta')=1} ||P_\theta \wedge P_{\theta'}||. \tag{1.87}$$

Proof of 1.85. Let $X_1, X_2, \ldots, X_n$ be i.i.d. $\mathcal{N}(0, \Sigma_p(\theta))$ with $\Sigma_\infty(\theta) \in \mathcal{F}_1$. Denote the joint distribution by P_θ. Applying Lemma 1.3.5 to the parameter space $\mathcal{F}_1$, we have

$$\inf_T \max_{\theta \in \{0,1\}^k} 4E ||T - \Sigma_p(\theta)||_2^2 \geq \min_{H(\theta,\theta') \geq 1} \frac{||\Sigma_p(\theta) - \Sigma_p(\theta')||_2^2}{H(\theta, \theta')}$$

$$\times \frac{k}{2} \min_{H(\theta,\theta')=1} ||P_\theta \wedge P_{\theta'}||. \tag{1.88}$$

Now it is easy to see that for some $C > 0$

$$\min_{H(\theta,\theta') \geq 1} \frac{||\Sigma_p(\theta) - \Sigma_p(\theta')||_2^2}{H(\theta, \theta')} \geq Cka^2 \text{ and } \min_{H(\theta,\theta')=1} ||P_\theta \wedge P_{\theta'}|| \geq C. \tag{1.89}$$

Therefore,

$$\inf_{T} \max_{\theta \in \{0,1\}^k} 4E\|T - \Sigma_p(\theta)\|_2^2 \geq Ck^2a^2 = Cn^{-\frac{2\alpha}{2\alpha+1}}. \tag{1.90}$$

This completes the proof of (1.85).

Construction of $\mathcal{F}_2$. Next, we define another collection of multivariate normal distributions whose covariance matrices are diagonal and belong to the following finite set.

$$\mathcal{F}_2 = \left\{\Sigma_m : \Sigma_m = I_p + \left(\sqrt{\frac{\tau}{n}\log p_1} I(i = j = m)\right)_{p \times p}, \ 0 \leq m \leq p_1\right\}, \tag{1.91}$$

where $p_1 = \max(p, e^{n/2})$ and $0 < \tau < \min\{(\epsilon^{-1} - 1)^2, (\alpha - 1)^2, 1\}$. Note that $m = 0$ in $\mathcal{F}_2$ indicates $\Sigma_0 = I_p$. Clearly,

$$\mathcal{F}_2 \subset \mathcal{U}(\epsilon, \alpha, C). \tag{1.92}$$

Now we shall show that for some $C_1 > 0$

$$\inf_{T} \sup_{\Sigma_\infty \in \mathcal{F}_2} E\|T - \Sigma_p\|_2^2 \geq C_1 \frac{\log p}{n}. \tag{1.93}$$

For this, we use Le Cam's lemma. Let X be an observation from a distribution in the collection $\{P_\theta : \theta \in \Theta\}$ where $\Theta = \{\theta_0, \theta_1, \theta_2, \ldots, \theta_{p_1}\}$. Using a two-point testing of hypothesis argument, the lemma provides a lower bound for the maximum estimation risk over the parameter set Θ. More specifically, let L be the loss function. Define

$$r(\theta_0, \theta_m) = \inf_{t}[L(t, \theta_0) + L(t, \theta_m)],$$

$$r_{\min} = \inf_{1 \leq m \leq p_1} r(\theta_0, \theta_m).$$

Denote $\bar{P} = \frac{1}{p_1} \sum_{m=1}^{p_1} P_{\theta_m}$.

Lemma 1.3.6. *(Le Cam's lemma, see Yu [1997]). Let T be an estimator of θ based on observations from a distribution in the collection $\{P_\theta : \theta \in \Theta = \{\theta_0, \theta_1, \theta_2, \ldots, \theta_{p_1}\}\}$, then*

$$\sup_{\theta} EL(T, \theta) \geq 0.5 \, r_{\min}\|P_{\theta_0} \wedge \bar{P}\|. \tag{1.94}$$

Proof of (1.93). For this we use Le Cam's lemma. Indeed for some $C > 0$,

$$r_{\min} > C_1 \frac{\log p}{n} \text{ and } \|P_{\theta_0} \wedge \bar{P}\| \geq C_1. \tag{1.95}$$

The first inequality in (1.95) holds as for some $C > 0$,

$$r(\Sigma_0, \Sigma_m) = \inf_{T}(E\|T - \Sigma_0\|_2^2 + E\|T - \Sigma_m\|_2^2)$$

$$\geq C\|\Sigma_0 - \Sigma_m\|_2^2$$

$$\geq C\frac{\log p_1}{n} \geq C\frac{\log p}{n}.$$

Therefore,

$$r_{\min} = \inf_{1 \leq m \leq p_1} r(\Sigma_0, \Sigma_m) \geq C \frac{\log p}{n}$$

and the first inequality of (1.95) holds.

To verify the second inequality in (1.95) note that P_{θ_m} corresponds to $\mathcal{N}_p(0, \Sigma_m)$. Let f_m be its density. The density of $\bar{P}$ is $\bar{f} = p_1^{-1} \sum_{m=1}^{p_1} f_m$. After some simplification, one can easily get

$$\int \frac{\bar{f}^2}{f_\theta} dx - 1 \to 0. \tag{1.96}$$

Therefore, by (1.86), $\|P_{\theta_0} \wedge \bar{P}\| \geq C$ for some $C > 0$ and this completes the proof of the second inequality in (1.95). Thus, (1.93) is established.

Now (1.85) and (1.93) together complete the proof of Theorem 1.3.1.   □

1.4   Toeplitz Σ_p

The covariance matrix Σ_p is said to be a Toeplitz matrix if it is of the form $((\sigma_{|i-j|}))$. Toeplitz covariance matrices arise naturally in the analysis of stationary stochastic processes. If $\{X_i\}$ is a stationary process then the covariance matrix of $X_1, \ldots, X_k$ (the *autocovariance matrix*) is a Toeplitz matrix. Toeplitz covariance matrices have a wide range of applications in many fields, including engineering, economics and biology. See, for instance, Franaszczuk et al. [1985], Fuhrmann [1991] and, Quah [2000] for specific applications.

In this section we focus on estimating large Toeplitz covariance matrices. For a univariate stationary process, it is known that the sample autocovariance matrix is not a consistent estimator. However, appropriate banding and tapering leads to consistency. See Wu and Pourahmadi [2003] and McMurry and Politis [2010] and also Basak et al. [2014] who prove consistency in a different sense.

However, in the high-dimensional setting, the standard estimators do not provide satisfactory performance and regularization is needed. The methods that we have discussed earlier now lend themselves to appropriate refinements that take advantage of the Toeplitz structure. We first describe appropriate parameter spaces that have been used in the literature. Then we present the results on banded and tapered estimators for different parameter spaces and finally we provide the minimax result.

1.4.1   Parameter space

Consider the natural parameter space defined in terms of the rate of decay of the covariance sequence $\{\sigma_m\}$.

$$\mathcal{G}_\beta(M) = \{\Sigma_\infty = ((\sigma_{|i-j|})) : |\sigma_m| \leq M(m+1)^{-\beta-1}, \Sigma_\infty > 0\} \tag{1.97}$$

where $0 < \beta, M < \infty$ and $\Sigma_\infty > 0$ denotes positive definiteness of Σ_p for all $p \geq 1$. It is easy to see that $\mathcal{G}_\beta(M) \subset \mathcal{X}(\beta, C)$ for some $C > 0$ depending on β and M. But $\Sigma_\infty \in \mathcal{G}_\beta(M)$ may not satisfy the condition $\inf_p \lambda_{\min}(\Sigma_p) > \epsilon > 0$ for some $\epsilon > 0$. Therefore, $\mathcal{G}_\beta(M)$ is not a subset of $\mathcal{U}(\epsilon, \alpha, C)$ and hence the results in Section 1.3 do not automatically hold for the class $\mathcal{G}_\beta(M)$.

The second parameter space that we shall consider is from Cai et al. [2013] and is defined in terms of the smoothness of the spectral density f. When $\sum_{i=1}^{\infty} |\sigma_i| < \infty$, its *spectral density* is defined as

$$f_{\Sigma_\infty}(x) = \frac{1}{2\pi}\left[\sigma_0 + 2\sum_{m=1}^{\infty} \sigma_i \cos(mx)\right], \quad \text{for } x \in [-\pi, \pi]. \tag{1.98}$$

For any positive number β, write $\beta = \alpha + \gamma$ where γ is the largest integer strictly less than β and $0 < \alpha \leq 1$. Let

$$\mathcal{F}_\beta(M_0, M) = \{\Sigma_\infty : \|f_{\Sigma_\infty}\|_\infty \leq M_0, \ \|f_{\Sigma_\infty}^{(\gamma)}(\cdot + h) - f_{\Sigma_\infty}^{(\cdot)}(\cdot)\|_\infty \leq Mh^\alpha\}. \tag{1.99}$$

In other words, $\mathcal{F}_\beta(M_0, M)$ contains the Toeplitz covariance matrices whose corresponding spectral density functions are *Hölder smooth* of order α.

For any $M > 0$ and non-integer $\beta > 0$, it can be shown that there exists some constants M_0 and M_1 depending on M such that

$$\mathcal{G}_\beta(M) \subset \mathcal{F}_\beta(M_0, M_1). \tag{1.100}$$

However, in general this is not true for integer β. See, for example Zygmund [2002]. Conversely, it is easy to see that for any $\Sigma_\infty \in \mathcal{F}_\beta(M_0, M_1)$, we have $|\sigma_m| \leq Mm^{-\beta}$ where M is some constant depending only on M_0 and M_1. Therefore

$$\mathcal{F}_\beta(M_0, M_1) \subset \mathcal{G}_{\beta-1}(M)$$

for some M depending on M_0 and M_1.

1.4.2 Estimation in $\mathcal{G}_\beta(M)$ or $\mathcal{F}_\beta(M_0, M)$

An immediate improvement of $\tilde{\Sigma}_{p,n}$ that takes advantage of the Toeplitz structure is to average the entries in the diagonals. Let

$$\bar{\sigma}_m = \frac{1}{p - m} \sum_{s-t=m} \tilde{\sigma}_{st} \tag{1.101}$$

and define the Toeplitz matrix $\bar{\Sigma}_{p,n} = ((\bar{\sigma}_{|s-t|}))_{1 \leq s,t \leq p}$. Note that $\bar{\Sigma}_{p,n}$ is unbiased for Σ_p.

Tapered estimator.

Consider the estimator $R_{g_{k_n,\beta},1}(\bar{\Sigma}_{p,n})$ with g_k as described in (1.78). Then we have the following theorem.

Theorem 1.4.1. *(Cai et al. [2013]) Suppose $\{C_{ip}\}$ are i.i.d. $\mathcal{N}(0, \Sigma_p)$ and $k_{n,\beta} \leq p/2$. Then for some $C > 0$,*

$$\sup_{\mathcal{F}_\beta(M_0, M) \cup \mathcal{G}_\beta(M)} E\|R_{g_{k_{n,\beta}},1}(\bar{\Sigma}_{p,n}) - \Sigma_p\|_2^2 \leq C \frac{k_{n,\beta} \log(np)}{np} + C k_{n,\beta}^{-2\beta}. \quad (1.102)$$

By setting an optimal choice $k_{n,\beta} = \left(\frac{np}{\log(np)}\right)^{\frac{1}{2\beta+1}} < p/2$, we have

$$\sup_{\Sigma_\infty \in \mathcal{F}_\beta(M_0, M) \cup \mathcal{G}_\beta(M)} E\|R_{g_{k_{n,\beta}},1}(\bar{\Sigma}_{p,n}) - \Sigma_p\|_2^2 \leq C \left(\frac{\log(np)}{np}\right)^{\frac{2\beta}{2\beta+1}}. \quad (1.103)$$

Outline of proof. We shall prove the result only for the class $\mathcal{G}_\beta(M)$. The proof for the class $\mathcal{F}_\beta(M_0, M)$ is significantly harder.

As in the proof of Theorems 1.3.1 and 1.3.2, we decompose the risk into bias and variance. Therefore,

$$\begin{aligned}
E\|R_{g_{k_{n,\beta}},1}(\bar{\Sigma}_{p,n}) - \Sigma_p\|_2^2 &\leq 2E\|R_{g_{k_{n,\beta}},1}(\bar{\Sigma}_{p,n}) - E(R_{g_{k_{n,\beta}},1}(\bar{\Sigma}_{p,n}))\|_2^2 \\
&\quad + 2\|E(R_{g_{k_{n,\beta}},1}(\bar{\Sigma}_{p,n})) - \Sigma_p\|_2^2. \quad (1.104)
\end{aligned}$$

Using Lemma 1.3.3 and similar arguments as in the proof of Theorem 1.3.2, it is easy to show that for some $C_1 > 0$

$$\sup_{\mathcal{G}_\beta(M)} E\|R_{g_{k_{n,\beta}},1}(\bar{\Sigma}_{p,n}) - E(R_{g_{k_{n,\beta}},1}(\bar{\Sigma}_{p,n}))\|_2^2 \leq C_1 \frac{k_{n,\beta} \log(np)}{np}. \quad (1.105)$$

Moreover, as $\Sigma_\infty \in \mathcal{G}_\beta(M)$ has polynomially decaying corner, it is obvious that for some $C_2 > 0$,

$$\sup_{\Sigma_\infty \in \mathcal{G}_\beta(M)} \|E(R_{g_{k_{n,\beta}},1}(\bar{\Sigma}_{p,n})) - \Sigma_p\|_2^2 \leq C_2 k_{n,\beta}^{-2\beta}. \quad (1.106)$$

Therefore, combining (1.105) and (1.106), for some $C > 0$,

$$\sup_{\Sigma_\infty \in \mathcal{G}_\beta(M)} E\|R_{g_{k_{n,\beta}},1}(\bar{\Sigma}_{p,n}) - \Sigma_p\|_2^2 \leq C \frac{k_{n,\beta} \log(np)}{np} + C k_{n,\beta}^{-2\beta}. \quad (1.107)$$

When we replace $\mathcal{G}_\beta(M)$ by $\mathcal{F}_\beta(M_0, M)$, relations similar to (1.105), (1.106) and (1.107) continue to hold. But the proofs are much harder and require some specific properties of tapered estimator and we omit them. The details are available in Section 6 of Cai et al. [2013] and use some results from harmonic analysis. A good reference for the latter is Zygmund [2002].

This completes the proof of Theorem 1.4.1. $\square$

Banded estimator.

We now turn to the performance of the banded estimator. First consider $\mathcal{F}_\beta(M_0, M)$. It is interesting to note that the best banding estimator is inferior to the optimal tapering estimator for estimating the Toeplitz covariance matrices in this class. The following theorem supports this.

Theorem 1.4.2. *(Cai et al. [2013]) Suppose $\{C_{ip}\}$ are i.i.d. $\mathcal{N}(0, \Sigma_p)$ and for some $\kappa < 2/5$, $(np \log(np))^{\frac{1}{2\beta+1}} = O(p^\kappa)$. Then*

$$\left(\frac{np}{\log(np)}\right)^{\frac{2\beta}{2\beta+1}} \inf_k \sup_{\Sigma_\infty \in \mathcal{F}_\beta(M_0, M)} E||B_k(\bar{\Sigma}_{p,n}) - \Sigma_p||_2^2 \to \infty. \qquad (1.108)$$

Proof of Theorem 1.4.2. It suffices to show that for each fixed pair (k, p) there exists some $\Sigma_\infty \in \mathcal{F}_\beta(M_0, M)$ such that

$$E||B_k(\bar{\Sigma}_{p,n}) - \Sigma_p||_2^2 \geq C\left(\frac{\log(np)}{np}\right)^{\frac{2\beta}{2\beta+1}} (\log(np))^{\frac{2}{2\beta+1}-\epsilon} \qquad (1.109)$$

for some $C > 0$ and $\epsilon < (2\beta+1)^{-1}$.

We shall again present an outline of the proof of (1.109). We state the following necessary lemmas. Proof of the first two lemmas are available in Section 7 of Cai et al. [2013]. Proof of the third lemma is similar to the proof in the tapered case.

Lemma 1.4.1. *The bias of the banded estimator $B_k(\bar{\Sigma}_{p,n})$ of the Toeplitz covariance matrix $\Sigma_\infty \in \mathcal{F}_\beta(M_0, M)$ with $k \leq p/2$ satisfies*

$$\sup_{\Sigma_\infty \in \mathcal{F}_\beta(M_0, M)} ||EB_k(\bar{\Sigma}_{p,n}) - \Sigma_p||_2^2 \geq Ck^{-2\beta}(\log k)^2 \text{ for some } C > 0. \quad (1.110)$$

Lemma 1.4.2. *Let Σ_∞ be the identity matrix. The banded estimator $B_k(\bar{\Sigma}_{p,n})$ with $k = O(p^\kappa)$ for some $\kappa < 2/5$ and $k \to \infty$ as $p \to \infty$ satisfies*

$$E||B_k(\bar{\Sigma}_{p,n}) - \Sigma_p||_2^2 \geq C\frac{p^\kappa \log p}{np} \text{ for some } C > 0. \qquad (1.111)$$

Lemma 1.4.3. *For some $C > 0$, $B_k(\bar{\Sigma}_{p,n})$ satisfies*

$$\sup_{\Sigma_\infty \in \mathcal{F}_\beta(M_0, M)} E||B_k(\bar{\Sigma}_{p,n}) - EB_k(\bar{\Sigma}_{p,n})||_2^2 \leq C\frac{k \log(np)}{np}. \qquad (1.112)$$

Now we continue the proof of (1.109) using above three lemmas.

First, we consider the banded estimator with $k < (np)^{\frac{1}{2\beta+1}} (\log(np))^{\frac{1}{2\beta+1}-\epsilon}$. It follows from Lemmas 1.4.1 and 1.4.3 that for some $C_1, C_2, C_3, C_4 > 0$,

$$
\begin{aligned}
E||B_k(\bar{\Sigma}_{p,n}) - \Sigma_p||_2^2 &\geq C_1||EB_k(\bar{\Sigma}_{p,n}) - \Sigma_p||_2^2 \\
&\qquad - E||B_k(\bar{\Sigma}_{p,n}) - EB_k(\bar{\Sigma}_{p,n})||_2^2] \\
&\geq C_2 k^{-2\beta}(\log k)^2 - C_3\frac{k \log(np)}{np} \\
&\geq C_4\left(\frac{\log(np)}{np}\right)^{\frac{2\beta}{2\beta+1}} (\log(np))^{\frac{2\beta}{2\beta+1}-\epsilon}. \quad (1.113)
\end{aligned}
$$

When $k \geq (np)^{\frac{1}{2\beta+1}}(\log np)^{\frac{1}{2\beta+1}-\epsilon} = O(p^\kappa)$ and Σ_∞ is the identity matrix, Lemma 1.4.2 implies that for some $C_1, C_2 > 0$,

$$
\begin{aligned}
E\|B_k(\bar{\Sigma}_{p,n}) - \Sigma_p\|_2^2 &\geq C_1 \frac{k \log k}{np} \\
&\geq C_2 \Big(\frac{\log(np)}{np}\Big)^{\frac{2\beta}{2\beta+1}} (\log(np))^{\frac{2\beta}{2\beta+1}-\epsilon}. \quad (1.114)
\end{aligned}
$$

The proof of (1.109) is now complete by combining (1.113) and (1.114). Therefore, Theorem 1.4.2 is proved. $\qquad\square$

Let us now consider $\mathcal{G}_\beta(M)$. In this parameter space the banded estimator achieves the same rate of convergence as the tapered estimator.

Theorem 1.4.3. *(Cai et al. [2013]) Suppose $\{C_{ip}\}$ are i.i.d. $\mathcal{N}(0, \Sigma_p)$ and $k_{n,\beta} \leq p/2$. Then*

$$
\sup_{\Sigma_\infty \in \mathcal{G}_\beta(M)} E\|B_{k_{n,\beta}}(\bar{\Sigma}_{p,n}) - \Sigma_p\|_2^2 \leq C \frac{k_{n,\beta} \log(np)}{np} + C k_{n,\beta}^{-2\beta} \quad (1.115)
$$

for some constant $C > 0$. By setting an optimal choice
$k_{n,\beta} = \Big(\frac{np}{\log(np)}\Big)^{\frac{1}{2\beta+1}} < p/2$, *we have*

$$
\sup_{\Sigma_\infty \in \mathcal{G}_\beta(M)} E\|B_{k_{n,\beta}}(\bar{\Sigma}_{p,n}) - \Sigma_p\|^2 \leq C \Big(\frac{\log(np)}{np}\Big)^{\frac{2\beta}{2\beta+1}}. \quad (1.116)
$$

Proof of the above theorem is similar to the proof of Theorem 1.4.1 and therefore we omit it.

1.4.3   Minimaxity

The proof of the following minimax theorem is similar to the proof Theorem 1.3.5. We leave it as an exercise.

Theorem 1.4.4. *(Cai et al. [2013]) Suppose $\{C_{ip}\}$ are i.i.d. $\mathcal{N}(0, \Sigma_p)$. Then for some $C > 0$,*

$$
\inf_{T} \sup_{\Sigma_\infty \in \mathcal{G}_\beta(M) \cup \mathcal{F}_\beta(M_0, M)} E\|T - \Sigma_p\|^2 \geq C \Big(\frac{\log(np)}{np}\Big)^{\frac{2\beta}{2\beta+1}}. \quad (1.117)
$$

Therefore with $k_{n,\beta} = \Big(\frac{np}{\log(np)}\Big)^{\frac{1}{2\beta+1}}$, both $B_{k_{n,\beta}}(\bar{\Sigma}_{p,n})$ and $R_{g_{k_{n,\beta}},1}(\bar{\Sigma}_{p,n})$ are optimal estimators in $\mathcal{G}_\beta(M)$. Moreover, $R_{g_{k_{n,\beta}},1}(\bar{\Sigma}_{p,n})$ with $k_{n,\beta} = \Big(\frac{np}{\log(np)}\Big)^{\frac{1}{2\beta+1}}$ is optimal in $\mathcal{F}_\beta(M_0, M)$. $B_k(\bar{\Sigma}_{p,n})$ is not optimal in $\mathcal{F}_\beta(M_0, M)$ for any $k \geq 1$.

1.5   Sparse Σ_p

For estimating bandable and Toeplitz covariance matrices, we have used an ordering of the variables. However, in many applications such as genomics, there may be no such order of variables. On the other hand, the covariance between most of the pairs of variables are often assumed to be insignificant. This leads us to the class of *sparse covariance matrices*. In this section we deal with such types of covariance matrices.

1.5.1   Parameter space

The idea of no natural ordering between indices of the variables and most of the variables being nearly uncorrelated can be captured as follows.

The *strong l_q ball* of radius c in $\mathbb{R}^p$ is defined by

$$\mathcal{B}_q^p(c) = \{\xi \in \mathbb{R}^p : \sum_{k=1}^p |\xi_k|^q \leq c\} \tag{1.118}$$

where ξ_k denotes the k-th element of the vector ξ.

Consider the following class of sparse covariance matrices. This class is invariant under permutations. See Bickel and Levina [2008b]. For $0 \leq q < 1$,

$$\mathcal{U}_\tau(q, C_0(p), M) = \{\Sigma : \sup_i \sigma_{ii} \leq M, \sum_{\substack{1 \leq j \leq p \\ j \neq i}} |\sigma_{ij}|^q \leq C_0(p) \ \forall i, p \geq 1\}. \tag{1.119}$$

The columns/rows of $\Sigma_p \in \mathcal{U}_\tau(q, C_0(p), M)$ are assumed to belong to a strong l_q ball in $\mathbb{R}^p$ for all $p \geq 1$. When $q = 0$,

$$\mathcal{U}_\tau(0, C_0(p), M) = \{\Sigma_\infty : \sup_i \sigma_{ii} \leq M, \sum_{j=1, j \neq i}^p I(\sigma_{ij} \neq 0) \leq C_0(p) \ \forall i\} \tag{1.120}$$

provides restriction on the number of non-zero entries in each column/row.

Cai and Zhou [2012] considered a broader class of covariance matrices where columns of Σ_p belong to a weak l_q ball in $\mathbb{R}^p$ for all $p \geq 1$. A *weak l_q ball* of radius c in $\mathbb{R}^p$ is given by

$$B_q^p(c) = \{\xi \in \mathbb{R}^p : |\xi|_{(k)}^q \leq ck^{-1} \ \forall 1 \leq k \leq p\} \tag{1.121}$$

where $|\xi|_{(k)}$ denotes the k-th largest element in magnitude of the vector ξ.

For a covariance matrix $\Sigma_p = ((\sigma_{ij}))_{1 \leq i, j \leq p}$, let $\sigma_{-j,j,p}$ be the j-th column of Σ_p with σ_{jj} removed. They assumed that $\sigma_{-j,j,p}$ is in a weak l_q ball for all $1 \leq j \leq p$. More specifically, for $0 \leq q < 1$, we have the following parameter space $G_q(C_{n,p})$ of covariance matrices:

$$G_q(C_{n,p}) = \{\Sigma_\infty : \sigma_{-j,j,p} \in B_q^{p-1}(C_{n,p}) \ \forall j, p \geq 1\}. \tag{1.122}$$

In the special case of $q = 0$, the matrix in $G_q(C_{n,p})$ has at most $C_{n,p}$ nonzero off diagonal elements in each column. Since a strong l_q ball is always contained in a weak l_q ball,

$$\mathcal{U}_\tau(q, C_0(p), M) \subset G_q(C_{n,p}) \tag{1.123}$$

for some $C_{n,p} > 0$ depending on $q, C_0(p)$ and M.

1.5.2 Estimation in $\mathcal{U}_\tau(q, C_0(p), M)$ or $G_q(C_{n,p})$

To estimate sparse covariance matrices, Bickel and Levina [2008b] and Cai and Zhou [2012] considered a *permutation invariant* covariance regularization, called *thresholding*. For a matrix $A = ((a_{ij}))$, its t_n-*thresholded* version is given by

$$T_{t_n}(A) = ((a_{ij}I(|a_{ij}| > t_n))). \tag{1.124}$$

The following theorem establishes the convergence rate of the thesholded version of $\tilde{\Sigma}_{p,n}$.

Theorem 1.5.1. *(Cai and Zhou [2012]) Suppose $\{C_{ip}\}$ are i.i.d. $\mathcal{N}(0, \Sigma_p)$. Then for $t_n = M\sqrt{n^{-1}\log p}$, $\log p = o(n)$, $C_{n,p} \le Mn^{\frac{1-q}{2}}(\log p)^{-\frac{3-q}{2}}$ and for some $C > 0$, we have*

$$\sup_{\Sigma_\infty \in G_q(C_{n,p})} E\|T_{t_n}(\tilde{\Sigma}_{p,n}) - \Sigma_p\|_2^2 \le C\Big[C_{n,p}^2\Big(\frac{\log p}{n}\Big)^{1-q} + \frac{\log p}{n}\Big]. \tag{1.125}$$

Remark 1.5.1. *It is easy to see that by (1.123) and Theorem 1.5.1, the same bound in (1.125) holds for supremum over $\Sigma_\infty \in \mathcal{U}_\tau(q, C_0(p), M)$. This was also independently established in Bickel and Levina [2008b].*

Proof of Theorem 1.5.1. Define the event

$$A_{ij} = \{|\hat{\sigma}_{ij} - \sigma_{ij}| \le 4\min\{|\sigma_{ij}|, t_n\}\}. \tag{1.126}$$

Using Lemma 1.3.3 one can show that for some $C > 0$

$$P(A_{ij}) \ge 1 - Cp^{-9/2}. \tag{1.127}$$

Let $I(A)$ be the indicator function for the set A and $D = (((\hat{\sigma}_{ij} - \sigma_{ij})(1 - I(A_{ij}))))$. Recall the $(1,1)$ norm from (1.26). Using (1.127), it is easy to see that for some $C > 0$

$$E\|D\|_{(1,1)}^2 \le Cn^{-1}. \tag{1.128}$$

Recall the ∞-norm from (1.34). By (1.43) and (1.58) of the proof of Theorem 1.3.1, we have

$$\|\tilde{\Sigma}_{p,n} - \Sigma_p\|_\infty^2 = O_P(n^{-1}\log p). \tag{1.129}$$

Also as $\Sigma_\infty \in G_q(C_{n,p})$, one can easily see that for some $C > 0$

$$\sum_{i\neq j} \min(|\sigma_{ij}|, t_n) \le CC_{n,p}\Big(\frac{\log p}{n}\Big)^{\frac{1-q}{2}}. \tag{1.130}$$

Therefore,

$$
\begin{aligned}
E||T_{t_n}(\tilde{\Sigma}_{p,n}) - \Sigma_p||_2^2 \;\le\;\; & 2E\Big[\max_j \sum_{i\neq j} |\hat{\sigma}_{ij} - \sigma_{ij}| I(A_{ij})\Big]^2 \\
& + 2E||D||_{(1,1)}^2 + C||\tilde{\Sigma}_{p,n} - \Sigma_p||_\infty^2 \\
\le\;\; & 2\Big(\max_j \sum_{i\neq j} \min(|\sigma_{ij}|, t_n)\Big)^2 \\
& + 2E||D||_{(1,1)}^2 + C||\tilde{\Sigma}_{p,n} - \Sigma_p||_\infty^2. \tag{1.131}
\end{aligned}
$$

Hence, Theorem 1.5.2 follows from (1.128)–(1.131).   □

1.5.3 *Minimaxity*

The following result establishes the minimax rate for the parameter spaces described in Section 1.5.1. Its proof is similar to the proof of Theorem 1.3.5. We leave it as an exercise.

Theorem 1.5.2. *(Cai and Zhou [2012]) Suppose $\{C_{ip}\}$ are i.i.d. $\mathcal{N}(0, \Sigma_p)$. Then for $t_n = M\sqrt{n^{-1}\log p}$, $\log p = o(n)$, $C_{n,p} \le Mn^{\frac{1-q}{2}}(\log p)^{-\frac{3-q}{2}}$ and for some $C > 0$, we have*

$$\inf_T \sup_{\Sigma_\infty \in \mathcal{U}_\tau(q, C_0(p), M)} E||T - \Sigma_p||_2^2 \ge C\Big[C_{n,p}^2\Big(\frac{\log p}{n}\Big)^{1-q} + \frac{\log p}{n}\Big]. \tag{1.132}$$

The same bound as in (1.132) follows for supremum over $\Sigma_\infty \in G_q(C_{n,p})$. This is easy to see from (1.123) and Theorem 1.5.2. Moreover, $T_{t_n}(\tilde{\Sigma}_{p,n})$ for $t_n = M\sqrt{n^{-1}\log p}$ is an (asymptotically) minimax estimator of $\Sigma_p \in \mathcal{U}_\tau(q, C_0(p), M)$ and $G_q(C_{n,p})$.

Exercises

1. Simulate observations from the models given in Example 1.1.1 and construct Table 1.

2. Let $\{X_t : 1 \le t \le n\}$ be a random sample of size n from $\mathcal{N}_p(0, I_p)$ and $\hat{\Sigma}_{p,n} = \frac{1}{n}\sum_{t=1}^p X_t X_t^*$. Find out $\lim \frac{1}{p} E\mathrm{Tr}(\hat{\Sigma}_{p,n} - I_p)^2$.

3. Give an example where banding destroys positive semi-definiteness.

4. Show that if A and B are positive semi-definite matrices then their Schur product is also positive semi-definite.

5. Give an example where thresholding destroys positive semi-definiteness.

6. Suppose $A = ((\theta^{|i-j|}))$ for some $\theta \in (0,1)$. Find the corner measure $T(A,k)$ and hence check whether A is in $\mathcal{X}(\alpha, C)$ for some $\alpha, C > 0$.

7. Suppose $A = (1-\rho)I + \rho J$. Find the corner measure $T(A,k)$ and hence check whether A is in $\mathcal{X}(\alpha, C)$ for some $\alpha, C > 0$.

8. Establish (1.30) and (1.31).

9. Give some examples of covariance matrices which belong to $\mathcal{U}(\epsilon, \alpha, C)$. Simulate Gaussian random variables from these matrices. Compute $k_{n,\alpha}^{\alpha}\|B_{k_{n,\alpha}}(\tilde{\Sigma}_{p,n}) - \Sigma_p\|_2$ for 500 replications, draw the histogram and comment on the rate of convergence.

10. Learn the proof of Lemmas 1.3.2 and 1.3.3.

11. Establish (1.67) and (1.72)

12. Learn the proof of (1.89) and (1.95)

13. Prove Theorems 1.3.3 and 1.3.4.

14. Establish (1.84), (1.92), and (1.96).

15. Show that $\mathcal{G}_\beta(M)$ is a subset of $\mathcal{X}(\alpha, C)$ for some $\alpha, C > 0$.

16. Give an example of $\Sigma_\infty \in \mathcal{G}_\beta(M)$ such that $\inf_p \lambda_{\min}(\Sigma_p) \to 0$ as $p \to \infty$.

17. Given an example of $\{\sigma_i\}$ such that $\sigma_i \neq 0 \; \forall i$ and its spectral density is Hölder continuous.

18. Show that for any $M > 0$ and non-integer $\beta > 0$, there exists some constants M_0 and M_1 depending on M such that $\mathcal{G}_\beta(M) \subset \mathcal{F}_\beta(M_0, M_1)$. Moreover, show that this result may not hold for integer β.

19. If $((\tilde{\sigma}_{st}))$ is positive semi-definite then check if $((\bar{\sigma}_{|s-t|}))$ is also so.

20. Prove (1.105) and (1.106).

21. Establish (1.105) and (1.106) when the class $\mathcal{G}_\beta(M)$ is replaced by $\mathcal{F}_\beta(M_0, M_1)$.

22. Learn the proofs of Lemmas 1.4.1–1.4.3.

23. Prove Theorem 1.4.3.

24. Prove Theorem 1.4.4.

25. Establish (1.123) and (1.127)–(1.130).

26. Prove Theorem 1.5.2.

Chapter 2

LARGE COVARIANCE MATRIX II

In the previous chapter we discussed estimation of the population covariance matrix when observations are independently and identically distributed. However, the independence is often inapplicable in practice and specific examples of a lack of independence may be found in the works of many researchers. So suppose that the p-dimensional (dependent) observations are marginally Gaussian with mean 0 and covariance matrix Σ_p. Consistent estimation of Σ_p under dependency has been developed very recently when it is either bandable or sparse. The goal of this chapter is to introduce some of these results.

One of the dependent structures is the so-called *cross-covariance structure* and we consider three different types of such structures. In the first, the restriction is on the growth of the powers of the trace of certain matrices derived from the cross-covariance structure. In the second, the dependence among any two observations weakens as the lag between them increases, and in the third case we assume weak dependence among the high-indexed observations.

In the first case, the convergence rate of the banded estimator is the same as in the i.i.d. case under a *trace condition*. For the other two structures too explicit rates of convergence for the banded estimators are provided. The tapered estimator also continues to remain consistent in operator norm in these dependent situations.

The literature on estimation of sparse Σ_p when dependency is present appears to be very scant. A natural *measure of stability*, based on the spectral density of the process captures the dependence in these cases. Under appropriate assumptions on this measure, the thresholded sample covariance matrix achieve consistency.

2.1 Bandable Σ_p

2.1.1 Models and examples

Efron [2009] proposed the matrix-variate normal distribution as a model for dependent observations. Recall the data matrix X and observations $\{C_{ip}\}$ respectively from (1.1) and (1.2). Denote

$$vec(X) = (C_{1p}^*, C_{2p}^*, \ldots, C_{np}^*)^*. \tag{2.1}$$

The matrix-variate normal model assumes that

$$vec(X) \quad \sim \quad \mathcal{N}_{np}(0, \Omega), \text{ where} \tag{2.2}$$

$$\Omega \quad = \quad \begin{pmatrix} \Sigma_p & \lambda_{21}\Sigma_p & \cdots & \lambda_{n1}\Sigma_p \\ \lambda_{21}\Sigma_p & \Sigma_p & \cdots & \lambda_{n2}\Sigma_p \\ \vdots & \vdots & \cdots & \vdots \\ \lambda_{n1}\Sigma_p & \lambda_{n2}\Sigma_p & \cdots & \Sigma_p \end{pmatrix}, \tag{2.3}$$

for some $n \times n$ real symmetric matrix $\Lambda_n = ((\lambda_{ij}))$, $\lambda_{ii} = 1$, $\forall i$. Here Λ_n and Σ_p are respectively called *column* and *row covariances*. Under this model, Allen and Tibshirani [2010] provided estimators of Σ_p using penalized log-likelihood method.

The main drawback of this model is that the correlation between rows is controlled without considering the effect of the columns; that is,

$$\frac{corr(X_{ki}, X_{lj})}{corr(X_{mi}, X_{mj})} = \frac{\sigma_{kl}}{\sqrt{\sigma_{ll}\sigma_{kk}}}, \quad \forall i, j = 1, 2, \ldots, p \text{ and } m = 1, 2, \ldots, n. \tag{2.4}$$

In Examples 2.1.2–2.1.4, we shall see that there are many models which are not accommodated by (2.2)–(2.3). A model which can overcome the limitation exhibited in (2.4) is the following from Bhattacharjee and Bose [2014a]:

Consider J_k, A_∞, $*$ and $vec(X)$ defined respectively in (1.10), (1.20), (1.18), and (2.1). Suppose

$$vec(X) \quad \sim \quad \mathcal{N}_{np}(0, \Delta_{np}), \text{ where} \tag{2.5}$$

$$\Delta_{np} \quad = \quad \begin{pmatrix} \Lambda_{11} * \Sigma_p & \Lambda_{21} * \Sigma_p & \cdots & \Lambda_{n1} * \Sigma_p \\ \Lambda_{12} * \Sigma_p & \Lambda_{22} * \Sigma_p & \cdots & \Lambda_{n2} * \Sigma_p \\ \vdots & \vdots & \cdots & \vdots \\ \Lambda_{1n} * \Sigma_p & \Lambda_{2n} * \Sigma_p & \cdots & \Lambda_{nn} * \Sigma_p \end{pmatrix}, \tag{2.6}$$

for some real $p \times p$ matrices $\{\Lambda_{ij}\}$ with $\Lambda_{ij} = \Lambda'_{ji}$, $\forall 1 \le j, i \le n$ and $\Lambda_{ii} = J_p$, $\forall 1 \le i \le n$.

We now recall a few notation from Chapter 1. Let Σ_∞ be the $\infty \times \infty$ extension of $\{\Sigma_p\}$ as defined in (1.20). For any $\epsilon > 0$, the set of all ϵ-well-conditioned covariance matrices is given by:

$$\mathcal{W}(\epsilon) = \{\Sigma_\infty : 0 < \epsilon < \inf_p \lambda_{\max}(\Sigma_p) \le \sup_p \lambda_{\max}(\Sigma_p) < \epsilon^{-1} < \infty\}. \tag{2.7}$$

Recall that a covariance matrix Σ_∞ has a polynomially decaying corner, if it belongs to

$$\mathcal{X}(\alpha, C) \quad = \quad \{A : T(A, k) \le Ck^{-\alpha}, \forall k \ge 1\}, \ \alpha, C > 0 \text{ where} \tag{2.8}$$

$$T(A, k) \quad = \quad \sup_j \sum_{i:|i-j|>k} |a_{ij}| \text{ is the corner meausre.} \tag{2.9}$$

Any $\Sigma_\infty \in \mathcal{X}(\alpha, C)$ is called a *bandable* covariance matrix. Recall

$$\mathcal{U}(\epsilon, \alpha, C) = \mathcal{W}(\epsilon) \cap \mathcal{X}(\alpha, C). \tag{2.10}$$

Let

$$\mathcal{V} = \{\Sigma_\infty = ((\sigma_{ij})) : \sigma_{ij} \neq 0, \ \forall i, j\}. \tag{2.11}$$

Note that $\Sigma_\infty \in \mathcal{V}$ is an *identifiability condition* since it allows to recover $\{\Lambda_{ij}\}$ uniquely when Δ_{np} and Σ_p are given. Recall the notation in (1.4). By (2.6), note that

$$\Delta((j-1)p + k, (i-1)p + l) = (\Lambda_{ij} * \Sigma_p)(k, l) = \Lambda_{ij}(k, l)\sigma_{kl}. \tag{2.12}$$

Therefore, when $\Sigma_\infty \in \mathcal{V}$, one can recover $\{\Lambda_{ij}\}$ from the matrices Δ_{np} and Σ_p by considering

$$\Lambda_{ij}(k, l) = \frac{\Delta((j-1)p + k, (i-1)p + l)}{\sigma_{kl}}, \quad 1 \leq i, j \leq n, \ 1 \leq k, l \leq p. \tag{2.13}$$

Example 2.1.1. Suppose $\{z_t\}$ are one-dimensional i.i.d. random variables with mean 0 and variance 1. Let

$$w_t = z_t + z_{t-1}, \text{ and } v_t = 0.5v_{t-1} + z_t, \ \forall t. \tag{2.14}$$

Then $\Sigma_\infty = \text{Var}(w_1, w_2, \ldots) \notin \mathcal{V}$ but $\Sigma_\infty = \text{Var}(v_1, v_2, \ldots) \in \mathcal{V}$.

The following examples provide models illustrating (2.2) and (2.5).

Example 2.1.2. Suppose

$$C_{ip} = A_p C_{(i-1)p} + Z_{ip}, \ \forall i = 0, \pm 1, \pm 2, \ldots, \tag{2.15}$$

where each Z_{ip} is a p-component column vector and i.i.d. with mean zero and $\text{Var}(Z_{ip}) = \tilde{\Sigma}_p$. Recall the operator norm defined in (1.8). Suppose A_p is a symmetric matrix of order p such that $\|A_p\|_2 < 1$ and $A_p\tilde{\Sigma}_p = \tilde{\Sigma}_p A_p$ for all p. From the properties of linear operators (see for example Bhatia [2009]), if $\|A_p\|_2 < 1$, then $(I - A_p)$ is invertible and

$$(I - A_p)^{-1} = (I + A_p + A_p^2 + \cdots).$$

Therefore, it is easy to see that for all $1 \leq i \neq j \leq n$,

$$\text{Var}(C_{ip}) = \Sigma_p = (I - A_p^2)^{-1}\tilde{\Sigma}_p, \text{ and} \tag{2.16}$$

$$\text{Cov}(C_{ip}, C_{jp}) = (I - A_p^2)^{-1}A_p^{|i-j|} = \Sigma_p A_p^{|i-j|}. \tag{2.17}$$

Hence,

$$\Delta_{np} = \text{Var}\left(vec(X_{p \times n})\right) = \left((\Sigma_p A_p^{|i-j|} I(i \neq j) + \Sigma_p I(i = j))\right)_{1 \leq i, j \leq n}.$$

Recall J_k defined in (1.10) and the notation in (1.4). As we have mentioned earlier, if $\Sigma_\infty \in \mathcal{V}$ then one can express Δ_{np} as

$$\Delta_{np} \;=\; ((\Sigma_p * \Lambda_{ij} I(i \neq j) + \Sigma_p * J_p I(i = j)))_{1 \leq i,j \leq n}, \quad \text{where} \quad (2.18)$$

$$\Lambda_{ij}(k,l) \;=\; \frac{(\Sigma_p A_p^{|i-j|})(k,l)}{((I - A_p^2)^{-1}\tilde\Sigma_p)(k,l)}, \quad 1 \leq i \neq j \leq n, \; 1 \leq k,l \leq p. \quad (2.19)$$

This satisfies the conditions (2.2)–(2.3) if for all $1 \leq k,l \leq p$, $1 \leq i,j \leq n$ and some $C_{ij} > 0$,

$$((I - A_p^2)^{-1}\tilde\Sigma_p A_p^{|i-j|})(k,l) = C_{ij}((I - A_p^2)^{-1}\tilde\Sigma_p)(k,l). \quad (2.20)$$

For example, if $A_p = \alpha I_p$ for some $0 < \alpha < 1$, then the model (2.15) satisfies (2.20). But in general, (2.20) may not hold always. Suppose, for some $0 < \alpha < 1$,

$$\tilde\Sigma_p = I_p \quad \text{and} \quad A_p = ((\alpha I(i + j = p))). \quad (2.21)$$

Then it is easy to see that $\Sigma_p = (1 - \alpha^2)^{-1} I_p$ and

$$(I - A_p^2)^{-1}\tilde\Sigma_p A_p^{|i-j|} = \begin{cases} (1 - \alpha^2)^{-1}\alpha^{|i-j|} I_p, & \text{if } |i - j| \text{ is even} \\ (1 - \alpha^2)^{-1}\alpha^{|i-j|} A_p, & \text{if } |i - j| \text{ is odd.} \end{cases} \quad (2.22)$$

Therefore, (2.20) does not hold when $|i - j|$ is odd. This shows that (2.5) is a more general model than (2.2)–(2.3).

Note that $\tilde\Sigma_p$ and A_p are also parameters of the model. The problem of their estimation will be addressed in details in Chapter 3.

Example 2.1.3. Suppose $\{Z_{ip}, \; i = 0, \pm 1, \pm 2, \dots\}$ is a sequence of random vectors such that $E(Z_{ip}) = 0 \; \forall i$ and $E(Z_{ip} Z_{jp}^*) = D_{|i-j|} \; \forall i,j$. Also, let Y_p be a mean zero p-component column random vector such that $\mathrm{Var}(Y_p) = \tilde\Sigma_p$ and which is independent of Z_{ip}'s. Recall the Schur product $*$ in (1.18). Define another sequence of p-component mean zero random vectors as

$$C_{ip} = Y_p * Z_{ip}, \; i = 0, 1, 2, \dots n. \quad (2.23)$$

Clearly, we have $\Delta_{np} = ((\tilde\Sigma_p * D_{|i-j|}))_{1 \leq i,j \leq n}$.

Suppose $\Sigma_p = \tilde\Sigma_p * D_0$. Then it is easy to see that

$$\Delta_{np} \;=\; ((\Sigma_p * \Lambda_{ij} I(i \neq j) + \Sigma_p * J_p I(i = j)))_{1 \leq i,j \leq n}, \quad \text{where} \quad (2.24)$$

$$\Lambda_{ij}(k,l) \;=\; \frac{D_{|i-j|}(k,l)}{D_0(k,l)}, \quad \forall 1 \leq i \neq j \leq n, \; 1 \leq k,l \leq p. \quad (2.25)$$

This satisfies the conditions (2.2)–(2.3) if for all $1 \leq k,l \leq p$, $i \geq 1$ and for some $C_i > 0$, we have

$$D_i(k,l) = C_i D_0(k,l). \quad (2.26)$$

But in general, (2.26) may not of course hold.

Suppose Z_{ip} satisfies (2.15). Then $D_i = (I - A_p)^{-1}\tilde{\Sigma}_p A_p^{|i|}$ for all $i = 0, \pm 1, \pm 2, \ldots$. Hence, as we have seen in Example 2.1.2, for the choice (2.21) of A_p and $\tilde{\Sigma}_p$, (2.26) is not satisfied and (2.2)–(2.3) is not applicable.

Example 2.1.4. Let

$$\Delta_{np} = \left(\left((B_p^{i+j}I(i \neq j) + (I - B_p^2)^{-1}I(i = j))\right)\right)_{1 \leq i,j \leq n} \quad (2.27)$$

where B_p is a symmetric $p \times p$ matrix and $||B_p||_2 < 1$ for all p. Then Δ_{np} is always positive semi-definite since

$$\Delta_{np} = (B_p \ldots B_p^n)'(B_p \ldots B_p^n) + Diag\left(I_p + \sum_{i=1}^{\infty} B_p^{2i} - B_p^{2k}, k \leq n\right)$$

where $Diag(A_i, \ i = 1, 2, \ldots, n)$ denotes the block-diagonal matrix with i-th diagonal block as A_i and I_p is the identity matrix of order p.

If $(I - B_p^2)^{-1}(k, l) \neq 0 \ \forall k, l$, then we can write

$$\Delta_{np} = \left(\left((I - B_p^2)^{-1} * \Lambda_{ij}I(i \neq j) + (I - B_p^2)^{-1} * J_pI(i = j)\right)\right)_{i,j \leq n},$$

$$\Lambda_{ij}(k, l) = \frac{(B_p^{i+j})(k, l)}{((I - B_p^2)^{-1})(k, l)}, \quad \forall 1 \leq i \neq j \leq n, \ 1 \leq k, l \leq p. \quad (2.28)$$

This satisfies the conditions (2.2)–(2.3) if for all $1 \leq k, l \leq p$, $1 \leq i, j \leq n$ and for some $C_{ij} > 0$,

$$(B_p^{i+j})(k, l) = C_{ij}((I - B_p^2)^{-1})(k, l). \quad (2.29)$$

For example, if $B_p = \alpha I_p$ for some $0 < \alpha < 1$, then Δ_{np} in (2.27) satisfies (2.29). But in general, (2.29) may not hold. Suppose, $B_p = A_p$ where A_p is as in (2.21). Then it is easy to see that $(I - B_p^2)^{-1} = (1 - \alpha^2)^{-1}I_p$ and for $i, j \geq 1$,

$$B_p^{i+j} = \begin{cases} \alpha^{i+j}I_p, & \text{if } i+j \text{ is even} \\ \alpha^{i+j}B_p, & \text{if } i+j \text{ is odd}. \end{cases} \quad (2.30)$$

Therefore, (2.29) does not hold when $i + j$ is odd.

2.1.2 Weak dependence

The dependence terms $\{\Lambda_{ij}\}$ can be separated out from Δ_{np} and the following matrix may be defined:

$$\nabla_{np} = \begin{bmatrix} J_p & \Lambda_{12} & \Lambda_{13} & \cdots & \Lambda_{1n} \\ \Lambda'_{12} & J_p & \Lambda_{23} & \cdots & \Lambda_{2n} \\ \Lambda'_{13} & \Lambda'_{23} & J_p & \cdots & \Lambda_{3n} \\ \vdots & \vdots & \vdots & \vdots & \vdots \\ \Lambda'_{1n} & \Lambda'_{2n} & \Lambda'_{3n} & \cdots & J_p \end{bmatrix}.$$

∇_{np} is called the *covariance structure* of the model (2.5). We can consider the following four different assumptions on ∇_{np} which provide feeble dependence among observations. In each of these cases a consistent estimator of Σ_p can be provided along with its convergence rate.

(1) A relevant question is under what restrictions on ∇_{np} and Σ_p, can one retain the consistency, preferably with the same convergence rate of the earlier estimators of Σ_p as in case of i.i.d. observations? In Theorem 2.1.1, we shall see that it is sufficient to assume that for some $M > 0$,

$$\sup_{n,j,k} n^{-1}\mathrm{Tr}\big((\Gamma_{\pm}^{jk})^r\big) \leq M^r \qquad (2.31)$$

where Γ_{+}^{jk} and Γ_{-}^{jk} are two $(n \times n)$ matrices defined by

$$\Gamma_{\pm}^{jk}(p,q) = \begin{cases} \dfrac{\Lambda_{pq}(jj)\pm(\Lambda_{pq}(jk)+\Lambda_{pq}(kj))\rho_{jk}+\Lambda_{pq}(kk)}{2(1\pm\rho_{jk})}, & p \neq q \\ 1, & p = q, \end{cases} \qquad (2.32)$$

$1 \leq j, k \leq p$ and $\rho_{jk} = \sigma_{jk}(\sigma_{jj}\sigma_{kk})^{-\frac{1}{2}}$.

Next, note that, in data where time is one of the latent variables which is responsible for the dependence, one may consider the Toeplitz (stationary) structure $\Lambda_{ij} = \Lambda_{|i-j|}$ for a suitable sequence of matrices $\{\Lambda_i\}$ and if $\Lambda_i = \Lambda^i$, then it yields the autoregressive structure of Example 2.1.2. Example 2.1.3 also has a Toeplitz structure. We have seen an example of the Hankel structure $\Lambda_{ij} = \Lambda_{i+j}$, $\forall i \neq j$ in Example 2.1.4.

Broadly speaking, "weak dependence" between columns can be modelled by assuming that Λ_{ij} is "small" when say both indices i and j are large or when $|i - j|$ or $i + j$ is large. While (2.31) demands control over all $\{\Lambda_{ij}\}$, the above assumption (weak dependence for large $|i - j|$ or $i + j$) has control on fewer $\{\Lambda_{ij}\}$. Therefore, results obtained under the assumption (2.31), do not hold true here and hence we need to discuss these cases separately. Below we mention the technical assumptions provided by Bhattacharjee and Bose [2014a] on these covariance structures, so that consistent estimators of Σ_p can be obtained.

Recall $|| \cdot ||_\infty$ defined in (1.34) and the notation in (1.20). Let $\{a_n\}_{n=1}^{\infty}$ be a sequence of non-negative integers such that $n^{-1}a_n < 1$, $\forall n \geq 1$.

(2) Weak dependence among the columns when i and j are large can be modelled as follows:

$$\mathcal{L}_n(a_n) = \Big\{\nabla_{np} : S'(a_n) := \max_{k,m\geq 1} ||\Lambda_{a_n+k,a_n+k+m}||_\infty = O\big(\frac{a_n}{n^2}\big)\Big\}, \qquad (2.33)$$

$$\mathcal{L}_\infty(\{a_n\}) \;=\; \{\nabla_\infty : \nabla_{np} \in \mathcal{L}_n(a_n)\}. \qquad (2.34)$$

(3) Weak dependence between i-th and j-th columns when $|i - j|$ is large is modelled as follows:

$$\mathcal{A}_n(a_n) = \{\nabla_{np} = ((\Lambda_{|i-j|})) : S(a_n) := \max_{a_n \le k \le n} ||\Lambda_k||_\infty = O(\frac{a_n}{n^2})\}, \quad (2.35)$$

$$\mathcal{A}_\infty(\{a_n\}) = \{\nabla_\infty : \nabla_{np} \in \mathcal{A}_n(a_n)\}. \quad (2.36)$$

(4) Finally, weak dependence among columns when $(i+j)$ is large, is modelled by:

$$\mathcal{H}_n(a_n) = \{\nabla_{np} = ((\Lambda_{i+j}I\,(i \ne j) + \Lambda_0 I\,(i = j))) : \max_{r \ge a_n} ||\Lambda_r||_\infty = O(\frac{a_n}{n^2})\},$$

$$\mathcal{H}_\infty(\{a_n\}) = \{\nabla_\infty : \nabla_{np} \in \mathcal{H}_n(a_n)\}. \quad (2.37)$$

Theorems 2.1.2 and 2.1.3 provide consistent estimator of Σ_p respectively for Cases (2) and (3). Case (4) does not have to be dealt with separately as all bounds for Case (2) will automatically hold for Case (4) due to the following Lemma.

Lemma 2.1.1. $\mathcal{H}_\infty(\{a_n\}) \subset \mathcal{L}_\infty(\{[2^{-1}a_n] + 2\})$, where $[x]$ is the largest integer contained in x.

Proof of the above lemma is immediate by observing that, for $b_n = [2^{-1}a_n] + 2$, $\forall n \ge 1$,

$$\max_{k \ge 1, m \ge 1} ||\Lambda_{b_n+k, b_n+k+m}||_\infty \le \sup_{r \ge a_n} ||\Lambda_r||_\infty. \quad (2.38)$$

Any model that satisfies (2.5) and any of the assumptions described in (1), (2), and (3), will be referred to as a *weak model*.

2.1.3 Estimation

In this section, we shall discuss consistent estimators of Σ_p under any of the following assumptions—(a) ∇_{np} and Σ_p satisfy (2.31), (b) $\nabla_\infty \in \mathcal{A}_\infty(\{a_n\})$ and (c) $\nabla_\infty \in \mathcal{L}_\infty(\{a_n\})$. Recall the banded and tapered version of a matrix in Section 1.2.2(a) and (b). Since we assume that $\Sigma_\infty \in \mathcal{U}(\epsilon, \alpha, C)$ for some $\epsilon, \alpha, C > 0$, from the experience of Section 1.3.2, we can expect that the banded or tapered version of $\hat{\Sigma}_{p,n}$ can serve our purpose. Let us first concentrate on the banded estimator. The tapered estimator will be discussed later.

Banding. Recall the k-banded version of a matrix $M = ((m_{ij}))$ given by

$$B_k(M) = ((m_{ij}I(|i - j| \le k))). \quad (2.39)$$

As discussed in the previous section, we are interested not only in just consistency but restrictions on ∇_{np} and Σ_p under which the convergence rate of the banded version of $\hat{\Sigma}_{p,n}$ remains the same as in case of i.i.d. observations dealt with in Theorem 1.3.1.

Recall U^i_{jh}, U_{ik}, V^i_{jk} and V_{jk} from (1.48) and (1.49). Note that in Remark 1.3.1, we pointed out that (1.52) may not hold for dependent $[C_{ip}]$ as then U^i_{jk} and V^i_{jk} are not independent over i. Recall $\Gamma^{jk}_{\pm}$ defined in (2.32). Under our model assumptions, $U_{jk} \sim \mathcal{N}_n(0, \Gamma^{jk}_+)$ and $V_{jk} \sim \mathcal{N}_n(0, \Gamma^{jk}_-)$. Hence, the problem boils down to finding conditions on $\Gamma^{jk}_{\pm}$, so that (1.53) follows directly from (1.51). In other words, if $U \sim \mathcal{N}_n(0, \Gamma_{n \times n})$, then we wish to claim that for any bounded t_n there exists $C_1, C_2, C_3 > 0$ so that

$$P(|U'U - n| \geq C_1 n t_n) \leq C_2 e^{-C_3 n t_n^2}.$$

To find out suitable conditions on $\Gamma_{n \times n}$ so that the above holds, we can use the following lemma on the large deviation rate of a random variable.

Lemma 2.1.2. *(Saulis and Statulevičius [1991]) Suppose $E\xi = 0$ and there exist $\gamma \geq 0$, $H > 0$ and $\overline{\Delta} > 0$ such that*

$$|\mathrm{Cum}_k(\xi)| \leq \left(\frac{k!}{2}\right)^{1+\gamma} \frac{H}{\overline{\Delta}^{k-2}}, \quad k = 2, 3, 4, \ldots, \tag{2.40}$$

where $|\mathrm{Cum}_k(\xi)| = |\frac{d^k}{dt^k}(\log E(e^{it\xi}))|_{t=0}$, is the k-th order cumulant of ξ. Then for all $x \geq 0$,

$$P[\pm \xi \geq x] \leq \exp\left\{ -\frac{x^2}{2}\left(H + x\overline{\Delta}^{\frac{-1}{2\gamma+1}}\right)^{-\frac{2\gamma+1}{\gamma+1}} \right\}.$$

Lemma 1.3.3 easily follows from Lemma 2.1.2 as for $\xi = (\chi_n^2 - n)$, we have $\gamma = 0$, $H = 4n$ and $\overline{\Delta} = \frac{1}{2}$.

Now to see whether (2.40) is satisfied by $U \sim \mathcal{N}_n(0, \Gamma_{n \times n})$, we need to calculate the characteristic function of $U'U$. For this purpose, the following lemma is useful.

Lemma 2.1.3. *Suppose A_k is a $k \times k$ positive definite matrix. Then*

$$\int_{\mathbb{R}^k} e^{-\frac{1}{2}y'(A_k - 2itI_k)y} dy = (2\pi)^{\frac{k}{2}}(\det(A_k - 2itI_k))^{-\frac{1}{2}}, \quad t \in \mathbb{R}.$$

Proof. Let $\lambda > 0$ be the minimum eigenvalue of A_k. Define f and g as

$$g(z) = (2\pi)^{\frac{k}{2}} [\det(A_k - 2zI)]^{-\frac{1}{2}}, \quad \mathcal{R}e\, z < \lambda,$$

$$f(z) = \int_{-\infty}^{\infty} e^{-y'(A_k - 2zI)y} dy, \quad \mathcal{R}e\, z < \lambda.$$

Note that both g and f are well defined. It is easy to check by direct integration that if $z = x \in (-\infty, \lambda)$, then $f(x) = g(x)$. It is also easy to check that both f and g are analytic functions on $\{z : \mathcal{R}e\, z < \lambda\}$. Since they agree on $\{z : z = x \in (-\infty, \lambda)\}$, they must be identical functions. Hence $f(it) = g(it)$, $t \in \mathbb{R}$ and the proof is complete. $\square$

The following lemma easily follows from the above two lemmas.

Lemma 2.1.4. $U \sim \mathcal{N}_n(0, \Gamma_{n \times n})$ *satisfies (2.40) if for some $M > 0$,*

$$\sup_n \frac{1}{n} \text{Tr}(\Gamma_{n \times n}^r) \leq M^r, \ \forall r \geq 1. \tag{2.41}$$

Proof. The characteristic function of $U'U$ is given by

$$
\begin{aligned}
E(e^{itU'U}) &= (2\pi)^{-n/2} \sqrt{\det(\Gamma_{n \times n}^{-1})} \int_{\mathbb{R}^k} e^{-\frac{1}{2} y' (\Gamma_{n \times n}^{-1} - 2itI_n) y} dy \\
&= \sqrt{\det(\Gamma_{n \times n}^{-1})} (\det(\Gamma_{n \times n}^{-1} - 2itI_n))^{-\frac{1}{2}}, \text{ by Lemma 2.1.3} \\
&= \left[\det(I_n - 2it\Gamma_{n \times n}) \right]^{-\frac{1}{2}}. \tag{2.42}
\end{aligned}
$$

Hence,

$$\frac{d^r}{dt^r} \log E(e^{itU'U}) = -\frac{1}{2} \sum_{u=1}^n \frac{d^r}{dt^r} \log(1 - 2it\lambda_u)$$

where λ_u, $1 \leq u \leq n$, are eigenvalues of $\Gamma_{n \times n}$. So, we have

$$
\begin{aligned}
|\text{Cum}_r(U'U - n)| &= |\text{Cum}_r(U'U)| \\
&= |\frac{d^r}{dt^r} \log E(e^{itU'U})|_{t=0} \\
&= \frac{1}{2} \sum_{u=1}^n (r-1)! 2^r (\lambda_u)^r.
\end{aligned}
$$

Hence using (2.41), we have

$$|\text{Cum}_r(U'U - n)| \leq \frac{r!}{2} \frac{4nM^2}{\left(\frac{1}{2M}\right)^{r-2}}.$$

Therefore, (2.40) is satisfied for $\gamma = 0, H = 4nM^2$ and $\bar{\Delta} = (2M)^{-1}$. Hence, using Lemma 2.1.2, for some $C_1, C_2, C_3 > 0$

$$
\begin{aligned}
P\big[|U'U - n| \geq C_1 n t_n\big] &\leq e^{-\frac{(C_1 n t_n)^2}{2}} \left(4nM^2 + 2C_1 M n t_n\right)^{-1} \\
&\leq C_2 e^{-C_3 n t_n^2}, \quad \text{provided } t_n \text{ is bounded,}
\end{aligned}
$$

and the proof of Lemma 2.1.4 is complete. $\square$

Lemma 2.1.4 motivates us to ask if under suitable restrictions on $\{\Gamma_{\pm}^{jk} : j, k \geq 1\}$, the banded sample covariance matrix can have the same rate of convergence as in the case of i.i.d. observations discussed in Theorem 1.3.1. To state such a result, recall the model assumptions in (2.5) and the classes of covariance matrices $\mathcal{U}(\epsilon, \alpha, C)$ and $\mathcal{V}$ respectively in (2.10) and (2.11).

Theorem 2.1.1. *(Bhattacharjee and Bose [2014a]) Suppose X satisfies the model assumption (2.5) and $\Sigma_\infty \in \mathcal{U}(\epsilon, \alpha, C) \cap \mathcal{V}$ for some $\epsilon, \alpha, C > 0$. Suppose (2.31) holds. Then for $k_{n,\alpha} \asymp (n^{-1}\log p)^{-\frac{1}{2(\alpha+1)}}$, we have $\|B_{k_{n,\alpha}}(\hat{\Sigma}_{p,n}) - \Sigma_p\|_2 = O_P(k_{n,\alpha}^{-\alpha})$.*

Proof. By Lemma 2.1.4 and (2.31), (1.51) implies (1.53). Therefore, exactly the same proof as for Theorem 1.3.1 goes through in this case also. $\qquad\square$

Condition (2.31) can be difficult to check. On the other hand, it is comparatively easy to bound $\|\Lambda_{ij}\|_\infty$ ($\|\cdot\|_\infty$ is defined in (1.34)). Now we provide some sufficient conditions for (2.31) to hold in terms of these quantities.

Sufficient conditions for (2.31). (a) Suppose, $\Sigma_\infty \in \mathcal{W}(\epsilon)$ for some $\epsilon > 0$ and $\{x_k\}$ is a sequence of non-negative real numbers such that $x_k = x_{-k}$ and $\|\Lambda_{ij}\|_\infty \le x_{i-j}$ $\forall i \ne j$, $1 \le i, j \le n$, then (2.31) holds if $\sum_1^\infty |x_k| < \infty$.

(b) Let $\Lambda_{ij} = 0$ $\forall |i-j| > k$. Then (2.31) holds if

$$\sum_{l=1}^k \left(\sup_{|i-j|=l} \|\Lambda_{ij}\|_\infty \right) < \infty. \tag{2.43}$$

(c) If $\Lambda_{ij} = \Lambda_{|i-j|}$ $\forall i, j$ and $\Lambda_r = 0$ $\forall r > k$, then (2.31) will hold if

$$\|\Lambda_r\|_\infty < \infty, \ \forall\, 1 \le r \le k. \tag{2.44}$$

Proof. (a) To prove this, we essentially show that

$$\frac{1}{n}\mathrm{Tr}((\Gamma_\pm^{jk})^r) \le \left(\sum_{i=1}^\infty |x_i| \right)^r, \ \forall\, 1 \le j, k \le p, \ r \ge 1. \tag{2.45}$$

Fix a $1 \le j, k \le p$ and $r \ge 1$. Note that

$$\frac{1}{n}\mathrm{Tr}((\Gamma_\pm^{jk})^r) \ \le \ \frac{1}{n}\sum |\Gamma_\pm^{jk}(u_1, u_2)\Gamma_\pm^{jk}(u_2, u_3) \cdots \Gamma_\pm^{jk}(u_r, u_1)|. \tag{2.46}$$

Now, by (2.32)

$$|\Gamma_\pm^{jk}(u, v)| \le \left(\frac{(1+|\rho_{jk}|)^2}{1 - \rho_{jk}^2} \right) \|\Lambda_{uv}\|_\infty. \tag{2.47}$$

Moreover, as $\Sigma_\infty \in \mathcal{W}(\epsilon)$

$$1 - \rho_{jk}^2 \ge \sqrt{\sigma_{jj}\sigma_{kk}} - \sigma_{jk}^2 = \det\begin{pmatrix} \sigma_{jj} & \sigma_{jk} \\ \sigma_{jk} & \sigma_{kk} \end{pmatrix} \ge (\inf_p \lambda_{\min}(\Sigma_p))^2 \ge \epsilon^2. \tag{2.48}$$

Now by (2.47), for some $C > 0$

$$|\Gamma_{\pm}^{jk}(u,v)| \leq C||\Lambda_{uv}||_{\infty} \leq Cx_{u-v}, \ \forall \ u,v. \tag{2.49}$$

Therefore, by (2.46)

$$\frac{1}{n}\text{Tr}((\Gamma_{\pm}^{jk})^r) \leq \frac{1}{n}C^r \sum_{u_1,\ldots,u_r} x_{u_1-u_2} x_{u_2-u_3} \cdots x_{u_r-u_1}$$

$$\leq C^r \sum_{k_1,\ldots,k_{m-1}=-(n-1)}^{n-1} x_{k_1} \cdots x_{k_{m-1}} x_{(-\sum_{j=1}^{m-1} k_j)}$$

$$\leq C^r \Big(\sum_{k=-\infty}^{\infty} |x_k| \Big)^r. \tag{2.50}$$

Hence, (2.45) is proved and the proof of (a) is complete.

(b) This immediately follows from (a) by observing that

$$||\Lambda_{ij}||_{\infty} \leq \sup_{|i-j|=l} ||\Lambda_{ij}||_{\infty}, \ \forall |i-j| = l, \ l \geq 1. \tag{2.51}$$

(c) This follows from (b) by observing that $\Lambda_{ij} = \Lambda_l, \ \forall |i-j| = l, \ l \geq 1$ and the sum in (2.43) reduces to $\sum_{l=1}^{k} ||\Lambda_l||_{\infty}$ and it is finite if (2.44) holds. $\square$

We now provide an example where (2.31) does not hold. Suppose, $g : [0, 2\pi] \to \mathbb{R}$ is a square integrable function. Then the Fourier coefficients of g are defined as

$$\hat{g}(k) = (2\pi)^{-1} \int_0^{2\pi} g(x)e^{-ikx}dx, \quad k = 0, \pm 1, \ldots$$

If g is symmetric (about π), then $\{\hat{g}(k)\}$ are real and $\hat{g}(k) = \hat{g}(-k) \ \forall k$. Let $T_{g,n}$ be the Toeplitz matrix defined by

$$T_{g,n} = ((\hat{g}(i-j)))_{1 \leq i,j \leq n}.$$

Example 2.1.5. Consider a function $g : [0, 2\pi] \to \mathbb{R}$ which is non-negative, symmetric (about π) and square integrable but is unbounded. Suppose $\Gamma_+^{jk} = T_{g,n}, \ \forall j, k$. Then (2.31) does not hold.

Proof. The proof is an application of *Szegö's theorem* (see Grenander and Szegö [1958]). Suppose if possible (2.31) holds. Let X_n be a random variable such that

$$P(X_n = \lambda_{in}) = n^{-1}, \quad i = 1, \ldots, n$$

where $\{\lambda_{1n}, \ldots, \lambda_{nn}\}$ are all the eigenvalues of $T_{g,n}$. By Szegö's theorem $X_n \xrightarrow{\mathcal{D}}$

$g(\mathcal{U})$ where $\mathcal{U}$ is a random variable distributed uniformly on $[0, 2\pi]$. Now from inequality (2.31) for all n,

$$EX_n^k = n^{-1} \sum_{i=1}^{n} \lambda_{in}^k = n^{-1} \mathrm{Tr}\left(T_{g,n}^k\right) \le M^k, k = 1, \ldots \qquad (2.52)$$

Thus, (2.52) implies that $\{X_n^{2k}\}$ is uniformly integrable for all $k = 1, 2, \ldots$. As a consequence

$$EX_n^{2k} \to E(g(\mathcal{U}))^{2k}, \quad k = 1, 2 \ldots$$

and using (2.52), $E(g(\mathcal{U}))^{2k} \le M^{2k}, k = 1, \ldots$. From this it is immediate that g is almost everywhere bounded. This contradicts our assumption that g is unbounded. Therefore (2.31) does not hold. $\qquad \square$

Recall that Λ_∞ is the $\infty \times \infty$ extension of the matrix Λ_{np} in the sense (1.20) and, the classes of cross-covariance structures $\mathcal{A}_\infty\left(\{a_n\}\right)$ and $\mathcal{L}_\infty\left(\{a_n\}\right)$ are respectively given in (2.36) and (2.34). As discussed in Section 2.1.2, if $\nabla_\infty \in \mathcal{L}_\infty\left(\{a_n\}\right)$ or $\mathcal{A}_\infty\left(\{a_n\}\right)$, then we cannot say whether (2.31) will hold or not. In these classes, we do not have any control over Λ_{ij} for $\min(i, j) < a_n$ or $|i - j| < a_n$ respectively and moreover $a_n \to \infty$. As the following theorems show, we have a slower rate of convergence for the two classes.

Theorem 2.1.2. *(Bhattacharjee and Bose [2014a]) Suppose X satisfies (2.5). If $\Sigma_\infty \in \mathcal{U}(\epsilon, \alpha, C) \cap \mathcal{V}$ for some $\epsilon, \alpha, C > 0$ and $\nabla_\infty \in \mathcal{L}_\infty(\{l_n\})$ for some non-decreasing sequence $\{l_n\}_{n \ge 1}$ of non-negative integers such that $n^{-1}l_n \log p \to 0$ as $n \to \infty$ and $\liminf n^{-1}l_n^2 \log p > 0$. Then with $k_{n,\alpha}^* \asymp \left(n^{-1}l_n \log p\right)^{-\frac{1}{1+\alpha}}$,*

$$||B_{k_{n,\alpha}^*}(\hat{\Sigma}_{p,n}) - \Sigma_p||_2 = O_P\left(k_{n,\alpha}^{*-\alpha}\right).$$

Proof. The relations (1.39)–(1.42) in the proof of Theorem 1.3.1, continue to hold when we replace $k_{n,\alpha}$ by $k_{n,\alpha}^*$. However, since the observations are not independent, the rate of convergence of $\hat{\Sigma}_{p,n}$ to Σ_p in $||\cdot||_\infty$ norm as mentioned in (1.43) of the proof of Theorem 1.3.1 does not hold. Instead here we prove

$$||\hat{\Sigma}_{p,n} - \Sigma_p||_\infty = O_P(l_n n^{-1} \log p). \qquad (2.53)$$

Note that once (2.53) holds, then steps similar to (1.61) and (1.62) in the proof of Theorem 1.3.1, imply

$$||B_{k_{n,\alpha}^*}(\hat{\Sigma}_{p,n}) - B_{k_{n,\alpha}^*}(\Sigma_p)||_2 = O_P(k_{n,\alpha}^* l_n n^{-1} \log p). \qquad (2.54)$$

Hence, to choose $k_{n,\alpha}^*$ appropriately, we set

$$k_{n,\alpha}^{*-\alpha} = k_{n,\alpha}^* l_n n^{-1} \log p \implies k_{n,\alpha}^* = \left(l_n n^{-1} \log p\right)^{-\frac{1}{1+\alpha}}. \qquad (2.55)$$

Now proof of Theorem 2.1.2 will be complete if (2.53) holds.

Recall U_{jk}^i, V_{jk}^i, U_{jk} and V_{jk} from (1.48) and (1.49). To prove (2.53), again note that the same calculations from (1.44) to (1.51) go through in this case also as the independence assumption is first used in (1.52). Therefore, we have

$$P\big(\|\hat{\Sigma}_{p,n} - \Sigma_p\|_\infty \geq t_n\big) \;\leq\; \sum_{j,k=1}^{p} \big(P|U'_{jk}U_{jk} - n| \geq C_1 n t_n\big)$$

$$+ \sum_{j,k=1}^{p} P\big(|V'_{jk}V_{jk} - n| \geq C_1 n t_n\big). \quad (2.56)$$

Let

$$U_{jkl_n} = (U_{jk}^{l_n+1}, U_{jk}^{l_n+2}, \ldots, U_{jk}^n)', \quad V_{jkl_n} = (V_{jk}^{l_n+1}, V_{jk}^{l_n+2}, \ldots, V_{jk}^n)'.$$

Suppose $\Gamma_\pm^{jkl_n}$ is the covariance matrices of U_{jkl_n} and U_{jkl_n}. Note that $\Gamma_\pm^{jkl_n}$ are symmetric positive semi-definite matrix constructed by deleting the first l_n rows and columns from $\Gamma_\pm^{jk}$. Let $C_\pm^{jkl_n} = I_{n-l_n} - (\Gamma_\pm^{jkl_n})^{-1}$. Then we can write

$$P[|U'_{jk}U_{jk} - n| \geq C_1 n t_n] \;\leq\; l_n P(|(U_{jk}^1)^2 - 1| \geq C_1 n l_n^{-1} t_n)$$

$$+ P(|U'_{jkl_n}(\Gamma_+^{jkl_n})^{-1}U_{jkl_n} - (n - l_n)| \geq C_1 n t_n/2)$$

$$+ P(|U'_{jkl_n} C_+^{jkl_n} U_{jkl_n}| \geq C_1 n t_n/2)$$

$$= \; T_1 + T_2 + T_3, \; \text{(say)}. \quad (2.57)$$

Let, $t_n = M l_n n^{-1} \log p$, for some fix constant $M > 0$. Later M will be chosen appropriately. Now, as $U_{jk}^i \sim \mathcal{N}(0,1)$ for all $i \geq 1$, by Lemma 1.3.3 and for some $C_2, C_3 > 0$, we have

$$T_1 \leq l_n P\big[|\chi_1^2 - 1| \geq C_1 n l_n^{-1} t_n\big] \leq 2 l_n C_2 e^{-C_3 M \log p}. \quad (2.58)$$

Then for some constants $C_4, C_5 > 0$, by Lemma 2.1.2

$$T_2 = P\big[|\chi_{(n-l_n)}^2 - (n - l_n)| \geq C_1 n t_n/2\big] \leq C_4 e^{-C_5 \frac{l_n^2}{n}(\log p)^2 M^2}. \quad (2.59)$$

Next, for some $C_6, C_7 > 0$, we have

$$T_3 \;=\; P[|(U^{jkl_n})'\big(C_+^{jkl_n}\big)(U^{jkl_n})| \geq C_1 n t_n/2]$$

$$\leq\; P\big[\max_{U \neq 0} \frac{|(U'C_+^{jkl_n}U)|}{U'U}(U^{jkl_n})'(U^{jkl_n}) \geq C_1 n t_n/2\big]$$

$$=\; P[\sqrt{\lambda_{\max}\big(C_+^{jkl_n}\big)'\big(C_+^{jkl_n}\big)}(U^{jkl_n})'(U^{jkl_n}) \geq C_1 n t_n/2]$$

$$=\; P[\|C_+^{jkl_n}\|_2 (U^{jkl_n})'(U^{jkl_n}) \geq C_1 n t_n/2]$$

$$\leq\; nP[\|C_+^{jkl_n}\|_2 \chi_1^2 \geq C_1 t_n/2]$$

$$\leq\; nC_6 \exp\{-C_7 \frac{t_n}{\|C_+^{jkl_n}\|_2}\}, \quad \text{by Lemma 1.3.3.}$$

Moreover, it is easy to see that for some $C_8 > 0$, $||C_+^{jkl_n}||_2 \leq n||C_+^{jkl_n}||_\infty \leq nC_8 S'(l_n)$. Hence, putting $t_n = Ml_n n^{-1} \log p$, for some constants C_9, $C_{10} > 0$,

$$P[|(U^{jkl_n})'(I - (\Gamma_+^{jkl_n})^{-1})U^{jkl_n}| \geq C_1 nt_n] \leq nC_6 e^{-C_9 \frac{t_n}{nS'(l_n)}}$$

$$\leq nC_6 e^{-C_{10}M \log p}. \quad (2.60)$$

Similar bound holds for V_{jk} also. By (2.56) to (2.60), for some $C_{11}, C_{12} > 0$ and for all sufficiently large n,

$$P[||\hat{\Sigma}_{p,n} - \Sigma_p||_\infty \geq Mn^{-1}l_n \log p]$$

$$\leq 2C_{11}p^2\left(l_n e^{-C_3 M \log p} + 2e^{-C_5 M^2 \frac{l_n^2}{n}(\log p)^2} + 2ne^{-C_{10}M \log p}\right)$$

$$= C_{12}\left(p^{3-C_3 M} + p^2 e^{-C_5 M^2 \frac{l_n^2}{n}(\log p)^2} + p^{3-C_{10}M}\right).$$

If $M > \max\{\frac{3}{C_3}, \frac{3}{C_{10}}\}$, then $p^{3-C_3 M} + p^{3-C_{10}M} \to 0$. The logarithm of the second term is

$$2\log p - C_5 M^2 n^{-1}l_n^2(\log p)^2 = \log p\left[2 - C_5 M^2 n^{-1}l_n^2(\log p)\right].$$

Now if $\liminf l_n^2 n^{-1}\log p > 0$ then the above expression is bounded away from zero by S (say).

So, if $M > \max\{\frac{3}{C_3}, \frac{3}{C_{10}}, \sqrt{\frac{2}{C_5 S}}\}$, then the second term also tends to zero.

This completes the proof of (2.53) and hence also of Theorem 2.1.2. $\qquad\square$

Remark 2.1.1. (i) If $l_n \approx (n^{-1}\log p)^{-\frac{1}{2}}$, then the rate of convergence will be same as that for the i.i.d. case. If $l_n \approx (n^{-1}\log p)^{-\beta}$ where β is more than $1/2$, then the rate is slower than the i.i.d. case. Note that $\beta < 1/2$ is not allowed in the theorem as $\liminf n^{-1}l_n^2 \log p > 0$.

(ii) Theorem 2.1.2 is not applicable in case the sequence $\{l_n\}$ is bounded above. This is because if $n^{-1}l_n \log p \to 0$ then $n^{-1}\log p \to 0$ and hence $n^{-1}l_n^2 \log p \to 0$. Recall $k_{n,\alpha}$ in Theorem 1.3.1. When $\{l_n\}$ is bounded by K, $C_{(K+1)p}$, $C_{(K+2)p}, \ldots$ will be an i.i.d. sample and we can construct the estimator on the basis of this i.i.d. sample i.e., we can consider the $k_{n,\alpha}$ banded version of $\dfrac{1}{n-K}\sum_{i=K+1}^{n} C_{ip}C_{ip}'$ with the same rate as the i.i.d. case.

The next theorem shows consistency of the banded sample covariance matrix when the cross-covariance structure $\nabla_\infty \in \mathcal{A}_\infty(\{a_n\})$.

Theorem 2.1.3. (Bhattacharjee and Bose [2014a]) Suppose X satisfies our model assumptions (2.5). If $\Sigma_\infty \in \mathcal{U}(\epsilon, \alpha, C) \cap \mathcal{V}$ for some $\epsilon, \alpha, C > 0$ and $\nabla_\infty \in \mathcal{A}_\infty(\{a_n\})$ for some non-decreasing sequence $\{a_n\}_{n\geq 1}$ of non-negative *integers such that $a_n\sqrt{n^{-1}\log p} \to 0$ and $a_n^{-1}\sqrt{n\log p} \to \infty$ as $n \to \infty$. Then with $k_{n,\alpha}^{**} \asymp \left(a_n n^{-\frac{1}{2}}\sqrt{\log p}\right)^{-\frac{1}{1+\alpha}}$,*

$$||B_{k_{n,\alpha}^{**}}(\hat{\Sigma}_{p,n}) - \Sigma_p||_2 = O_P(k_{n,\alpha}^{**-\alpha}).$$

Proof. As before, note that in the proof of Theorem 1.3.1, (1.39)–(1.42) hold true when we replace $k_{n,\alpha}$ by $k_{n,\alpha}^{**}$. As observations are not independent, the rate of convergence of $\hat{\Sigma}_{p,n}$ to Σ_p in $||\cdot||_\infty$ norm as mentioned in (1.43) of the proof of Theorem 1.3.1 does not hold. Instead here we will prove

$$||\hat{\Sigma}_{p,n} - \Sigma_p||_\infty = O_P(a_n\sqrt{n^{-1}\log p}). \tag{2.61}$$

Note that if (2.61) is true, then steps similar to (1.61) and (1.62) in the proof of Theorem 1.3.1, will yield

$$||B_{k_{n,\alpha}^{**}}(\hat{\Sigma}_{p,n}) - B_{k_{n,\alpha}^{**}}(\Sigma_p)||_2 = O_P(k_{n,\alpha}^{**}a_n\sqrt{n^{-1}\log p}). \tag{2.62}$$

Then the appropriate choice of $k_{n,\alpha}^{**}$ is obtained by setting

$$k_{n,\alpha}^{**-\alpha} = k_{n,\alpha}^{**}a_n\sqrt{n^{-1}\log p} \text{ that is } k_{n,\alpha}^{**} = \left(a_n\sqrt{n^{-1}\log p}\right)^{-\frac{1}{1+\alpha}}. \tag{2.63}$$

Hence proof of Theorem 2.1.3 will be complete if we can show (2.61) holds.

Recall U_{jk}^i, V_{jk}^i, U_{jk} and V_{jk} in (1.48) and (1.49). To prove (2.53), again note that the calculations from (1.44) to (1.51) go through in this case also as the independence assumption is first used in (1.52). Therefore, we have

$$P(||\hat{\Sigma}_{p,n} - \Sigma_p||_\infty \geq t_n) \leq \sum_{j,k=1}^{p} P\left(|U_{jk}'U_{jk} - n| \geq C_1 n t_n\right)$$

$$+ \sum_{j,k=1}^{p} P\left(|V_{jk}'V_{jk} - n| \geq C_1 n t_n\right). \tag{2.64}$$

Let, for all $1 \leq r \leq a_n$,

$$A_{r,a_n} = \{i \in \mathbb{Z}^+ \cup \{0\} : ia_n + r \leq n\} \text{ and } C_{r,a_n} = \text{cardinality of } A_{r,a_n}.$$

Let

$$U_{jkra_n} = vec\left(U_{jk}^i : i \in A_{r,a_n}\right) \text{ and } V_{jkra_n} = vec\left(V_{jk}^i : i \in A_{r,a_n}\right).$$

Now, by (2.64), we have,

$$P(||\hat{\Sigma}_{p,n} - \Sigma_p||_\infty \geq t_n) \leq \sum_{j,k}\sum_{r=1}^{a_n} P\left[|(U_{jkra_n})'U_{jkra_n} - C_{r,a_n}| \geq \frac{C_1 n t_n}{a_n}\right]$$

$$+ \sum_{j,k}\sum_{r=1}^{a_n} P\left[|(V_{jkra_n})'V_{jkra_n} - C_{r,a_n}| \geq \frac{C_1 n t_n}{a_n}\right]. \tag{2.65}$$

Recall $\Gamma_\pm^{jk}$ in (2.32). Note that for each $1 \le r \le a_n$,

$$U_{jkra_n} \sim \mathcal{N}_{C_{\cdot,a_n}}(0, \Gamma_+^{jkra_n}) \text{ and } V_{jkra_n} \sim \mathcal{N}_{C_{r,a_n}}(0, \Gamma_-^{jkra_n}), \qquad (2.66)$$

where $\Gamma_\pm^{jkra_n}$ is nothing but the sub-matrix consisting of the A_{r,a_n}-th rows and columns from $\Gamma_\pm^{jk}$. Let $C_\pm^{jkra_n} = I_{C_{r,a_n}} - (\Gamma_\pm^{jkar_n})^{-1}$ $\forall r \ge 1$. Therefore, for some $C_2 > 0$

$$P\big[\|\hat{\Sigma}_{p,n} - \Sigma_p\|_\infty \ge t_n\big] \le 2a_n p^2 P\big[|\chi^2_{C_{r,a_n}} - C_{r,a_n}| \ge C_2 n a_n^{-1} t_n\big]$$

$$+ \sum_{j,k} \sum_{r=1}^{a_n} P\big[|U_{jkra_n})' C_+^{jkra_n} U_{jkra_n}| \ge C_2 n a_n^{-1} t_n\big]$$

$$+ \sum_{j,k} \sum_{r=1}^{a_n} P\big[|(V_{jkra_n})' C_-^{jkra_n} V_{jkra_n}| \ge C_2 n a_n^{-1} t_n\big].$$

Again, by Lemma 1.3.3, for $t_n = M a_n (n^{-1} \log p)^{\frac{1}{2}}$ and some $C_3, C_4 > 0$, as $n^{-1}\log p \to 0$ we have

$$P\big[|\chi^2_{C_{r,a_n}} - C_{r,a_n}| \ge C_2 n a_n^{-1} t_n\big] \le C_3 e^{-C_4 M \log p}. \qquad (2.67)$$

Now, as in the proof of Theorem 2.1.2, for some $C_5, C_6 > 0$,

$$P\big[|(U^{jkra_n})' C_+^{jkra_n} U^{jkra_n}| \ge C_2 n a_n^{-1} t_n\big] \le n C_5 e^{-C_6 t_n a_n^{-1} \|C_+^{jkan}\|_2^{-1}}.$$

Since $\nabla_\infty \in \mathcal{A}_\infty(\{a_n\})$, for some $C_7 > 0$ we have

$$\|C_+^{jkra_n}\|_2 \le C_7 n S(a_n).$$

Therefore, putting $t_n = M a_n (n^{-1}\log p)^{\frac{1}{2}}$, we have, for some constants $C_8, C_9 > 0$,

$$P\big[|(U^{jkra_n})' C_+^{jkra_n} U^{jkra_n}| \ge C_2 a_n^{-1} n t_n\big] \le n C_8 e^{-C_9 M a_n^{-1} \sqrt{n \log p}}.$$

Similarly, for some constants $C_{10}, C_{11} > 0$,

$$P\big[|(V_{jkra_n})' C_-^{jkra_n} V_{jkra_n}| \ge C_2 n a_n^{-1} t_n\big] \le n C_{10} e^{-C_{11} M a_n^{-1} \sqrt{n \log p}}.$$

Hence, for some constants $C_{12}, C_{13}, C_{14} > 0$, we have

$$P\big[\|\hat{\Sigma}_{p,n} - \Sigma_p\|_\infty \ge t_n\big] \le C_{12}(p^3 e^{-C_{13} M \log p} + p^4 e^{-C_{14} M \frac{\sqrt{n \log p}}{a_n}}).$$

Clearly, the first term $\to 0$ as $n \to \infty$ if $M > \frac{3}{C_1}$. Now, since $a_n \sqrt{n^{-1}\log p} \to 0$ and $a_n^{-1}\sqrt{n \log p} \to \infty$, we have, for some constants C_{15} and $C_{16} > 0$,

$$p^4 e^{-C_{14} M \frac{\sqrt{n \log p}}{a_n}} = e^{C_{15} \frac{\sqrt{n \log p}}{a_n} \left(a_n \sqrt{\frac{\log p}{n}} - C_{16} M\right)} \to 0.$$

Hence (2.61) is proved and proof of Theorem 2.1.3 is complete. $\qquad \square$

Observe that if $\{a_n\}$ is bounded above, then the rate of convergence reduces to the rate for i.i.d. sample as given in Theorem 1.3.1. This completes our discussion on banded estimators of Σ_p.

Tapering. Recall tapering of a matrix from Section 1.2.2(b). For any matrix M,

$$R_{g,\tau_{n,\alpha}}(M) = ((m_{ij}g(|i-j|/\tau_{n,\alpha}))) \qquad (2.68)$$

where the tapering function g satisfies the following assumptions.

(T1) $g : \mathbb{R}^+ \cup \{0\} \to \mathbb{R}^+ \cup \{0\}$ is positive semi-definite, continuous, non-increasing such that

$$g(0) = 1 \text{ and } \lim_{x\to\infty} g(x) = 0.$$

(T2) $\int_0^\infty g(x) < \infty$ and $1 - g(x) = O(x^\gamma)$ for some $\gamma \geq 1$ in some neighborhood of zero.

We have the following theorem.

Theorem 2.1.4. *(Bhattacharjee and Bose [2014a]) Suppose (T1) and (T2) hold.*

(a) If conditions of Theorem 2.1.1 hold, and $\tau_{n,\alpha} \asymp \left(\dfrac{\log p}{n}\right)^{-\frac{1}{2(1+\gamma)}\left[\frac{\gamma}{1+\alpha}+1\right]}$, then

$$\|R_{g,\tau_{n,\alpha}}(\hat{\Sigma}_{p,n}) - \Sigma_p\|_2 = O_P\big[(n^{-1}\log p)^{\frac{\gamma\alpha}{2(1+\alpha)(1+\gamma)}}\big].$$

(b) If conditions of Theorem 2.1.2 hold and $\tau_{n,\alpha} \asymp \left(\dfrac{l_n}{n}\log p\right)^{-\frac{1}{2(1+\gamma)}\left[\frac{1}{1+\alpha}+1\right]}$, then

$$\|R_{g,\tau_{n,\alpha}}(\hat{\Sigma}_{p,n}) - \Sigma_p\|_2 = O_P\big[(l_n n^{-1}\log p)^{\frac{\gamma\alpha}{(1+\alpha)(1+\gamma)}}\big].$$

(c) If conditions of Theorem 2.1.3 hold, and $\tau_{n,\alpha} \asymp \left(\dfrac{a_n}{\sqrt{n}}\sqrt{\log p}\right)^{-\frac{1}{2(1+\gamma)}\left[\frac{1}{1+\alpha}+1\right]}$, then

$$\|R_{g,\tau_{n,\alpha}}(\hat{\Sigma}_{p,n}) - \Sigma_p\|_2 = O_P\big[(a_n\sqrt{n^{-1}\log p})^{\frac{\gamma\alpha}{(1+\alpha)(1+\gamma)}}\big].$$

Proof. By Lemma 1.3.2 and triangle inequality,

$$\|R_{g,\tau_{n,\alpha}}(\hat{\Sigma}_{p,n}) - \Sigma_p\|_2 \; < \; \|R_{g,\tau_{n,\alpha}}(\hat{\Sigma}_{p,n}) - R_{g,\tau_{n,\alpha}}(\Sigma_p)\|_{(1,1)}$$
$$+ \|R_{g,\tau_{n,\alpha}}(\Sigma_p) - \Sigma_p\|_{(1,1)}. \qquad (2.69)$$

Now, for some constant $C_1 > 0$,

$$\|R_{g,\tau_{n,\alpha}}(\hat{\Sigma}_{p,n}) - R_{g,\tau_{n,\alpha}}(\Sigma_p)\|_{(1,1)} \; \leq \; \|\hat{\Sigma}_{p,n} - \Sigma_p\|_\infty \Big(2\sum_{l=0}^{p} g\big(\frac{l}{\tau_{n,\alpha}}\big)\Big)$$

$$\leq \; \tau_{n,\alpha}\|\hat{\Sigma}_p - \Sigma_p\|_\infty C_1 \int_0^\infty g(x)dx. \qquad (2.70)$$

As before, $\quad \|\hat{\Sigma}_{p,n} - \Sigma_p\|_\infty = \begin{cases} O_P\left(\sqrt{n^{-1}\log p}\right), & \text{in Theorem 2.1.1} \\ O_P\left(l_n n^{-1}\log p\right), & \text{in Theorem 2.1.2} \\ O_P\left(u_n\sqrt{n^{-1}\log p}\right) & \text{in Theorem 2.1.3} \end{cases}$

Again, by triangle inequality

$$\|R_{g,\tau_{n,\alpha}}(\Sigma_p) - \Sigma_p\|_{(1,1)} \;\leq\; \|R_{g,\tau_{n,\alpha}}(\Sigma_p) - B_{k'_{n,\alpha}}[R_{g,\tau_{n,\alpha}}(\Sigma_p)]\|_{(1,1)}$$
$$+ \|B_{k'_{n,\alpha}}[R_{g,\tau_{n,\alpha}}(\Sigma_p)] - B_{k'_{n,\alpha}}(\Sigma_p)\|_{(1,1)} \quad (2.71)$$
$$+ \|B_{k'_{n,\alpha}}(\Sigma_p) - \Sigma_p\|_{(1,1)}.$$

By Lemma 1.3.1, we have

$$\|B_{k'_{n,\alpha}}(\Sigma_p) - \Sigma_p\|_{(1,1)} = O((k'_{n,\alpha})^{-\alpha}) \quad \text{and}$$

$$\|R_{g,\tau_{n,\alpha}}(\Sigma_p) - B_{k'_{n,\alpha}}[R_{g,\tau_{n,\alpha}}(\Sigma_p)]\|_{(1,1)} = O((k'_{n,\alpha})^{-\alpha}).$$

Now, for some constants $C_2, C_3 > 0$, as σ_{ij}'s are bounded, for sufficiently large n,

$$\|B_{k'_{n,\alpha}}[R_{g,\tau_{n,\alpha}}(\Sigma_p)] - B_{k'_{n,\alpha}}(\Sigma_p)\|_{(1,1)}$$

$$\leq \max_i \sum_{j:|i-j|\leq k'_{n,\alpha}} \left(1 - g\left(\frac{|i-j|}{\tau'_{n,\alpha}}\right)\right)|\sigma_{ij}|$$

$$\leq C_2 \sum_{l=-k'_{n,\alpha}}^{k'_{n,\alpha}} \left(1 - g\left(\frac{l}{\tau_{n,\alpha}}\right)\right) \leq C_3 \left(\frac{k'_{n,\alpha}}{\tau_{n,\alpha}}\right)^\gamma k'_{n,\alpha}.$$

Now, consider $\quad k'_{n,\alpha} = \begin{cases} \left(n^{-1}\log p\right)^{-\frac{\gamma}{2(1+\gamma)(1+\alpha)}}, & \text{for (a)}, \\ \left(l_n n^{-1}\log p\right)^{-\frac{\gamma}{(1+\gamma)(1+\alpha)}}, & \text{for (b)}, \\ \left(a_n n^{-1/2}\sqrt{\log p}\right)^{-\frac{\gamma}{(1+\gamma)(1+\alpha)}}, & \text{for (c)}. \end{cases}$

This completes the proof.   □

2.2   Sparse Σ_p

Let us assume that

(E1) $\{C_{ip}\}$ is stationary Gaussian with mean 0 and variance Σ_p.

As mentioned in Section 1.5, sparse covariance matrices arise in many applications of biological and medical sciences where there is no natural ordering among the components of $\{C_{ip}\}$ and covariance between most of the components is insignificant. Recall the following class of sparse covariance matrices from Section 1.5.1:

$$\mathcal{U}_\tau(q, C_0(p), M) = \left\{\Sigma : \sup_i \sigma_{ii} \leq M, \ \sup_i \sum_{\substack{1\leq j\leq p \\ j\neq i}} |\sigma_{ij}|^q \leq C_0(p) \ p \geq 1\right\} \quad (2.72)$$

for $q \in [0,1)$. Thresholding of the sample covariance matrix $\hat{\Sigma}_{p,n}$ is necessary to achieve consistency for $\Sigma_p \in \mathcal{U}_\tau(q, C_0(p), M)$. Recall that for a matrix $A = ((a_{ij}))$, its t_n-threshold version is given by

$$T_{t_n}(A) = ((a_{ij}I(|a_{ij}| > t_n))). \tag{2.73}$$

Bickel and Levina [2008b] and Cai and Zhou [2012] independently proved that $T_{t_n}(\hat{\Sigma}_{p,n})$ with $t_n = M\sqrt{n^{-1}\log p}$ for sufficiently large M, is consistent for $\Sigma_p \in \mathcal{U}_\tau(q, C_0(p), M)$ if $\log p = o(n)$, $C_0(p) \leq Mn^{\frac{1-q}{2}}(\log p)^{-\frac{3-q}{2}}$ and observations $\{C_{ip}\}$ are independent.

Basu and Michailidis [2015] appears to be the only work that has dealt with estimation of sparse covariance matrices in dependent models. In finite-dimensional time series, temporal dependence is usually quantified by some mixing conditions on the underlying stochastic process. For example, a widely used mixing condition is the functional dependence measure introduced by Wu [2005]. Wu and Wu [2014] and Chen et al. [2013] investigated the asymptotic properties of *lasso* and *covariance thresholding* in finite-dimensional time series, assuming a specific rate of decay for this functional dependence measure.

This route is hard to follow in the high-dimensional context, even for simple linear processes. For instance, consider the high-dimensional AR(1) process $X_t = AX_{t-1} + \varepsilon_t$. Then the above functional dependence measure boils down to some restrictions on the spectral radius $\rho(A) := |\lambda_{\max}(A)|$ of A. Simulation results on high-dimensional AR(1) process, given in Basu and Michailidis [2015], provides evidence that dependence in the data is not completely captured by $\rho(A)$ and can affect the convergence rates of estimates in a more intricate manner.

This is a motivation to introduce a different mixing condition to quantify dependence in the high-dimensional setting.

Measure of stability. Under Assumption (E1), the covariance matrix

$$\mathrm{Cov}(C_{ip}, C_{jp}) = C(|i-j|)$$

depends only on the lag $|i-j|$. Note that $C(h)$ are all $p \times p$ matrices and moreover $C(0) = \Sigma_p$. Consider the following assumption.

(E2) The *matrix* spectral density function

$$f(\theta) := \frac{1}{2\pi} \sum_{h=-\infty}^{\infty} C(h)e^{-ih\theta}, \quad \theta \in [-\pi, \pi]$$

exists, and

$$\mathcal{M}(f) := \mathrm{ess\ sup}_{\theta \in [-\pi,\pi]} \lambda_{\max}(f(\theta)) < \infty.$$

Existence of the spectral density is guaranteed if $\sum_{h=0}^{\infty} \|C(h)\|_2^2 < \infty$. Further, it also implies that the spectral density is bounded, continuous and essential supremum in the definition of $\mathcal{M}(f)$ is actually the maximum. Assumption (E2) is satisfied by a large class of general linear processes, including causal and invertible ARMA processes. Moreover, the spectral density has a closed form expression for these processes. For example, for the causal invertible ARMA(d, l) process

$$X_t = A_1 X_{t-1} + \cdots + A_d X_{t-d} + \varepsilon_t - B_1 \varepsilon_{t-1} - \cdots - B_l \varepsilon_{t-l} \qquad (2.74)$$

with $\mathrm{Var}(\varepsilon_t) = \Sigma_p(\varepsilon)$, the spectral density takes the form

$$f(\theta) = \frac{1}{2\pi} \mathcal{A}^{-1}(e^{-i\theta}) \mathcal{B}(e^{-i\theta}) \Sigma_p(\varepsilon) \mathcal{B}^*(e^{-i\theta}) (\mathcal{A}^{-1}(e^{-i\theta}))^*, \qquad (2.75)$$

where $-\pi \leq \theta \leq \pi$ and $\mathcal{A}^{-1}(z)$ and $\mathcal{B}^{-1}(z)$ are the usual matrix inverses of

$$\mathcal{A}(z) = I_p - \sum_{t=1}^{d} A_t z^t \text{ and } \mathcal{B}(z) = I_p - \sum_{t=1}^{l} B_t z^t.$$

When the spectral density exists, we have

$$C(h) = \int_{-\pi}^{\pi} f(\theta) e^{ih\theta} d\theta \;\; \forall h = 0, \pm 1, \pm 2, \ldots.$$

Since $\{C(h)\}$ or $f(\theta)$ uniquely characterizes a centered Gaussian process, it can be used to quantify the temporal and cross-sectional dependence for Gaussian processes.

The quantity $\mathcal{M}(f)$ may be called a *measure of stability* of the process. For any subset $\{i_1, i_2\}$ of $\{1, 2, \ldots, p\}$, we can similarly define the measure of stability $\mathcal{M}(f, (i_1, i_2))$ of the two-dimensional sub-process that consists of the i_1-th and i_2-th components of C_{ip}. The combined stability measure of all 2-dimensional sub-processes of $\{C_{ip}\}$ can then be defined as

$$\mathcal{M}(f, 2) = \max_{\{i_1, i_2\} \subset \{1, 2, \ldots, p\}} \mathcal{M}(f, (i_1, i_2)). \qquad (2.76)$$

Clearly, $\mathcal{M}(f, 2) \leq \mathcal{M}(f)$. Therefore, (E2) implies $\mathcal{M}(f, 2) < \infty$.

Then we have the following theorem whose proof is left as an exercise.

Theorem 2.2.1. *(Basu and Michailidis [2015]) Suppose $\{C_{ip}\}$ satisfies (E1) and (E2). For sufficiently large M, let $u_n = \mathcal{M}(f, 2) M \sqrt{\log p/n}$ and $n \geq \mathcal{M}^2(f, 2) \log p$. Then uniformly on $\mathcal{U}_\tau(q, C_0(p), M)$,*

$$\|T_{u_n}(\hat{\Sigma}_{p,n}) - \Sigma_p\|_2 = O_P\big(C_0(p)\big(\mathcal{M}^2(f, 2) \frac{\log p}{n}\big)^{(1-q)/2}\big).$$

Therefore, if $\mathcal{M}(f, 2)$ is bounded, then the above threshold estimator recovers the rate of convergence given in Theorem 1.5.1 for the i.i.d. setting.

We conclude by noting that, estimation of population autocovariance matrices of different orders is very important in the analysis of a stationary time series model. The population autocovariance matrix of order 0 is nothing but the population covariance matrix. Therefore, some very specific situations of estimation of autocovariance matrices can be handled by the results of this chapter. The next chapter deals with estimation of autocovariance matrices in details.

Exercises

1. Consider the model (2.2)–(2.3). Discuss the performance of banded and tapered sample covariance matrix to estimate Σ_p. In this case rewrite the parameter spaces given in (2.34), (2.36), and (2.37).

2. Provide an example where the cross covariance structure does not obey any of (2.34), (2.36), and (2.37).

3. Consider the model (2.5). Check whether an appropriately banded sample covariance matrix is consistent for $\Sigma_p = ((\theta^{|i-j|}))$ $(\theta \in (0,1))$ and establish its convergence rate. In this context, also state sufficient conditions on the cross covariance structure ∇_{np}. What can one say about the tapered estimator?

4. Consider the model (2.5) with

$$
\Sigma_p = \begin{bmatrix} ((\theta_1^{|i-j|}))_{r \times r} & ((\theta_2^{|i-j|+r}))_{r \times p-r} \\ ((\theta_2^{|i-j|+r}))_{p-r \times r} & ((\theta_3^{|i-j|}))_{p-r \times p-r} \end{bmatrix}
$$

and $\theta_1, \theta_2, \theta_3 \in (0,1)$. Discuss sufficient conditions on $\theta_1, \theta_2, \theta_3$ and ∇_{np} so that an appropriately banded and tapered sample covariance matrix is consistent for Σ_p. Also establish its rate of convergence.

5. Learn a proof of Lemma 2.1.2.

6. Simulate Gaussian random vectors so that $\Sigma_p \in \mathcal{U}(\epsilon, \alpha, C) \cap \mathcal{V}$ and $\nabla_{np} \in \mathcal{L}_n(a_n)$. Compute $k_{n,\alpha}^\alpha \| B_{k_{n,\alpha}}(\hat{\Sigma}_{p,n}) - \Sigma_p \|_2$ for $k_{n,\alpha} = (n^{-1} a_n \log p)^{-\frac{1}{(\alpha+1)}}$ and 500 replications, draw its histogram and comment on the rate of convergence.

7. Establish the form (2.75) of the spectral density of the ARMA(d, l) process (2.74).

8. Show that (E2) implies $\mathcal{M}(f, 2) < \infty$.

9. Provide an example where (E2) does not hold.

10. Learn the proof of Theorem 2.2.1.

Chapter 3

LARGE AUTOCOVARIANCE MATRIX

In Chapters 1 and 2 we have seen examples of high-dimensional time series data. The most common assumptions made in modelling such data is stationarity.

Let $\{X_{t,p} : t = 0, \pm 1, \pm 2, \ldots\}$ be p-dimensional random vectors with $E(X_{t,p}) = 0$ for all t. It is called *weak or covariance stationary* if and only if, for all $u \geq 0$, the $p \times p$ matrix

$$\Gamma_{u,p} = E(X_{t,p} X_{t+u,p}^*) \tag{3.1}$$

does not depend on t and is a function of only u. The matrix $\Gamma_{u,p}$ is called the *(population) autocovariance matrix* of order u. Note that $\Gamma_{0,p}$ is the covariance matrix of $X_{t,p}$.

Both p and n grow so that we are in a high-dimensional setting. In this chapter, the interest is on estimating $\{\Gamma_{u,p}\}$ based on a sample $\{X_{t,p} : 1 \leq t \leq n\}$ from a linear infinite-dimensional moving average process of order ∞ (MA(∞)) (see (3.3)). Under some causality conditions, this model includes, infinite-dimensional IID processes (see (3.6)), infinite-dimensional finite-order moving average processes MA(r) (see (3.8)) and, infinite-dimensional vector autoregressive processes IVAR(r) with *i.i.d.* innovations (see (3.10)).

A method of moment estimator of $\Gamma_{u,p}$ is given by the $p \times p$ matrix

$$\hat{\Gamma}_{u,p,n} = \frac{1}{n} \sum_{t=1}^{n-u} X_{t,p} X_{t+u,p}^*. \tag{3.2}$$

$\hat{\Gamma}_{u,p,n}$ is called the *sample autocovariance matrix*. Just like the sample variance-covariance matrices in Chapters 1 and 2 turned out to be inconsistent, the matrices $\hat{\Gamma}_{u,p,n}$ also fail to consistently estimate $\Gamma_{u,p}$.

Taking a cue from Chapter 2, we provide consistency results for banded and tapered $\hat{\Gamma}_{0,p,n}$, under restrictions on the parameter space. However, these restrictions are cumbersome and may be difficult to check. Also the approach from Chapter 2 does not provide any direction on how to estimate the entire autocovariance sequence $\{\Gamma_{u,p}\}$ consistently. We discuss simpler restrictions on the parameter spaces for the particular cases of MA(r) and IVAR(r).

Then we deal with the consistent estimation of $\Gamma_{u,p}$ by using banding and tapering, with an appropriate parameter space, under the Gaussian assumption on the driving process. Upper bound for the convergence rate of these estimators are also demonstrated. We also show how to obtain consistent estimators for the parameter matrices of IVAR(r).

For applications, the Gaussian assumption may be deemed too strong. We argue how one can replace this by an appropriate condition on the moment generating function. Most of the earlier results continue to hold under this condition.

3.1 Models and examples

As mentioned in the previous section, a very general high-dimensional linear time series model is the MA(∞) process given by

$$X_{t,p}^{(n)} = \sum_{j=0}^{\infty} \psi_{j,p}\varepsilon_{t-j}, \quad t, n \geq 1 \text{ (almost surely)}, \tag{3.3}$$

where $\{X_{t,p}^{(n)}\}$ and $\{\varepsilon_t\}$ are both p-dimensional random vectors, $\{\varepsilon_t\}$ are i.i.d. with mean 0 and $p \times p$ variance-covariance matrix Σ_p, $\{\psi_{j,p}\}$ are $p \times p$ *parameter matrices*. Appropriate conditions are always assumed on $\{\psi_{j,p}\}$ so that the above infinite series is meaningful. The dimension $p = p(n) \to \infty$ as the sample size $n \to \infty$. It may be noted that the dimension of $X_{t,p}$ is not infinite. However, since $p \to \infty$, it has become customary to refer to such models as "infinite dimensional". This is a weakly stationary time series and the *population autocovariance matrix* of order u is given by

$$\Gamma_{u,p} = \sum_{j=0}^{\infty} \psi_{j,p}\Sigma_p\psi_{j+u,p}^{*}, \text{ for all } u \geq 0. \tag{3.4}$$

Clearly in high-dimensional setting, the size of the coefficient matrices $\{\psi_{j,p}\}$ increases as p increases and consequently as we move from the n-th stage to the $(n+1)$-th stage, all the components of $X_{t,p}^{(n)}$ get changed. Hence, in the high-dimensional setting, we have the following triangular sequence:

$$\begin{aligned}
&X_{1,p(1)}^{(1)} \\
&X_{1,p(2)}^{(2)}, X_{2,p(2)}^{(2)} \\
&X_{1,p(3)}^{(3)}, X_{2,p(3)}^{(3)}, X_{3,p(3)}^{(3)} \\
&\quad\vdots \\
&X_{1,p(n)}^{(n)}, X_{2,p(n)}^{(n)}, X_{3,p(n)}^{(n)}, \ldots, X_{n,p(n)}^{(n)} \\
&\quad\vdots
\end{aligned} \tag{3.5}$$

and the sample at the n-th stage is the n-th row of this triangular sequence. For convenience, we usually write $X_{t,p}$ for $X_{t,p}^{(n)}$.

Example 3.1.1. The infinite-dimensional IID process is given by

$$X_{t,p} = \varepsilon_t, \text{ for all } t \tag{3.6}$$

where $\{\varepsilon_t\}$ is a sequence of i.i.d. p-dimensional random vectors with mean 0 and $p \times p$ variance-covariance matrix Σ_p.

The process in (3.6) is a weak stationary time series process with

$$\Gamma_{u,p} = \begin{cases} \Sigma_p, & \text{if } u = 0, \\ 0, & \text{otherwise.} \end{cases} \tag{3.7}$$

Note that if $\psi_{0,p} = I_p$ and $\psi_{j,p} = 0$, for $j \geq 1$, then (3.3) reduces to (3.6).

Example 3.1.2. The infinite-dimensional moving average process MA(r) of order r is given by

$$X_{t,p} = \sum_{i=0}^{r} M_{i,p} \varepsilon_{t-i}, \ t \geq 1 \tag{3.8}$$

where $\{\varepsilon_t\}$ is as in Example 3.1.1, $M_{i,p}, i = 0, 1, 2, \ldots, r$ are square matrices of order p and the parameter matrices with $M_{0,p} = I_p$.

It is easy to see that MA(r) is a weak stationary model and

$$\Gamma_{u,p} = \begin{cases} \sum_{i=0}^{r-u} M_{i,p} \Sigma_p M_{i+u,p}^*, & \text{for } 0 \leq u \leq r, \\ 0, & \text{otherwise.} \end{cases} \tag{3.9}$$

For $r = 0$, (3.8) is same as the IID process given in Example 3.1.1. If $\psi_{j,p} = M_{j,p} I (0 \leq j \leq r)$, then (3.3) reduces to (3.8).

Example 3.1.3. The infinite-dimensional vector autoregressive process IVAR(r) of order r is given by

$$X_{t,p} = \sum_{i=1}^{r} A_{i,p} X_{t-i,p} + \varepsilon_t, \ t \geq 1 \tag{3.10}$$

where $\{\varepsilon_t\}$ is as in Example 3.1.1. The $p \times p$ matrices $\{A_{i,p}\}$ are called the *parameter matrices*.

Suppose I_p is the identity matrix of order p and $\mathbb{C}$ is the set of all complex numbers. If for some $\epsilon > 0$, $\{A_{i,p}\}$ satisfy the *causality condition*

$$det(I_p - A_{1,p}z - A_{2,p}z^2 - \cdots - A_{r,p}z^r) \neq 0, \text{ for all } z \in \mathbb{C}, \ |z| \leq 1 + \epsilon \tag{3.11}$$

then (3.10) is a weak stationary process and has the representation,

$$X_{t,p} = \sum_{j=0}^{\infty} \phi_{j,p}\varepsilon_{t-j}, \quad l \geq 1 \quad \text{(almost surely)}, \tag{3.12}$$

where

$$\phi_{0,p} = I_p \text{ and } \phi_{j,p} = \sum_{i=1}^{j} A_{i,p}\phi_{j-i,p}, \quad j \geq 1. \tag{3.13}$$

Let $\{z_{i,p} : 1 \leq i \leq r\}$ be the r roots of the equation

$$det(I_{p(n)} - A_{1,p(n)}z - A_{2,p(n)}z^2 - \cdots - A_{r,p(n)}z^r) = 0, \quad z \in \mathbb{C}.$$

Let

$$\alpha_p = \min\{|z_{i,p}| : 1 \leq i \leq r\}.$$

By Theorem 11.3.1 in Brockwell and Davis [2009], for each fixed p, (3.10) can be represented as (3.12) with the coefficient matrices (3.13), if

$$\alpha_p > 1. \tag{3.14}$$

Note that (3.11) implies (3.14) for all $p \geq 1$. Further, its autocovariance matrices are given by

$$\Gamma_{u,p} = \sum_{j=0}^{\infty} \phi_{j,p}\Sigma_p\phi_{j+u,p}^*, \quad \text{for all } u \geq 0. \tag{3.15}$$

3.2   Estimation of $\Gamma_{0,p}$

Note that $\Gamma_{0,p}$ is the covariance matrix of $\{X_{t,p}\}$. As in the covariance matrix estimation in Chapters 1 and 2, to get a consistent estimator of $\{\Gamma_{0,p}\}$, we need suitable restrictions on $\{\psi_{j,p}\}$ and on Σ_p. Further we also need appropriate modification such as banding or tapering of $\{\hat{\Gamma}_{0,p,n}\}$. This section provides a class of such restrictions so that the appropriately banded and tapered $\hat{\Gamma}_{0,p,n}$ become consistent. These restrictions are directly borrowed from the developments of Chapter 2.

We recall some notions from Chapters 1 and 2. The $\infty \times \infty$ extension Σ_∞ of $\{\Sigma_p\}$ is defined in the sense (1.20). The class of well-conditioned and bandable dispersion matrices as defined in (1.28) is given by

$$\mathcal{U}(\epsilon, \alpha, C) = \mathcal{W}(\epsilon) \cap \mathcal{X}(\alpha, C) \tag{3.16}$$

where the class of well-conditioned dispersion matrices is

$$\mathcal{W}(\epsilon) = \{\Sigma_\infty : 0 < \epsilon < \inf_p \lambda_{\max}(\Sigma_p) \leq \sup_p \lambda_{\max}(\Sigma_p) < \epsilon^{-1} < \infty\} \tag{3.17}$$

and the class of bandable dispersion matrices is given by

$$\mathcal{X}(\alpha, C) = \{A : T(A, k) \leq Ck^{-\alpha}, \forall k \geq 1\}, \text{ for } \alpha, C > 0 \text{ with} \quad (3.18)$$

$$T(A, k) = \sup_{j} \sum_{i:|i-j|>k} |a_{ij}|. \quad (3.19)$$

Recall the k-banded version of a matrix $M = ((m_{ij}))$ as defined in Section 1.2.2(a):

$$B_k(M) = ((m_{ij}I(|i-j| \leq k))). \quad (3.20)$$

Also recall tapering from Section 1.2.2(b). For any matrix M,

$$R_{g,\tau_{n,\alpha}}(M) = ((m_{ij}g(|i-j|/\tau_{n,\alpha}))) \quad (3.21)$$

where the tapering function g satisfies the following assumption.

(T1) $g : \mathbb{R}^+ \cup \{0\} \rightarrow \mathbb{R}^+ \cup \{0\}$ is continuous, non-increasing such that

$$g(0) = 1 \text{ and } \lim_{x \to \infty} g(x) = 0.$$

Also recall

$$\Delta_{g,\tau_{n,\alpha}} = \sum_{j=0}^{n-1} g\left(\frac{j}{\tau_{n,\alpha}}\right).$$

The following theorem on infinite-dimensional IID or MA(0) process is a restatement of Theorems 1.3.1 and 1.3.3.

Theorem 3.2.1. *Consider the model (3.6). Suppose $\varepsilon_t \sim \mathcal{N}_p(0, \Sigma_p)$, for all t and $\Sigma_\infty \in \mathcal{U}(\epsilon, \alpha, C)$ for some $\epsilon, \alpha, C > 0$. Then*

(a) for $k_{n,\alpha} \asymp (n^{-1}\log p)^{-\frac{1}{2(1+\alpha)}}$, we have

$$||B_{k_{n,\alpha}}(\hat{\Gamma}_{0,p,n}) - \Gamma_{0,p}||_2 = O_P(k_{n,\alpha}^{-\alpha}). \quad (3.22)$$

(b) Further suppose (T1) holds. Then for $\Delta_{g,\tau_{n,\alpha}} \asymp (n^{-1}\log p)^{-\frac{1}{2(1+\alpha)}}$, we have

$$||R_{g,\tau_{n,\alpha}}(\hat{\Gamma}_{0,p,n}) - \Gamma_{0,p}||_2 = O_P(\Delta_{g,\tau_{n,\alpha}}^{-\alpha}). \quad (3.23)$$

Recall the class of dispersion matrices $\mathcal{V}$ as defined in (2.11) is given by

$$\mathcal{V} = \{\Sigma_\infty = ((\sigma_{ij})) : \sigma_{ij} \neq 0, \forall i, j\}. \quad (3.24)$$

Additionally consider the following assumption on the tapered function g.

(T2) $\int_0^\infty g(x) < \infty$ and $1 - g(x) = O(x^\gamma)$ for some $\gamma \geq 1$ in some neighborhood of zero.

Then we have the following theorem for the infinite-dimensional MA(∞) process.

Theorem 3.2.2. *Consider the model (3.3) where $\varepsilon_t \sim \mathcal{N}_p(0, \Sigma_p)$. Suppose that for some $\alpha, \epsilon, C > 0$*

$$\sum_{j=0}^{\infty} \psi_{j,p} \Sigma_p \psi_{j,p}^* \in \mathcal{U}(\epsilon, \alpha, C) \cap \mathcal{V} \tag{3.25}$$

and

$$\max_{a_n \leq u} \max_{v,w} \frac{\left|\left(\sum_{j=0}^{\infty} \psi_{j,p} \Sigma_p \psi_{j+u,p}^*\right)(v,w)\right|}{\left|\left(\sum_{j=0}^{\infty} \psi_{j,p} \Sigma_p \psi_{j,p}^*\right)(v,w)\right|} = O(n^{-2} a_n), \tag{3.26}$$

for some a_n such that $a_n \sqrt{n^{-1} \log p} \to 0$ and $a_n^{-1} \sqrt{n \log p} \to \infty$ as $n \to \infty$. Then

(a) for $k_{n,\alpha} \asymp (a_n \sqrt{n^{-1} \log p})^{-\frac{1}{1+\alpha}}$, we have

$$\|B_{k_{n,\alpha}}(\hat{\Gamma}_{0,p,n}) - \Gamma_{0,p}\|_2 = O_P(k_{n,\alpha}^{-\alpha}). \tag{3.27}$$

(b) Additionally suppose (T1) and (T2) hold. Then for $\tau_{n,\alpha} \asymp (a_n \sqrt{n^{-1} \log p})^{-\frac{1}{2(1+\alpha)}[\frac{\gamma}{1+\alpha}+1]}$, we have

$$\|R_{g,\tau_{n,\alpha}}(\hat{\Gamma}_{0,p,n}) - \Gamma_{0,p}\|_2 = O_P\left[\left(a_n \sqrt{n^{-1} \log p}\right)^{\frac{\gamma\alpha}{(1+\alpha)(1+\gamma)}}\right]. \tag{3.28}$$

Proof. To prove the above theorem, we use Theorems 2.1.3 and 2.1.4(c). There the approach was to separate out the cross covariance structure ∇_{np}, an $np \times np$ matrix consisting of n^2-many $p \times p$ matrices $\{\Lambda_{ij} : 1 \leq i, j, \leq n\}$. Note that, by (2.6) and weak stationarity of (3.3), for all $1 \leq i, j \leq n$, $1 \leq v, w \leq p$ we have

$$\Lambda_{ij}(v,w) = \Lambda_{|i-j|}(v,w) = \frac{E(X_{i,p} X_{j,p}^*)(v,w)}{E(X_{i,p} X_{i,p}^*)(v,w)} = \frac{\Gamma_{i-j,p}(v,w)}{\Gamma_{0,p}(v,w)}, \tag{3.29}$$

provided $\Gamma_{0,p}(v,w) \neq 0$, for all v, w. Now, by (3.4), for all $1 \leq v, w \leq p$ we have

$$\Lambda_u(v,w) = \frac{\left(\sum_{j=0}^{\infty} \psi_{j,p} \Sigma_p \psi_{j+u,p}^*\right)(v,w)}{\left(\sum_{j=0}^{\infty} \psi_{j,p} \Sigma_p \psi_{j,p}^*\right)(v,w)}, \quad \text{for all } u \geq 0 \tag{3.30}$$

provided $(\sum_{j=0}^{\infty} \psi_{j,p} \Sigma_p \psi_{j,p}^*)(v,w) \neq 0$, for all v, w. Then Theorem 3.2.2 follows from Theorems 2.1.3 and 2.1.4(c) provided (3.25) and (3.26) hold. □

Conditions (3.25) and (3.26) are cumbersome and difficult to check in general unless there is some additional structure in the model. It is not at all clear what conditions on the parameter matrices are needed for general $MA(\infty)$ models so that these are satisfied. Here are some specific $MA(\infty)$ models.

Example 3.2.1. Consider the model (3.3), with $\psi_{j,p} = \theta^j A_p$ for any $0 < \theta < 1$ and $p \times p$ matrix A_p such that all elements of $A_p \Sigma_p A_p^*$ are non-zero. Then for all $1 \le v, w \le p$ and $u \ge 1$, we have

$$
\begin{aligned}
\Lambda_u(v, w) &= \frac{\left| \left(\sum_{j=0}^{\infty} \psi_{j,p} \Sigma_p \psi_{j+u,p}^* \right)(v, w) \right|}{\left| \left(\sum_{j=0}^{\infty} \psi_{j,p} \Sigma_p \psi_{j,p}^* \right)(v, w) \right|} \\
&= \frac{\theta^u \left| \left(A_p \Sigma_p A_p^* \right)(v, w) \right| \left(\sum_{j=0}^{\infty} \theta^{2j} \right)}{\left| \left(A_p \Sigma_p A_p^* \right)(v, w) \right| \left(\sum_{j=0}^{\infty} \theta^{2j} \right)} = \theta^u.
\end{aligned}
$$

Therefore, $\sup_{u \ge a_n} ||\Lambda_u||_\infty = \theta^{a_n}$ and (3.26) holds.

The *cross covariance structure* model used in Theorem 3.2.2 is meaningful if and only if all elements of the matrix $\Gamma_{0,p}$ are non-zero (see (3.25)). There are of course many processes where this may not be the case. Here are two simple examples.

Example 3.2.2. $\psi_{j,p} = \theta^j I_p$ for all j with at least one zero element in Σ_p. Then $\Gamma_{0,p}$ has some zero entries.

Example 3.2.3. Suppose $\Sigma_p = I_p$ and $\psi_{j,p}$'s are such that $\psi_{j,p} \psi_{j,p}^*$'s are diagonal. For example, one can think of $\psi_{j,p}$ to be any asymmetric Toeplitz matrix made of $\{t_i\}_{i=-\infty}^{\infty}$ with $t_i = 0$ for all i except one. Then again $\Gamma_{0,p}$ has some zero entries.

3.3   Estimation of $\Gamma_{u,p}$

Having shown how to estimate $\Gamma_{0,p}$, we now move to the problem of estimating $\{\Gamma_{u,p}\}$, $u \ge 1$. It is not clear from Theorems 3.2.1 and 3.2.2, how to estimate the *cross covariances* in general. Moreover, the assumptions of Theorem 3.2.2 do not offer any control over the first few cross covariances. These restrictions are not sufficient for consistent estimation of the autocovariance matrices. Thus, the first task is to identify appropriate parameter spaces.

3.3.1   Parameter spaces

There are two kinds of parameters in the model (3.3)· One is Σ_p and the other is the set of coefficient matrices $\{\psi_{j,p}\}$. From the experience of Chapter 1 (see (1.40), (1.54), and (1.55)), the following two conditions should be a minimal requirement:

(i) $\sup_p ||\Gamma_{u,p}||_\infty < \infty$, for all $u \ge 0$ where $|| \cdot ||_\infty$ in (1.34).

(ii) $\Gamma_{u,p} \in \mathcal{X}(\alpha, C)$ for some $\alpha, C > 0$ where $\mathcal{X}(\alpha, C)$ is the class of matrices having polynomially decaying corner as given in (3.18).

Below we discuss appropriate restrictions on both type of parameters so that

$\{\Gamma_{u,p}\}$ satisfy (i) and (ii) above and as a consequence, consistent estimators of $\{\Gamma_{u,p}\}$ can be obtained.

Restrictions on Σ_p. Recall the notation (1.20) and let Σ_∞ be the $\infty \times \infty$ extension of $\{\Sigma_p\}$. Since infinite-dimensional IID process is a particular case of the model (3.3), it is justifiable to continue to assume that

$$\Sigma_\infty \in \mathcal{U}(\epsilon, \alpha, C), \text{ for some } \epsilon, \alpha, C > 0 \tag{3.31}$$

where $\mathcal{U}(\epsilon, \alpha, C)$ is as given in (3.16).

Restrictions on $\{\psi_{j,p}\}$. For each $j \geq 0$, consider the $\infty \times \infty$ extension of the sequence of matrices $\{\psi_{j,p(n)}\}_{n \geq 1}$ as $\psi_{j,\infty}$ (in the sense (1.20)). Recall the $||\cdot||_{(1,1)}$ norm and the corner measure $T(\cdot, \cdot)$ respectively from (1.26) and (3.19).

(i) **Time lag criterion:** We ensure that the dependence decreases appropriately with the lag. For this purpose, define

$$\max(||\psi_{j,\infty}||_{(1,1)}, ||\psi_{j,\infty}^*||_{(1,1)}) = r_j, \text{ for all } j \geq 0. \tag{3.32}$$

We define the following class $\Im(\beta, \lambda)$ of sequence of matrices $\{\psi_{j,\infty}\}_{j=0}^\infty$ for some $0 < \beta < 1$ and $\lambda \geq 0$.

$$\Im(\beta, \lambda) = \{\{\psi_{j,\infty}\}_{j=0}^\infty : \sum_{j=0}^\infty r_j^\beta < \infty, \sum_{j=0}^\infty r_j^{2(1-\beta)} j^\lambda < \infty\}. \tag{3.33}$$

Note that the summability above implies that the decay of r_j cannot be slower than at a polynomial rate.

(ii) **Spatial lag criterion:** For any $1 \leq i \leq p$, let $X_{t,p.i}$ be the i-th component of the vector $X_{t,p}$. Here we ensure that for any $t_1 < t$ and $k > 0$, the dependence between $X_{t_1,p.(i \pm k)}$ and $X_{t,p.i}$ grows weaker as the lag k increases. We achieve this by putting restrictions over $\{T(\psi_{j,\infty}, k) : j = 0, 1, 2, \dots\}$ for all $k > 0$. Consider the following class $\mathcal{G}(C, \alpha, \eta, \nu)$ for some $C, \alpha, \nu > 0$ and $0 < \eta < 1$ as

$$\mathcal{G}(C, \alpha, \eta, \nu) = \{\{\psi_{j,\infty}\} : T(\psi_{j,\infty}, t \sum_{u=0}^j \eta^u) < Ct^{-\alpha} r_j j^\nu \sum_{u=0}^j \eta^{-u\alpha},$$

$$\sum_{j=0}^\infty \frac{r_j r_{j+u} j^\nu}{\eta^{\alpha j}} < \infty \text{ for all } u \geq 0\}. \tag{3.34}$$

Recall the two conditions (i) and (ii) described at the beginning of this section and the class of well conditioned dispersion matrices $\mathcal{W}(\epsilon)$ defined in (3.17). In the following theorems we provide sufficient conditions for (i) and (ii) to hold.

Theorem 3.3.1. *(Bhattacharjee and Bose [2014b]) Consider the model (3.3). Suppose $\Sigma_\infty \in \mathcal{W}(\epsilon)$ and $\{\psi_{j,\infty}\} \in \mathfrak{S}(\beta, \lambda)$ for some $\epsilon > 0$, $\lambda \geq 0$ and $0 < \beta < 1$. Then*

$$\sup_p \|\Gamma_{u,p}\|_\infty < \infty, \text{ for all } u \geq 0.$$

To prove the above theorem we need the following lemma.

Lemma 3.3.1. *For any two square matrices A and B of same order,*

$$\|AB\|_\infty \leq \min\{\|A\|_\infty\|B\|_{(1,1)}, \|B\|_\infty\|A^*\|_{(1,1)}\}.$$

Proof. Recall the notation (1.4). Then

$$
\begin{aligned}
\|AB\|_\infty &= \max_{i,j}|AB(i,j)| \leq \max_{i,j}\sum_k |A(i,k)B(k,j)| \\
&\leq \max_{i,k}|A(i,k)|\max_j\sum_k |B(k,j)| = \|A\|_\infty\|B\|_{(1,1)}.
\end{aligned}
$$

Similarly, one can show that $\|AB\|_\infty \leq \|B\|_\infty\|A^*\|_{(1,1)}$. This completes the proof. $\square$

Proof of Theorem 3.3.1. Note that as $\Sigma_\infty = ((\sigma_{ij})) \in \mathcal{W}(\epsilon)$, we have

$$|\sigma_{ij}| \leq \sqrt{\sigma_{ii}\sigma_{jj}} \leq \lambda_{\max}(\Sigma_p) < \epsilon^{-1}, \text{ for all } i,j.$$

Therefore,

$$\sup_p \|\Sigma_p\|_\infty = \|\Sigma_\infty\|_\infty < \epsilon^{-1}. \tag{3.35}$$

Also for the model (3.3), $\Gamma_{u,p} = \sum_{j=0}^\infty \psi_{j,p}\Sigma_p\psi^*_{j+u,p}$, for all $u \geq 0$.

Therefore, by (3.35) and the repeated use of Lemma 3.3.1, we have

$$
\begin{aligned}
\sup_p \|\Gamma_{u,p}\|_\infty &\leq \sup_p \sum_{j=0}^\infty \|\psi_{j,p}\Sigma_p\psi^*_{j+u,p}\|_\infty \\
&\leq \sup_p \sum_{j=0}^\infty \|\psi^*_{j,p}\|_{(1,1)}\|\Sigma_p\psi^*_{j+u,p}\|_\infty \\
&\leq \sup_p \sum_{j=0}^\infty \|\psi^*_{j,p}\|_{(1,1)}\|\Sigma_p\|_\infty\|\psi^*_{j+u,p}\|_{(1,1)} \\
&\leq \epsilon^{-1}\sum_{j=0}^\infty r_j r_{j+u} < \infty,
\end{aligned}
$$

as $\{\psi_{j,\infty}\} \in \mathfrak{S}(\beta, \lambda)$ for some $\lambda \geq 0$ and $0 < \beta < 1$. $\square$

Theorem 3.3.2. *(Bhattacharjee and Bose [2014b]) Consider the model (3.3). If $\Sigma_\infty \in \mathcal{X}(\alpha, C)$ and $\{\psi_{j,\infty}\}_{j=0}^\infty \in \mathcal{G}(C, \alpha, \eta, \nu)$ for some $C, \alpha, \nu > 0$ and $0 < \eta < 1$, then for all $t > 0$ and some $c' > 0$,*

$$T(\Gamma_{u,p(n)}, t) < c' t^{-\alpha} \|\Sigma_{p(n)}\|_{(1,1)}, \text{ for all } u \geq 0, \ n \geq 1.$$

Moreover, if $\|\Sigma_\infty\|_{(1,1)} < \infty$ then $\Gamma_{u,p(n)} \in \mathcal{X}(\alpha, c')$, for all $u \geq 0$, $n \geq 1$.

To prove the above theorem, we need the following lemma on the corner measure $T(\cdot, \cdot)$ and $\|\cdot\|_{(1,1)}$ norm of a square matrix.

Lemma 3.3.2. *Let A and B be two $r \times r$ matrices. Then,*

(i) $T(A, k) \leq T(A, k')$, *for all* $0 < k' < k < \infty$.

(ii) $\|AB\|_{(1,1)} \leq \|A\|_{(1,1)} \|B\|_{(1,1)}$,

(iii) $T(AB, (\alpha + \beta)t) \leq \|A\|_{(1,1)} T(B, \alpha t) + \|B\|_{(1,1)} T(A, \beta t)$, *for any* $\alpha, \beta, t > 0$.

Proof. Proofs of (i) and (ii) are trivial. To prove (iii), consider the following steps.

$$T(AB, (\alpha + \beta)t)$$

$$\leq \max_k \sum_{j:|j-k|>(\alpha+\beta)t} \sum_{l=1}^\infty |a_{jl}b_{lk}|$$

$$\leq \max_k \sum_{j:|j-k|>(\alpha+\beta)t} \sum_{l:|l-k|\leq\alpha t} |a_{jl}b_{lk}| + \max_k \sum_{j:|j-k|>(\alpha+\beta)t} \sum_{l:|l-k|>\alpha t} |a_{jl}b_{lk}|$$

$$\leq \max_k \sum_{j:|j-l|>\beta t, \ l:|l-k|\leq\alpha t} |a_{jl}b_{lk}| + \max_k \sum_{j:|j-k|>(\alpha+\beta)t, \ l:|l-k|>\alpha t} |a_{jl}b_{lk}|$$

$$\leq \Big(\max_l \sum_{j:|j-l|>\beta t} |a_{jl}|\Big)\Big(\max_k \sum_{l=1}^\infty |b_{lk}|\Big) + \Big(\max_k \sum_{l:|l-k|>\alpha t} |b_{lk}|\Big)\Big(\max_l \sum_{j=1}^\infty |a_{jl}|\Big)$$

$$\leq \|B\|_{(1,1)} T(A, \beta t) + \|A\|_{(1,1)} T(B, \alpha t).$$

This completes the proof. $\square$

Now we are ready to prove Theorem 3.3.2.

Proof of Theorem 3.3.2. Let $\delta_p = \|\Sigma_p\|_{(1,1)}$. From the properties of $\mathcal{X}(\alpha, C)$ and $\mathcal{G}(C, \alpha, \eta, \nu)$ and, by Lemma 3.3.2 (iii), for some $C_1 > 0$,

$$T(\psi_{j,p}\Sigma_p, (\sum_{k=0}^j \eta^k)t) \leq \delta_p C t^{-\alpha}(\sum_{k=0}^j \eta^{-k\alpha})r_j j^\nu + r_j C t^{-\alpha} \eta^{-(j+1)\alpha}$$

$$\leq C_1 t^{-\alpha}\delta_p(\sum_{k=0}^{j+1} \eta^{-k\alpha})r_j j^\nu.$$

Again, by Lemma 3.3.2 (ii), (iii) and for some $C_2 > 0$,

$$T(\psi_{j,p}\Sigma_p\psi_{j+u,p}^*, 2(\sum_{k=0}^{j+1}\eta^k)t) \leq r_{j+u}C_1 t^{-\alpha}\delta_p(1 + \sum_{k=0}^{j+1}\eta^{-k\alpha})r_j j^\nu$$

$$+\delta_p r_j C t^{-\alpha}(\sum_{k=0}^{j+1}\eta^{-k\alpha})r_{j+u}j^\nu$$

$$\leq C_2 t^{-\alpha}\delta_p(\sum_{k=0}^{j+1}\eta^{-k\alpha})r_j r_{j+u}j^\nu.$$

Hence, as $\{\psi_{j,\infty}\} \in \mathcal{G}(C, \alpha, \eta, \nu)$ and by Lemma 3.3.2 (i), for some $C_3, C_4 > 0$, we have

$$T(\Gamma_u, \frac{2}{1-\eta}t) = T(\sum_{j=0}^{\infty}\psi_{j,p}\Sigma_p\psi_{j+u,p}^*, \frac{2}{1-\eta}t)$$

$$< C_3 t^{-\alpha}\delta_p \sum_{j=0}^{\infty}\frac{r_j r_{j+u}}{\eta^{\alpha j}}j^\nu$$

$$< C_4 t^{-\alpha}\delta_p.$$

Hence, the proof of Theorem 3.3.2 is complete. $\square$

Thus, by Theorems 3.3.1 and 3.3.2, it is clear that we need to assume $\Sigma_\infty \in \mathcal{U}(\epsilon, \alpha, C)$ and $\{\psi_{j,\infty}\} \in \mathfrak{S}(\beta, \lambda) \cap \mathcal{G}(C, \alpha, \eta, \nu)$ for some $\lambda \geq 0$, $C, \alpha, \epsilon, \nu > 0$ and $0 < \beta, \eta < 1$, to guarantee $\{\Gamma_{u,p}\}$ have polynomially decaying corners and $\sup_p ||\Gamma_{u,p}||_\infty < \infty$, for all $u \geq 0$. As mentioned at the beginning of this section, these two conditions will be crucially used when we deal with the banded and tapered sample autocovariance matrices in the next section.

3.3.2 Estimation

We are now ready to show that appropriate banded and tapered version of $\hat{\Gamma}_{u,p,n}$ are consistent for $\Gamma_{u,p}$ in the sense of (1.14). *Throughout this section, we assume $p = p(n) \to \infty$ as $n \to \infty$ in such a way that $n^{-1}\log p(n) \to 0$.*

Recall that for any matrix M of order p, its k-*banded* version is as in (3.20). Also recall the tapered version of a matrix as in (3.21) and Assumptions (T1) and (T2) stated respectively before Theorems 3.2.1 and 3.2.2 on the tapering function. Then we have the following theorem.

Theorem 3.3.3. *(Bhattacharjee and Bose [2014b]) Consider the model (3.3). Suppose the driving process $\epsilon_t \sim \mathcal{N}_p(0, \Sigma_p)$, for all t, $\Sigma_\infty \in \mathcal{U}(\epsilon, \alpha, C)$ and $\{\psi_{j,\infty}\} \in \mathfrak{S}(\beta, \lambda) \cap \mathcal{G}(C, \alpha, \eta, \nu)$ for some $C, \epsilon, \alpha, \mu > 0$, $\lambda \geq 0$ and $0 < \beta, \eta < 1$. Then for $k_{n,\alpha} \asymp (n^{-1}\log p)^{-\frac{1}{2(\alpha+1)}}$ and $u \geq 0$, we have*

$$||B_{k_{n,\alpha}}(\hat{\Gamma}_{u,p,n}) - \Gamma_{u,p}||_2 = O_P(k_{n,\alpha}^{-\alpha}||\Sigma_p||_{(1,1)}). \tag{3.36}$$

Further suppose (T1) and (T2) hold. Then for $u \geq 0$ and
$$\tau_{n,\alpha} \asymp \left(n^{-1}\log p\right)^{-\frac{1}{2(1+\gamma)}\left[\frac{\gamma}{1+\alpha}+1\right]},$$

$$\|R_{g,\tau_{n,\alpha}}(\hat{\Gamma}_{u,p,n}) - \Gamma_{u,p}\|_2 = O_P\left[\left(n^{-1}\log p\right)^{\frac{\gamma\alpha}{2(1+\alpha)(1+\gamma)}}\|\Sigma_p\|_{(1,1)})\right]. \quad (3.37)$$

To prove the above theorem, we need the following two lemmas. Lemma 3.3.3 provides the rate for convergence of the sample autocovariance matrices to their corresponding population autocovariance matrices in $\|\cdot\|_\infty$ norm for the infinite-dimensional IID process. This turns out to be useful since the model (3.3) is driven by an infinite-dimensional IID process. Lemma 3.3.4 provides a summability condition which is useful to establish an upper bound to the rate of convergences involved in Theorem 3.3.3.

Lemma 3.3.3. *Suppose $\{\varepsilon_t\}$ are i.i.d. $\mathcal{N}_p(0,\Sigma_p)$. Then*

$$(i) \quad \|\frac{1}{n}\sum_{t=1}^{n}\varepsilon_t\varepsilon_t^* - \Sigma_p\|_\infty = O_P\left(\sqrt{n^{-1}\log p}\right)$$

$$(ii) \quad \|\frac{1}{n}\sum_{t=1}^{n-u}\varepsilon_t\varepsilon_{t+u}^*\|_\infty = O_P\left(\sqrt{n^{-1}\log p}\right), \quad \text{for all } u \geq 1.$$

Proof. (i) follows from (1.43). For (ii) let, $z_{t,i} = \dfrac{\varepsilon_{t,i}}{\sqrt{\Gamma_{0,p.ii}}}$ for all i,t, where $\varepsilon_{t,i}$ is the i-th component of ε_t and $\Gamma_{0,p.ii}$ is the (i,i)th entry of $\Gamma_{0,p}$. Then for some $c_1 > 0$,

$$P[\|\frac{1}{n}\sum_{t=1}^{n-u}\varepsilon_t\varepsilon_{t+u}^*\|_\infty > t] \leq \sum_{l,m}P[|\sum_{t=1}^{n-u}\{\frac{(z_{t,l}+z_{t+u,m})^2}{2} - 1\}| > c_1 nt]$$

$$+ \sum_{l,m}P[|\sum_{t=1}^{n-u}\{\frac{(z_{t,l}-z_{t+u,m})^2}{2} - 1\}| > c_2 nt].$$

Since, $\dfrac{(z_{t,l} \pm z_{t+u,m})^2}{2}, t \geq 1$ are all independent χ_1^2 variables, by Lemma 1.35, for some $c_2, c_3 > 0$,

$$P[\|\frac{1}{n}\sum_{t=1}^{n-u}\varepsilon_t\varepsilon_{t+u}^*\|_\infty > t] \leq c_3 p^2 e^{-c_2 nt^2} \to 0 \quad \text{as } n \to \infty \quad (3.38)$$

for $t = M\sqrt{n^{-1}\log p}$ and an appropriate $M > 0$. Hence, (ii) is proved. $\quad\square$

Lemma 3.3.4. *Let $\{a_j\}_{j=0}^{\infty}$ be any sequence of positive real members such that $\displaystyle\sum_{j=0}^{\infty}a_j^{\beta} < \infty$ and $\displaystyle\sum_{j=0}^{\infty}a_j^{2(1-\beta)}j^\lambda < \infty$, for some $\lambda > 0$ and $0 < \beta < 1$.*

Then for an appropriately chosen $M > 0$,

$$p^2 \sum_{1\leq i,j<\infty} p^{-\frac{M}{(a_i a_j)^{2(1-\beta)}}} \to 0 \quad as \ p \to \infty.$$

Proof. Let $\nu = 2(1-\beta)$. As $\sum_{j=1}^{\infty} a_j^\nu j^\lambda < \infty$, we have $\frac{1}{a_j^\nu} > j^\lambda$ for all $j > N$ and some $N \geq 1$. Now

$$p^2 \sum_{i,j} p^{-\frac{M}{(a_i a_j)\nu}} \leq \sum_{1\leq i,j\leq N} p^{-\frac{M}{(a_i a_j)\nu}+2} + p^2 \sum_{\{1\leq i,j\leq N\}^c} p^{-M(ij)^\lambda}.$$

In the first sum, as we have finitely many terms, it tends to 0 as $p \to \infty$. Now,

$$p^2 \sum_{\{1\leq i,j\leq N\}^c} p^{-M(ij)^\lambda} \leq p^2 \sum_{k=N}^\infty (k-N+1)p^{-Mk^\lambda} \leq C_1 p^2 \sum_{r=R}^\infty r^{\frac{1}{\lambda}} p^{-Mr}$$

$$\leq C_1 p^{2-MR} \sum_{r=0}^\infty (r+R)^{\frac{1}{\lambda}} p^{-Mr} \leq C_2 p^{2-MR}$$

for some C_1, C_2 and $R > 0$. This tends to 0 for an appropriately chosen large $M > 0$. Hence, the proof is complete. $\square$

Now we are ready to prove Theorem 3.3.3.

Proof of Theorem 3.3.3

Proof of (3.36). By Lemma 3.5.1, $\|B_{k_{n,\alpha}}(\hat{\Gamma}_{u,p,n}) - \Gamma_{u,p}\|_2$ is bounded above by

$$\sqrt{\|B_{k_{n,\alpha}}(\hat{\Gamma}_{u,p,n}) - \Gamma_{u,p}\|_{(1,1)}\|B_{k_{n\alpha}}(\hat{\Gamma}^*_{u,p,n}) - \Gamma^*_{u,p}\|_{(1,1)}}. \quad (3.39)$$

First, we shall show that

$$\|B_{k_{n,\alpha}}(\hat{\Gamma}_{u,p,n}) - \Gamma_{u,p}\|_{(1,1)} = O_P(k_{n,\alpha}^{-\alpha}\|\Sigma_p\|_{(1,1)}). \quad (3.40)$$

Using triangle inequality and by Lemma 1.3.2, the left side, say L, of the above display satisfies

$$L \leq \|B_{k_{n,\alpha}}(\hat{\Gamma}_{u,p,n}) - B_{k_{n,\alpha}}(\Gamma_{u,p})\|_{(1,1)} + T(\Gamma_{u,p}, k_{n,\alpha})$$
$$\leq (2k_{n,\alpha}+1)\|\hat{\Gamma}_{u,p,n} - \Gamma_{u,p}\|_\infty + T(\Gamma_{u,p}, k_{n,\alpha}) \quad (3.41)$$

By Theorem 3.3.2, we have

$$T(\Gamma_{u,p}, k_{n,\alpha}) = O(k_{n,\alpha}^{-\alpha}\|\Sigma_p\|_{(1,1)}). \quad (3.42)$$

Using the model (3.3) and Lemma 3.3.2 (ii),

$$\|\hat{\Gamma}_{u,p,n} - \Gamma_{u,p}\|_\infty \leq \sum_{j=0}^\infty \sum_{i=0}^\infty r_i r_j \|\frac{1}{n}\sum_{t=1}^{n-u} \varepsilon_{t,j}\varepsilon^*_{(t+u),i} - E_{ij}\|_\infty$$

where $E_{ij} = E\varepsilon_{t,j}\varepsilon^*_{(t+u),i}$ for all i, j. Hence, for some $C_1 > 0$,

$$P[||\hat{\Gamma}_{u,p,n} - \Gamma_{u,p}||_\infty > t]$$

$$\leq P\Big[\sum_j\sum_i r_i r_j ||\frac{1}{n}\sum_{t=1}^{n-u}\varepsilon_{t,j}\varepsilon^*_{(t+u),i} - E_{ij}||_\infty > \sum_j\sum_i \frac{C_1 t}{r_i^{-\beta}r_j^{-\beta}}\Big]$$

$$\leq \sum_j\sum_i P[||\frac{1}{n}\sum_{t=1}^{n-u}\varepsilon_{t,j}\varepsilon^*_{(t+u),i} - E_{ij}||_\infty > \frac{C_1 t}{r_i^{1-\beta}r_j^{1-\beta}}].$$

Now, by (1.53) and (3.38), for $t = M\sqrt{n^{-1}\log p}$,

$$P[||\hat{\Gamma}_{u,p,n} - \Gamma_{u,p}||_\infty > t] \leq p^2 \sum_{i,j} p^{-\frac{M}{(r_i r_j)^{2(1-\beta)}}}.$$

By Lemma 3.3.4, this tends to zero as $n \to \infty$. Hence,

$$||\hat{\Gamma}_{u,p,n} - \Gamma_{u,p}||_\infty = O_P(\sqrt{n^{-1}\log p}), \quad \text{for all } u \geq 0. \tag{3.43}$$

Now by (3.41) and (3.42), the proof of (3.40) is complete.

Similarly, one can show that

$$||B_{k_{n,\alpha}}(\hat{\Gamma}_{u,p,n})^* - \Gamma^*_{u,p}||_{(1,1)} = O_P(k_{n,\alpha}^{-\alpha}||\Sigma_p||_{(1,1)}). \tag{3.44}$$

Therefore, putting together (3.39), (3.40), (3.43), and (3.44), proof of (3.36) is complete.

Proof of (3.37). By Lemma 3.5.1, $||R_{g,\tau_{n,\alpha}}(\hat{\Gamma}_{u,p,n}) - \Gamma_{u,p}||_2$ is bounded above by

$$\sqrt{||R_{g,\tau_{n,\alpha}}(\hat{\Gamma}_{u,p,n}) - \Gamma_{u,p}||_{(1,1)}||R_{g,\tau_{n,\alpha}}(\hat{\Gamma}^*_{u,p,n}) - \Gamma^*_{u,p}||_{(1,1)}}. \tag{3.45}$$

First, we shall show that

$$||R_{g,\tau_{n,\alpha}}(\hat{\Gamma}_{u,p,n}) - \Gamma_{u,p}||_{(1,1)} = O_P\Big[(n^{-1}\log p)^{\frac{\gamma\alpha}{2(1+\alpha)(1+\gamma)}}||\Sigma_p||_{(1,1)})\Big]. \tag{3.46}$$

Using triangle inequality,

$$||R_{g,\tau_{n,\alpha}}(\hat{\Gamma}_{u,p,n}) - \Gamma_{u,p}||_{(1,1)} \leq ||R_{g,\tau_{n,\alpha}}(\hat{\Gamma}_{u,p,n}) - R_{g,\tau_{n,\alpha}}(\Gamma_{u,p})||_{(1,1)}$$
$$+ ||R_{g,\tau_{n,\alpha}}(\Gamma_{u,p}) - \Gamma_{u,p}||_{(1,1)}. \tag{3.47}$$

Now, for some constant $C_1 > 0$,

$$||R_{g,\tau_{n,\alpha}}(\hat{\Gamma}_{u,p,n}) - R_{g,\tau_{n,\alpha}}(\Gamma_{u,p})||_{(1,1)} \leq ||\hat{\Gamma}_{u,p,n} - \Gamma_{u,p}||_\infty 2\sum_{l=0}^p g\Big(\frac{l}{\tau_{n,\alpha}}\Big)$$

$$\leq \tau_{n,\alpha}||\hat{\Gamma}_{u,p,n} - \Gamma_{u,p}||_\infty C_1 \int_0^\infty g(x)dx.$$

Therefore, by (3.43), we have

$$\|R_{g,\tau_{n,\alpha}}\left(\hat{\Gamma}_{u,p,n}\right) - R_{g,\tau_{n,\alpha}}\left(\Gamma_{u,p}\right)\|_{(1,1)} = O_P(\tau_{n,\alpha}\sqrt{n^{-1}\log p}). \qquad (3.48)$$

Again, by triangle inequality

$$\begin{aligned}
\|R_{g,\tau_{n,\alpha}}(\Gamma_{u,p}) - \Gamma_{u,p}\|_{(1,1)} &\leq \quad \|R_{g,\tau_{n,\alpha}}(\Gamma_{u,p}) - B_{k'_{n,\alpha}}[R_{g,\tau_{n,\alpha}}(\Gamma_{u,p})]\|_{(1,1)} \\
&\quad + \|B_{k'_{n,\alpha}}[R_{g,\tau_{n,\alpha}}(\Gamma_{u,p})] - B_{k'_{n,\alpha}}(\Gamma_{u,p})\|_{(1,1)} \\
&\quad + \|B_{k'_{n,\alpha}}(\Gamma_{u,p}) - \Gamma_{u,p}\|_{(1,1)}. \qquad (3.49)
\end{aligned}$$

By Lemma 1.3.1 and Theorem 3.3.2, we have

$$\|B_{k'_{n,\alpha}}(\Gamma_{u,p}) - \Gamma_{u,p}\|_{(1,1)} = O((k'_{n,\alpha})^{-\alpha}\|\Sigma_p\|_{(1,1)}) \qquad (3.50)$$

and

$$\|R_{g,\tau_{n,\alpha}}(\Gamma_{u,p}) - B_{k'_{n,\alpha}}[R_{\tau_{n,\alpha}}(\Gamma_{u,p})]\|_{(1,1)} = O((k'_{n,\alpha})^{-\alpha}\|\Sigma_p\|_{(1,1)}). \qquad (3.51)$$

Now, by Theorem 3.3.1, for some constant $C_2, C_3 > 0$ and for sufficiently large n,

$$\begin{aligned}
&\|B_{k'_{n,\alpha}}[R_{g,\tau_{n,\alpha}}(\Gamma_{u,p})] - B_{k'_{n,\alpha}}(\Gamma_{u,p})\|_{(1,1)} \\
&\leq \max_i \sum_{j:|i-j|\leq k'_{n,\alpha}} \left(1 - g(\frac{|i-j|}{\tau'_{n,\alpha}})\right)\left(\sup_p \|\Gamma_{u,p}\|_\infty\right) \\
&\leq C_2 \sum_{l=-k'_{n,\alpha}}^{k'_{n,\alpha}} \left(1 - g(\frac{l}{\tau_{n,\alpha}})\right) \leq C_3\left(\frac{k'_{n,\alpha}}{\tau_{n,\alpha}}\right)^\gamma k'_{n,\alpha}. \qquad (3.52)
\end{aligned}$$

Now, consider $k'_{n,\alpha} = \left(n^{-1}\log p\right)^{-\frac{\gamma}{2(1+\gamma)(1+\alpha)}}$. Therefore, by (3.47)–(3.52), the proof of (3.46) is complete.

Similarly, one can show that

$$\|R_{g,\tau_{n,\alpha}}(\hat{\Gamma}^*_{u,p,n}) - \Gamma^*_{u,p}\|_{(1,1)} = O_P\left[(n^{-1}\log p)^{\frac{\gamma\alpha}{2(1+\alpha)(1+\gamma)}}\|\Sigma_p\|_{(1,1)})\right]. \qquad (3.53)$$

Hence, by (3.45), (3.46), and (3.53), the proof of (3.37) is complete. Therefore, Theorem 3.3.3 is proved. $\qquad\square$

Remark 3.3.1. *The rate of convergence depends not only on the class of coefficient matrices and the covariance matrix but also on $\|\Sigma_p\|_{(1,1)}$. This is to be expected since we are considering linear regression type models. Moreover, if $\|\Sigma_p\|_{(1,1)}$ is bounded, then the rate of convergence for $\Gamma_{0,p}$ is same as that for infinite-dimensional IID process as given in Theorem 1.3.1.*

As we have seen, the infinite-dimensional MA(r) processes and IVAR(r) processes defined respectively in Examples 3.1.2 and 3.1.3, are all particular cases of the model (3.3). Therefore, the obvious curiosity is under what condition on $\{M_{i,p}\}$ in Examples 3.1.2 and $\{A_{i,p}\}$ in Examples 3.1.3, would the corresponding coefficient matrices be in $\Im(\beta, \lambda) \cap \mathcal{G}(C, \alpha, \eta, \nu)$ for some $\lambda \geq 0$, $C, \alpha, \nu > 0$ and $0 < \beta, \eta < 1$ so that consistent estimation is possible in these models? We deal with these two models in the next two sections.

3.4   Estimation in MA(r)

Parameter space. Consider the model (3.8) and its $p \times p$ parameter matrices $\{M_{i,p} : 1 \leq i \leq r\}$ with $M_{0,p} = I_p$. For each $0 \leq i \leq r$, let $M_{i,\infty}$ be the $\infty \times \infty$ extension of the matrices $\{M_{i,p(n)}\}_{n \geq 1}$ in the sense (1.20). The following theorem provides a simplified condition on $\{M_{i,\infty} : 0 \leq i \leq r\}$ so that they belong to $\Im(\beta, \lambda) \cap \mathcal{G}(C, \alpha, \eta, \nu)$ for some $\lambda \geq 0$, $C, \alpha, \nu > 0$ and $0 < \beta, \eta < 1$. Recall the class of matrices having polynomially decaying corners, denoted by $\mathcal{X}(\alpha, C)$ for some $\alpha, C > 0$, in (3.18).

Theorem 3.4.1. *(Bhattacharjee and Bose [2014b]) Suppose $\|M_{i,\infty}\|_{(1,1)} < \infty$ and $M_{i,\infty} \in \mathcal{X}(\alpha, C)$ for some $\alpha, C > 0$ and for all $1 \leq i \leq r$. Then*

$$\{M_{i,\infty} : i \geq 0\} \in \left(\bigcap_{\substack{0 < \beta < 1 \\ \lambda \geq 0}} \Im(\beta, \lambda) \right) \cap \left(\bigcap_{\substack{0 < \eta < 1 \\ \nu > 0}} \mathcal{G}(Cm^{-1}, \alpha, \eta, \nu) \right),$$

where $m = \min\{\|M_{i,\infty}\|_{(1,1)} : 1 \leq i \leq r\}$.

Proof. Note that in the model (3.8), r_j as in (3.32) is given by

$$r_j = \begin{cases} \|I_\infty\|_{(1,1)} = 1, & \text{if } j = 0, \\ \|\psi_{j,\infty}\|_{(1,1)} = \|M_{j,\infty}\|_{(1,1)}, & \text{if } 1 \leq j \leq r, \\ 0, & \text{if } j > r. \end{cases}$$

Therefore, as there are only finitely many non-zero r_j's, all the summability conditions on $\{r_j\}$ in $\Im(\beta, \lambda)$ and $\mathcal{G}(C, \alpha, \eta, \nu)$ are satisfied for all $\lambda \geq 0$, $\alpha, \nu > 0$ and $0 < \beta, \eta < 1$.

Next, for all $j \geq 0$, $\nu > 0$ and $0 < \eta < 1$, we have

$$T\left(M_{j,\infty}, t \sum_{u=0}^{j} \eta^u\right) \quad < \quad Ct^{-\alpha}\left(\sum_{u=0}^{j} \eta^u\right)^{-\alpha} \tag{3.54}$$

$$< \quad (Cm^{-1})r_j t^{-\alpha} j^\nu j^{-\alpha}\left(j^{-1}\sum_{u=0}^{j} \eta^u\right)^{-\alpha}$$

$$< \quad (Cm^{-1})r_j t^{-\alpha} j^\nu j^{-\alpha-1}\sum_{u=0}^{j} \eta^{-u\alpha} \tag{3.55}$$

$$< \quad (Cm^{-1})r_j t^{-\alpha} j^\nu \sum_{u=0}^{j} \eta^{-u\alpha}.$$

This completes the proof. $\square$

Estimation. We now specialize Theorem 3.3.3 to the infinite-dimensional MA(r) processes. The next theorem follows directly from Theorem 3.3.3 once we invoke Theorem 3.4.1

Theorem 3.4.2. *(Bhattacharjee and Bose [2014b]) Consider the model (3.8). Suppose the driving process $\varepsilon_t \sim \mathcal{N}_p(0, \Sigma_p)$, for all t. Also suppose $\Sigma_\infty \in \mathcal{U}(\epsilon, \alpha, C)$, $0 < \|M_{i,\infty}\|_{(1,1)} < \infty$ and $M_{i,\infty} \in \mathcal{X}(\alpha, C)$ for some $\epsilon, \alpha, C > 0$ and for all $1 \leq i \leq r$. Then for $k_{n,\alpha} \asymp (n^{-1}\log p)^{-\frac{1}{2(\alpha+1)}}$, we have*

$$\|B_{k_{n,\alpha}}(\hat{\Gamma}_{u,p,n}) - \Gamma_{u,p}\|_2 \;=\; O_P(k_{n,\alpha}^{-\alpha}\|\Sigma_p\|_{(1,1)}). \tag{3.56}$$

Additionally suppose (T1) and (T2) hold. Then for $u \geq 0$ and
$\tau_{n,\alpha} \asymp (n^{-1}\log p)^{-\frac{1}{2(1+\gamma)}\left[\frac{\gamma}{1+\alpha}+1\right]}$,

$$\|R_{g,\tau_{n,\alpha}}(\hat{\Gamma}_{u,p,n}) - \Gamma_{u,p}\|_2 = O_P\left[(n^{-1}\log p)^{\frac{\gamma\alpha}{2(1+\alpha)(1+\gamma)}}\|\Sigma_p\|_{(1,1)}\right]. \tag{3.57}$$

3.5   Estimation in IVAR(r)

Parameter space Consider the model (3.10) and its representation (3.12). For each $i \geq 0$, let $\phi_{i,\infty}$ be the $\infty \times \infty$ extension of the sequence of matrices $\{\phi_{i,p(n)}\}_{n\geq 1}$ in the sense (1.20).

Theorem 3.5.1 provides direct conditions on the parameter matrices $\{A_{i,p}\}$ so that the corresponding coefficient matrices $\{\phi_{i,\infty}\} \in \Im(\beta, \lambda) \cap \mathcal{G}(C, \alpha, \eta, \nu)$ for some $\lambda \geq 0$, $C, \alpha, \nu > 0$ and $0 < \beta, \eta < 1$. To state the theorem, we need some preparation.

Let

$$\|A_{i,p}\|_{(1,1)} = \theta_{i,n} \quad \text{and} \quad \|A_{i,p}^*\|_{(1,1)} = \theta'_{i,n}, 1 \leq i \leq r. \tag{3.58}$$

Also let $\{\alpha_{i,n} : i = 1, 2, \ldots, r\}$ and $\{\alpha'_{i,n} : i = 1, 2, \ldots, r\}$ respectively be the roots of the following polynomials.

$$1 - \theta_{1,n}z - \theta_{2,n}z^2 - \cdots - \theta_{r,n}z^r \;=\; 0,$$
$$1 - \theta'_{1,n}z - \theta'_{2,n}z^2 - \cdots - \theta'_{r,n}z^r \;=\; 0.$$

For each $1 \leq i \leq r$, let $A_{i,\infty}$ be the $\infty \times \infty$ extension of the sequence of matrices $\{A_{i,p(n)}\}_{n\geq 1}$. Consider the parameter space $\mathcal{P}(C, \alpha, \epsilon)$ for $\{A_{i,\infty}\}_{i=1}^r$ defined as,

$$\left\{\{A_{i,\infty}\}_{i=1}^r : \inf_p \min_{1 \leq i \leq r}(|\alpha_{i,p}|, |\alpha'_{i,p}|) > 1 + \epsilon, \; A_{i,\infty} \in \mathcal{X}(\alpha, C) \; \forall \; i\right\} \tag{3.59}$$

for some $C, \epsilon, \alpha > 0$. Now, we are prepared to state the following theorem.

Theorem 3.5.1. *(Bhattacharjee and Bose [2014b]) If $\{A_{i,\infty}\}_{i=1}^r \in \mathcal{P}(C, \alpha, \varepsilon)$, then (3.12) holds. Also, $\{\phi_{i,\infty}\}_{i=0}^\infty \in \Im(\beta, 0) \cap \mathcal{G}(C, \alpha, \eta, 1)$ for any $0 < \beta < 1$ and some $0 < \eta < 1$.*

To prove the above theorem, we need the following two lemmas. Lemma 3.5.1 provides an inequality on matrix norms and Lemma 3.5.2 describes an important property of stationary univariate autoregressive processes.

Lemma 3.5.1. *(see Golub and van Loan [1996]) Let M be a square matrix. Then*

$$||M||_2 \leq \sqrt{||M||_{(1,1)}||M^*||_{(1,1)}}.$$

Note that Lemma 3.5.1 implies Lemma 1.3.2, when M is a symmetric matrix.

Lemma 3.5.2. *(see Brockwell and Davis [2009]) Consider a univariate autoregressive process of order r:*

$$x_t = b_1 x_{t-1} + b_2 x_{t-2} + \cdots + b_r x_{t-r} + e_t, \text{ for all } t, \qquad (3.60)$$

where $\{e_t\}$ are i.i.d. with mean 0 and variance σ^2. If $\{b_i\}$ satisfies

$$1 - b_1 z - b_2 z^2 - \cdots - b_r z^r \neq 0, \text{ for all } z \in \mathbb{C}, |z| < 1, \qquad (3.61)$$

then we have the following representation

$$x_t = \sum_{i=0}^{\infty} d_i e_{t-i}, \text{ where } d_0 = 1, \ d_j = \sum_{i=1}^{j} b_i d_{j-i}, \text{ for all } j \geq 1.$$

and moreover, there exists a $0 < \delta < 1$ and $c > 0$ such that $|d_i| < c\delta^i$.

Now, we are ready to prove Theorem 3.5.1.

Proof of Theorem 3.5.1. This proof involves the following three steps.

Step 1: *Proof that (3.12) holds.* Note that, we need to show that if $\{A_{i,\infty}\}_{i=1}^r \in \mathcal{P}(C, \alpha, \varepsilon)$, then it will satisfy condition (3.11). Define the polynomials

$$\begin{aligned} p_1(x) &= \theta_{1,n}x + \theta_{2,n}x^2 + \cdots + \theta_{r,n}x^r, \\ p_2(x) &= \theta'_{1,n}x + \theta'_{2,n}x^2 + \cdots + \theta'_{r,n}x^r. \end{aligned}$$

Note that $p_1(0) = p_2(0) = 0$ and both of them are increasing functions of x. Also, as $(1 - p_1(x))$ and $(1 - p_2(x))$ have all their roots strictly greater than $(1 + \varepsilon)$, $p_i(1 + \varepsilon) < 1$ for all $i = 1, 2$. Let us write I and A_i respectively for I_p and $A_{i,p}$ for all $i \geq 1$. Now, for any $|z| \leq 1 + \varepsilon$ and any $x \neq 0$, by Lemma 3.5.1

$$|\sum_{k=1}^{r} x'A_k x z^k| \leq \sum_{i=1}^{r} \sqrt{\theta_i \theta'_i}|z|^i \leq \frac{1}{2}(p_1(1 + \varepsilon) + p_2(1 + \varepsilon)) < 1.$$

Hence, there exists no $x \neq 0$ such that $(I - A_1 z - A_2 z^2 \cdots - A_r z^r)x = 0$. Therefore, (3.11) is satisfied.

Step 2: *Proof that $\{\phi_{i,\infty}\}_{i=0}^{\infty} \in \Im(\beta, 0)$ for any $0 < \beta < 1$ holds.* Consider the autoregressive processes

$$\begin{aligned} y_t &= \theta_1 y_{t-1} + \theta_2 y_{t-2} + \cdots + \theta_r y_{t-r} + e_t, \\ z_t &= \theta'_1 z_{t-1} + \theta'_2 z_{t-2} + \cdots + \theta'_r z_{t-r} + e_t \end{aligned}$$

where e_t, $t = 1, 2, \ldots$ are independently distributed with mean 0 and variance σ^2 and for all $1 \leq i \leq r$, $\theta_i = ||A_{i,\infty}||_{(1,1)}$ and $\theta_i' = ||A_{i,\infty}^*||_{(1,1)}$. If $\{A_{i,\infty}\}_{i=1}^r \in \mathcal{P}(C, \alpha, \varepsilon)$, then by Lemma 3.5.2, we have the representations,

$$y_t = \sum_{i=0}^{\infty} \alpha_i e_{t-i} \quad \text{and} \quad z_t = \sum_{i=0}^{\infty} \beta_i e_{t-i} \text{ for all } t$$

where

$$\alpha_0 = 1, \ \alpha_j = \sum_{i=1}^{j} \theta_i \alpha_{j-i} \quad \text{and} \quad \beta_0 = 1, \ \beta_j = \sum_{i=1}^{j} \theta_i' \beta_{j-i} \text{ for all } j \geq 1$$

and there exist $0 < \delta < 1$, $c > 0$, such that

$$\max(\alpha_i, \beta_i) < c\delta^i \text{ for all } i.$$

Therefore, using Lemma 3.3.2(i) repeatedly, we have $||\phi_{i,\infty}||_{(1,1)} < \alpha_i$ and hence

$$||\phi_{i,\infty}||_{(1,1)} < c\delta^i \text{ for all } i \text{ for some } c > 0, \ 0 < \delta < 1. \tag{3.62}$$

Therefore, $\{\phi_{i,\infty}\}_{i=0}^{\infty} \in \mathfrak{S}(\beta, \lambda)$ for any $0 < \beta < 1$ and $\lambda = 0$.

Step 3: *Proof that $\{\phi_{i,\infty}\} \in \mathcal{G}(C, \alpha, \eta, 1)$ for some $0 < \eta < 1$ holds.* By (3.62), the summability condition on $\{||\phi_{i,\infty}||_{(1,1)}\}$ in $\mathcal{G}(C, \alpha, \eta, 1)$ is satisfied. Therefore, it remains to justify the condition on $T(\cdot, \cdot)$ in $\mathcal{G}(C, \alpha, \eta, 1)$.

Now consider any $i \leq r$. Then

$$||\phi_{i,\infty}||_{(1,1)} \leq \sum_{j=1}^{i} ||A_{j,\infty}||_{(1,1)} ||\phi_{i-j,\infty}||_{(1,1)} < c\delta^i.$$

Hence,

$$||A_{i,\infty}||_{(1,1)} < c\delta^i, \ 1 \leq i \leq r.$$

Since $A_{i,\infty} \in \mathcal{X}(\alpha, C)$ for all $i \leq r$, we have $T(A_{i,\infty}, t) < C_1 \delta^i t^{-\alpha}$, for some $C_1 > 0$. Note that

$$T(\phi_{1,\infty}, (1 + \eta)t) < ct^{-\alpha} \delta(1 + \eta^{-\alpha}).$$

We now apply induction. Suppose,

$$T(\phi_{j,\infty}, \sum_{k=0}^{j} \eta^k t) < ct^{-\alpha} (\sum_{k=0}^{j} \eta^{-k\alpha}) \delta^j j.$$

Then, by Lemma 3.3.2(iii), for all $j > k$,

$$T(A_{k,\infty} \phi_{j-k,\infty}, \sum_{k=0}^{j} \eta^k t) \leq \delta^k ct^{-\alpha} (\sum_{s=0}^{j-k} \eta^{-s\alpha}) \delta^{j-k} j + \delta^{j-k} ct^{-\alpha} \eta^{-(j-k+1)\alpha} \delta^k$$

$$\leq ct^{-\alpha} \delta^j j (\sum_{k=0}^{j} \eta^{-k\alpha}).$$

Since, $\phi_{j+1,\infty} = \sum_{i=0}^{j+1} A_{i,\infty}\phi_{j-i+1,\infty}$, for some $C' > 0$,

$$T\left(\phi_{j+1,\infty}, \sum_{k=0}^{j+1} \eta^k t\right) \leq C' t^{-\alpha} \delta^j j^2 \sum_{k=0}^{j+1} \eta^{-k\alpha}.$$

Hence, the proof of the theorem is complete. □

Estimation of autocovariance matrices. We now specialize Theorem 3.3.3 to the IVAR(r) processes. The next theorem follows directly from Theorem 3.3.3 once we invoke Theorem 3.5.1. We omit its proof.

Theorem 3.5.2. *(Bhattacharjee and Bose [2014b]) Consider the model (3.10). Suppose the driving process $\varepsilon_t \sim \mathcal{N}_p(0, \Sigma_p)$, for all t. Also suppose $\Sigma_\infty \in \mathcal{U}(\epsilon, \alpha, C)$ and $\{A_{i,\infty}\}_{i=1}^r \in \mathcal{P}(C, \alpha, \varepsilon)$ for some $\alpha, \epsilon, C > 0$ and for all $1 \leq i \leq r$. Then for $k_{n,\alpha} \asymp (n^{-1}\log p)^{-\frac{1}{2(\alpha+1)}}$, we have*

$$||B_{k_{n,\alpha}}(\hat{\Gamma}_{u,p,n}) - \Gamma_{u,p}||_2 \quad = \quad O_P(k_{n,\alpha}^{-\alpha}||\Sigma_p||_{(1,1)}). \tag{3.63}$$

Further suppose (T1) and (T2) hold. Then for $u \geq 0$ and $\tau_{n,\alpha} \asymp (n^{-1}\log p)^{-\frac{1}{2(1+\gamma)}\left[\frac{\gamma}{1+\alpha}+1\right]}$,

$$||R_{g,\tau_{n,\alpha}}(\hat{\Gamma}_{u,p,n}) - \Gamma_{u,p}||_2 = O_P\left[(n^{-1}\log p)^{\frac{\gamma\alpha}{2(1+\alpha)(1+\gamma)}}||\Sigma_p||_{(1,1)}\right]. \tag{3.64}$$

Estimation of parameter matrices. The next task is to consistently estimate the parameter matrices $\{A_{i,p} : 1 \leq i \leq r\}$ and the covariance matrix of the driving process $\{\varepsilon_{t,p}\}$, i.e., Σ_p for the IVAR(r) process. By right multiplying both sides of (3.10) with $X_{t-k,p}^*$, $k = 1, 2, \ldots, r$ successively and then taking expectation, we have

$$\begin{aligned}
\Gamma_{1,p}^* &= A_{1,p}\Gamma_{0,p} + A_{2,p}\Gamma_{1,p} + \cdots + A_{r,p}\Gamma_{r-1,p} \tag{3.65}\\
\Gamma_{2,p}^* &= A_{1,p}\Gamma_{1,p}^* + A_{2,p}\Gamma_{0,p} + \cdots + A_{r,p}\Gamma_{r-2,p}
\end{aligned}$$

$$\vdots$$

$$\Gamma_{r,p}^* = A_{1,p}\Gamma_{r-1,p}^* + A_{2,p}\Gamma_{r-2,p}^* + \cdots + A_{r,p}\Gamma_{0,p}.$$

Let

$$\mathcal{Y}_{r,n} = (\Gamma_{1,p}, \Gamma_{2,p}, \ldots, \Gamma_{r,p})^*, \quad \mathcal{A}_{r,n} = (A_{1,p}^*, A_{2,p}^*, \ldots, A_{r,p}^*)^*$$

and let $G_{r,n}$ be a block matrix with r^2 many $p \times p$ blocks

$$G_{r,n}(i,j) = \Gamma_{|i-j|,p}I(i < j) + \Gamma_{|i-j|,p}^* I(i \geq j), \ 1 \leq i,j \leq r. \tag{3.66}$$

Then from (3.65) we have,

$$\mathcal{Y}_{r,n} = G_{r,n}\mathcal{A}_{r,n}. \tag{3.67}$$

This is analogous to the Yule–Walker equations for a finite-dimensional AR process. The following lemma implies the invertibility of the matrix $G_{r,n}$ for all $n \geq 1$. Recall the definition of $\lambda_{\min}$ in (1.7).

Lemma 3.5.3. *Fix any $n \geq 1$. If $\lambda_{\min}(\Gamma_{0,p(n)}) > 0$ and $||\Gamma_{h,p(n)}||_2 \to 0$ as $h \to \infty$, then $G_{r,n}$ is non-singular.*

Proof. Suppose that $G_{q,p}$ is non-singular but $G_{q+1,p}$ is singular. Then there exist $a, a_1, a_2, \ldots, a_q$ such that

$$a^* X_{q+1,p} = \sum_{j=1}^{q} a_j^* X_{j,p} \quad \text{a.s.}.$$

By stationarity,

$$a^* X_{q+h+1,p} = \sum_{j=1}^{q} a_j^* X_{h+j,p} \quad \text{for all } h \geq 1 \text{ a.s.}.$$

So, for all $K \geq q + 1$, there exists $a_1^{(K)}, a_2^{(K)}, \ldots, a_q^{(K)}$ such that $A^{(K)} = (a_1^{(K)*}, a_2^{(K)*}, \ldots, a_q^{(K)*})$, $Y_{q,p} = (X_{1,p}^*, X_{2,p}^* \ldots, X_{q,p}^*)^*$ and $a^* X_{K,p} = A^{(K)} Y_{q,p}$. Hence,

$$a^* \Gamma_{0,p} a = A^{(K)} G_{q,p} A^{(K)*} \geq \lambda_1 A^{(K)} A^{(K)*} = \lambda_1 \sum_{i=1}^{q} ||a_i^{(K)}||_2$$

where λ_1 is the smallest eigenvalue of $G_{q,p}$. Therefore, $||a_i^{(K)}||_2$ are bounded function of K for each i. Again,

$$a^* X_{K,p} = A^{(K)} Y_{q,p} \Rightarrow a^* X_{K,p} X_{K,p}^* a - A^{(K)} Y_{q,p} X_{K,p}^* a$$
$$\Rightarrow a^* \Gamma_{0,p} a = \sum_{j=1}^{q} a_j^{(K)*} \Gamma_{K-j,p} a.$$

Hence,

$$|a^* \Gamma_{0,p} a| \leq \sum_{j=1}^{q} ||a_j^{(K)}||_2 ||\Gamma_{K-j,p}||_2 ||a||_2 \leq C \sum_{j=1}^{q} ||\Gamma_{K-j,p}||_2$$

for some $C > 0$ and tends to zero as $K \to \infty$. So, $a^* \Gamma_{0,p} a = 0$ for some $a \neq 0$. This contradicts the assumption $\lambda_{\min}(\Gamma_{0,p}) > 0$. Hence, the result holds as $G_{1,p} = \Gamma_{0,p}$ is non-singular. □

Recall the class of dispersion matrices $\mathcal{W}(\epsilon)$ in (3.17). It is easy to see that, for the model (3.10), if $\{A_{i,\infty}\} \in \mathcal{P}(C, \alpha, \epsilon)$ and $\Sigma_\infty \in \mathcal{W}(\epsilon)$ for some $C, \alpha, \epsilon > 0$, then $||\Gamma_{h,p}||_2 \to 0$ as $h \to \infty$ and for all $n \geq 1$. The above statement follows because, by Theorem 3.5.1, (3.10) can be represented in the form (3.12), (3.62) holds and

$$\Gamma_{u,p} = \sum_{j=0}^{\infty} \phi_{j,p} \Sigma_p \phi_{j+u,p}^*, \quad \text{for all } u \geq 0. \tag{3.68}$$

Therefore, by (3.62), Lemma 3.5.1 and as $\Sigma_\infty \in \mathcal{W}(\epsilon)$, for some $C_1 > 0$, we have

$$\|\Gamma_{u,p}\|_2 \leq \sum_{j=0}^{\infty} \|\phi_{j,p}\|_2 \|\Sigma_p\|_2 \|\phi^*_{j+u,p}\|_2$$

$$\leq \epsilon^{-1} \sum_{j=0}^{\infty} \sqrt{\|\phi_{j,p}\|_{(1,1)} \|\phi^*_{j,p}\|_{(1,1)}} \sqrt{\|\phi_{j,p}\|_{(1,1)} \|\phi^*_{j+u,p}\|_{(1,1)}}$$

$$\leq C_1 \epsilon^{-1} \delta^u \Big(\sum_{j=0}^{\infty} \delta^{2j} \Big), \quad 0 < \delta < 1.$$

$$\to \quad 0, \quad \text{as } u \to \infty \text{ and for all } n \geq 1. \tag{3.69}$$

Hence, for the model (3.10), if $\Gamma_{0,p}$ is non-singular for each n, then

$$\mathcal{A}_{r,n} = G_{r,n}^{-1} \mathcal{Y}_{r,n} \tag{3.70}$$

i.e., each $A_{i,p}$ is the finite sum of the finite products of $\{\Gamma_{u,p}, \Gamma_{u,p}^{-1}, \ 1 \leq u \leq r\}$. Hence, (3.70) provides consistent estimates of A_i, once we replace the population autocovariance matrices by their consistent estimates. We illustrate this by the IVAR(1) model. Similar result also holds for estimating the parameter matrices of other finite-order IVAR processes.

For $\alpha, C > 0$ and $0 < \delta < 1$, define the parameter space $\mathcal{A}(\delta, C, \alpha)$ as

$$\big\{ A_\infty : \max(\|A_\infty\|_{(1,1)}, \|A^*_\infty\|_{(1,1)}) < (1 - \delta), \ A_\infty, \ A^*_\infty \in \mathcal{X}(\alpha, C) \big\}. \tag{3.71}$$

Theorem 3.5.3. *(Bhattacharjee and Bose [2014b]) Consider the model (3.10) for $r = 1$. Suppose $\varepsilon_t \sim \mathcal{N}_p(0, \Sigma_p)$, $\Sigma_\infty \in \mathcal{U}(\epsilon, \alpha, C)$ and $A_{1,\infty} \in \mathcal{A}(\delta, C, \alpha)$ for some $\epsilon, \alpha, C > 0$ and $0 < \delta < 1$. Also suppose (T1) and (T2) hold. Assume all the inverses below exist. Then for*

$k_{n,\alpha} \asymp (n^{-1} \log p)^{-\frac{1}{2(\alpha+1)}}$ and $\tau_{n,\alpha} \asymp (n^{-1} \log p)^{-\frac{1}{2(1+\gamma)} \left[\frac{\gamma}{1+\alpha} + 1 \right]}$,

(i) $\|B_{k_{n,\alpha}}(\hat{\Gamma}_{1,p,n})(B_{k_{n,\alpha}}(\hat{\Gamma}_{0,p,n}))^{-1} - A_{1,p}\|_2 = O_P(k_{n,\alpha}^{-\alpha} \|\Sigma_p\|_{(1,1)})$,

(ii) $\|R_{g,\tau_{n,\alpha}}(\hat{\Gamma}_{1,p,n})(R_{g,\tau_{n,\alpha}}(\hat{\Gamma}_{0,p,n}))^{-1} - A_{1,p}\|_2$ *is of order*

$$O_P \Big[\Big(\frac{\log p}{n} \Big)^{\frac{\gamma\alpha}{2(1+\alpha)(1+\gamma)}} \|\Sigma_p\|_{(1,1)} \Big],$$

(iii) $\|\hat{\Sigma}_{p,n,\alpha} - \Sigma_p\|_2 = O_P(\|\Sigma_p\|_{(1,1)} k_{n,\alpha}^{-\alpha})$ *and*

(iv) $\|\hat{\tilde{\Sigma}}_{p,n,\alpha} - \Sigma_p\|_2 = O_P \big[(n^{-1} \log p)^{\frac{\gamma\alpha}{2(1+\alpha)(1+\gamma)}} \|\Sigma_p\|_{(1,1)} \big]$,

where

$$\hat{\Sigma}_{p,n,\alpha} = B_{k_{n,\alpha}}(\hat{\Gamma}_{0,p}) - B_{k_{n,\alpha}}(\hat{\Gamma}_{1,p})(B_{k_{n,\alpha}}(\hat{\Gamma}_{0,p}))^{-1} B_{k_{n,\alpha}}(\hat{\Gamma}^*_{1,p}),$$

$$\hat{\tilde{\Sigma}}_{p,n,\alpha} = R_{g,\tau_{n,\alpha}}(\hat{\Gamma}_{0,p}) - R_{g,\tau_{n,\alpha}}(\hat{\Gamma}_{1,p})(R_{g,\tau_{n,\alpha}}(\hat{\Gamma}_{0,p}))^{-1} R_{g,\tau_{n,\alpha}}(\hat{\Gamma}^*_{1,p}).$$

To prove the above theorem, we need the following lemma.

Lemma 3.5.4. *(see Bhatia [2009]) If A and B are invertible and $||A - B||_2 \leq ||A^{-1}||_2^{-1}$, then*

$$||B^{-1} - A^{-1}||_2 \leq \frac{||A^{-1}||_2^2 ||A - B||_2}{1 - ||A^{-1}||_2 ||A - B||_2}. \tag{3.72}$$

Proof of Theorem 3.5.3. It is easy to see that, $A_{1,\infty} \in \mathcal{A}(\delta, C, \alpha)$ implies $A_{1,\infty} \in \mathcal{P}(C, \alpha, \delta(1-\delta)^{-1})$. Therefore, the conclusions of Theorem 3.5.2 hold. Using Lemma 3.5.4, for large n,

$$||(B_{k_{n,\alpha}}(\hat{\Gamma}_{0,p,n}))^{-1} - \Gamma_{0,p}^{-1}||_2 \leq \frac{||\Gamma_{0,p}^{-1}||_2^2 ||B_{k_{n,\alpha}}(\hat{\Gamma}_{0,p,n}) - \Gamma_{0,p}||_2}{1 - ||\Gamma_{0,p}^{-1}||_2 ||B_{k_{n,\alpha}}(\hat{\Gamma}_{0,p,n}) - \Gamma_{0,p}||_2},$$

$$||(R_{g,\tau_{n,\alpha}}(\hat{\Gamma}_{0,p,n}))^{-1} - \Gamma_{0,p}^{-1}||_2 \leq \frac{||\Gamma_{0,p}^{-1}||_2^2 ||R_{g,\tau_{n,\alpha}}(\hat{\Gamma}_{0,p,n}) - \Gamma_{0,p}||_2}{1 - ||\Gamma_{0,p}^{-1}||_2 ||R_{g,\tau_{n,\alpha}}(\hat{\Gamma}_{0,p,n}) - \Gamma_{0,p}||_2}.$$

If $n^{-1}\log p \to 0$, then for some $C > 0$ and for sufficiently large n,

$$||(B_{k_{n,\alpha}}(\hat{\Gamma}_{0,p,n}))^{-1} - \Gamma_{0,p}^{-1}||_2 \leq C||B_{k_{n,\alpha}}(\hat{\Gamma}_{0,p,n}) - \Gamma_{0,p}||_2,$$

$$||(R_{g,\tau_{n,\alpha}}(\hat{\Gamma}_{0,p,n}))^{-1} - \Gamma_{0,p}^{-1}||_2 \leq C||R_{g,\tau_{n,\alpha}}(\hat{\Gamma}_{0,p,n}) - \Gamma_{0,p}||_2. \tag{3.73}$$

Therefore, by Theorem 3.5.2

$$||(B_{k_{n,\alpha}}(\hat{\Gamma}_{0,p,n}))^{-1} - \Gamma_{0,p}^{-1}||_2 = O_P(||\Sigma_p||_{(1,1)} k_{n,\alpha}^{-\alpha}) \tag{3.74}$$

and

$$||(R_{g,\tau_{n,\alpha}}(\hat{\Gamma}_{0,p,n}))^{-1} - \Gamma_{0,p}^{-1}||_2 = O_P\left[(n^{-1}\log p)^{\frac{\gamma\alpha}{2(1+\alpha)(1+\gamma)}} ||\Sigma_p||_{(1,1)} \right]. \tag{3.75}$$

Again, by the fact

$$||AB - CD||_2 \leq ||A - C||_2 ||B - D||_2 + ||A - C||_2 ||D||_2 + ||C||_2 ||B - D||_2$$

and using $A_{1,p} = \Gamma_{1,p}\Gamma_{0,p}^{-1}$, (i) and (ii) follow.

Results (iii) and (iv) are immediate from the relation

$$\Gamma_{0,p} - A_{1,p}\Gamma_{1,p}^* - \Gamma_{1,p}A_{1,p}^* + A_{1,p}\Gamma_{0,p}A_{1,p}^* = \Sigma_p.$$

This completes the proof of Theorem 3.5.3. $\qquad\square$

Next we shall relax the Gaussian assumption on the driving process $\{\varepsilon_{t,p}\}$ in Theorems 3.3.3, 3.4.2, 3.5.2, and 3.5.3 and Lemma 3.3.3.

3.6 Gaussian assumption

The Gaussian assumption made so far (see Theorems 3.4.2, 3.5.2, and 3.5.3) may seem to be a very strong restriction. However, note that in the proofs of these theorems, the Gaussian assumption is used only while invoking Theorem 3.3.3. Moreover, the proof of Theorem 3.3.3 uses the Gaussian assumption only via application of Lemma 3.3.3. Our goal is to now replace the Gaussian assumption by a suitable weaker assumption in Lemma 3.3.3.

We borrow an idea from Theorem 1.3.1 which is quoted from Bickel and Levina [2008a]. They first proved the consistency of the covariance matrix $\hat{\Sigma}_p$ for the IID process under the assumption $\varepsilon_t \sim \mathcal{N}_p(0, \Sigma_p)$. Later they relaxed this assumption and proved (1.32) under the weaker assumption that,

$$\sup_{j \geq 1} E(e^{\lambda \varepsilon_{t,j}}) < \infty \quad \text{for all } |\lambda| < \lambda_0 \text{ and some } \lambda_0 > 0, \qquad (3.76)$$

where $\varepsilon_{t,j}$ is the j-th element of ε_t.

As a prelude we need the following lemma. For $n \geq 1$, let $U_1, U_2, \ldots, U_n$ be independent random variables with

$$EU_j = 0 \quad \text{and} \quad \sigma_j^2 = Var(U_j) > 0, j = 1, 2, \ldots.$$

Set

$$S_n = \sum_{j=1}^{n} U_j \quad \text{and} \quad B_n^2 = \sum_{j=1}^{n} \sigma_j^2, \quad Z_n = \frac{S_n}{B_n}.$$

We say that $\{U_j\}$ *satisfies condition (P)*, if there exist positive constants $A, C, C_1, C_2, \ldots$ such that for all $|z| < A$ and $j = 1, 2, \ldots$,

$$\left| \frac{\ln E(e^{zU_j})}{z^2} \right| \leq C_j^2 \quad \text{and} \quad \lim_{n \to \infty} \frac{1}{B_n^2} \sum_{j=1}^{n} C_j^2 \leq C. \qquad (3.77)$$

Lemma 3.6.1. *Suppose a sequence of random variables $\{U_j\}$ with $EU_j = 0$ and $\sigma_j^2 = Var(U_j) > 0$ satisfies condition (P). Then there exist some A, $C > 0$ such that*

$$Cum_k(Z_n)| \leq \frac{k!C}{(AB_n)^{k-2}} \quad \text{for all } k \geq 3. \qquad (3.78)$$

Hence, the conclusion of Lemma 2.1.2 holds for $\xi = Z_n$, with $\nu = 0, H = 2C, \bar{\Delta} = AB_n$. In particular, if U_i are i.i.d. then, (3.77) holds if

$$\left| \frac{\ln E(e^{zU_1})}{z^2} \right| \leq C, \quad \text{for all } |z| < A, \quad \text{for some } A, C > 0. \qquad (3.79)$$

Also for a random variable U_1 with $EU_1 = 0$, if there exists A', $C' > 0$ such that $E(e^{\lambda U_1}) \leq C'$ for all $|\lambda| < A'$, then (3.79) holds.

Proof. Equation (3.78) easy to show and the proof is given in Saulis and Statulevičius [1991]. Hence, we prove only the last statement. The cumulants $\{K_n\}$ of a random variable U_1 are defined by the cumulant generating function

$$g(z) = \log(E(e^{zU_1})) = \sum_{n=1}^{\infty} K_n \frac{z^n}{n!}. \qquad (3.80)$$

Note that the series in (3.80) converges absolutely for $|z| < A$. The cumulants are related to the moments $\{\mu'_n = E(U^n)\}$ by the following recursion formula

$$K_n = \mu'_n - \sum_{m=1}^{n-1} \binom{n-1}{m-1} K_m \mu'_{n-m}.$$

As all the moments of U_1 exist, $K_n, n = 1, 2, \ldots$ are finite. Moreover, $K_1 = \mu'_1 = 0$ and $K_2 = \mu'_2 - \mu'_1 = \mu'_2$. Hence,

$$\left| \frac{g(z)}{z^2} \right| \leq \frac{\mu'_2}{2!} + \sum_{n=3}^{\infty} |K_n| \frac{|A|^{n-2}}{n!} < \infty.$$

This completes the proof.   □

Lemma 3.6.2. *Let $\{\varepsilon_t\}$ be i.i.d. with mean 0 and covariance matrix Σ_p. Suppose (3.76) holds. Then (i) and (ii) of Lemma 3.3.3 hold.*

Proof. (i) follows from Bickel and Levina [2008a]. For (ii), using Lemma 3.6.1, we need the existence of the moment generating function of $\frac{(Z_{t,j} \pm Z_{(t+u),l})^2}{2} - 1$ for all j, l in some neighborhood of zero. This existence follows from the fact that $(x+y)^2 < 2(x^2 + y^2)$.   □

Thus, the conclusions of Theorems 3.3.3, 3.4.2, 3.5.2, 3.5.3, and Lemma 3.3.3 hold true if we assume (3.76) instead of $\varepsilon_t \sim \mathcal{N}_p(0, \Sigma_p)$.

3.7   Simulations

Consider the IVAR model (3.10) for $r = 1$. In this section we show some simulations for this model with two different choices of the parameter matrix $A_{1,p}$ which have the Toeplitz structure. As we move away from the main diagonal, in one case the entries decrease exponentially and in the other case, they decrease polynomially. The following simulations show that the convergence rate obtained in Theorem 3.5.3 is quite sharp. Establishing the exact rate appears to be a very difficult open problem.

Example 3.7.1. Exponentially decaying corners: Consider the IVAR(1) model with $A_{1,\infty} = (((-0.5)^{|i-j|}))$. Note that

$$\|A_{1,\infty}\|_{(1,1)} = \|A^*_{1,\infty}\|_{(1,1)} \leq 1 + 2 \sum_{u=1}^{\infty} (-0.5)^u = 1 - 2/3,$$

$$T(A_{1,\infty}, k) \leq 2 \sum_{u=k+1}^{\infty} (-0.5)^u \leq (2/3)(0.5)^k < (2/3)k^{-1}.$$

Therefore, $A_{1,\infty} \in \mathcal{A}(2/3, 2/3, 1)$ and the conclusion of Theorem 3.5.3 hold.

Example 3.7.2. Polynomially decaying corners: Consider the IVAR(1) model with $A_{1,\infty} = (((-1)^{|i-j|}(|i-j| + 1)^{-\beta}))$, for some $\beta > 1$. Then the following two relations hold.

$$||A_{1,\infty}||_{(1,1)} = ||A_{1,\infty}^*||_{(1,1)} \leq 1 + 2 \sum_{u=1}^{\infty} (-1)^u (u+1)^{-\beta} \leq 1 - 2(2^{-\beta} - 3^{-\beta})$$

$$T(A_{1,\infty}, k) \leq 2 \sum_{u=k+1}^{\infty} (-1)^u (u+1)^{-\beta} \leq 2 \int_k^{\infty} x^{-\beta} dx = 2(\beta - 1)^{-1} k^{-(\beta-1)}.$$

Therefore, $A_{1,\infty} \in \mathcal{A}(2(2^{-\beta} - 3^{-\beta}), 2(\beta - 1)^{-1}, \beta - 1)$ and the conclusion of Theorem 3.5.3 holds. For the following simulations, we chose $\beta = 1.01, 1.1, 1.2$, and 1.5.

Recall I_k from (1.9). We let $\varepsilon_t \sim \mathcal{N}_p(0, I_p)$, for all t. In each case, we draw the histogram for the values of $||B_{k_{n,\alpha}}(\hat{\Gamma}_{1,p,n})(B_{k_{n,\alpha}}(\hat{\Gamma}_{0,p,n}))^{-1} - A_{1,p}||_2$ using $R = 300$ replications. We consider two combinations of n and p, namely $n = 20$, $p = e^{\sqrt{n}} \sim 87$ and $n = 40$, $p = e^{\sqrt{n}} \sim 558$.

Note that most of the mass is concentrated near zero. Expectedly, the accuracy is sharper in Example 3.7.1 than in Example 3.7.2. Moreover, as β increases, the histogram has more mass near zero and there is some mass in the high values of the tail. Some stray values beyond the range given in the figures were observed over the different sets of simulations but overall most of the mass was concentrated in the range $(0, 600)$. This indicates that the rates of convergence are probably quite sharp. No results on the exact rate of convergence are currently known.

Exercises

1. Establish the autocovariances given in (3.7), (3.9), and (3.15).

2. Learn the proof of Theorem 11.3.1 of Brockwell and Davis [2009].

3. Give an example of IVAR(2) process which is not causal.

4. Establish Theorems 3.2.1 and 3.2.2.

5. Consider the IVAR process (3.10) with $r = 2$. State and prove a rate of convergence result for banded and tapered estimators of parameter matrices $A_{1,p}, A_{2,p}$ and the dispersion matrix Σ_p of $\{\varepsilon_t\}$.

6. Let $X_t = AX_{t-1} + B\varepsilon_{t-1} + \varepsilon_t$ where $\varepsilon_t \overset{\text{i.i.d.}}{\sim} \mathcal{N}(0, I_p)$. Obtain consistent estimators of Γ_1 and Γ_2. Find their convergence rate. In this context also state sufficient conditions on A and B.

7. Learn the proof of the first part of Lemma 3.6.1.

8. Learn the proof of Lemma 3.6.2(a) from Bickel and Levina [2008a].

9. Let $X_t = \varepsilon_t \overset{\text{i.i.d.}}{\sim} \mathcal{N}(0, I_p)$. Show that $\frac{1}{p}\text{Tr}(\hat{\Gamma}_1)$ does not converge in probability to $\frac{1}{p}\text{Tr}(\Gamma_1)$.

10. Let $X_t = \varepsilon_{t-1} + \varepsilon_t$ where $\varepsilon_t \overset{\text{i.i.d.}}{\sim} \mathcal{N}(0, I_p)$. Find $\lim E \frac{1}{p} \text{Tr}(\hat{\Gamma}_u)$ for $u = 0, 1, 2, 3, 4, 5$.

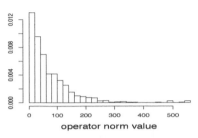

Example 3.7.1 $n = 20$

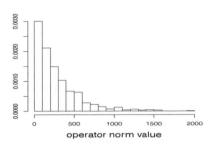

Example 3.7.1 $n = 40$

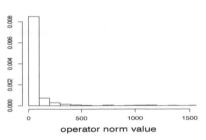

Example 3.7.2 $\beta = 1.01$, $n = 20$

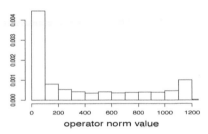

Example 3.7.2 $\beta = 1.01$, $n = 40$

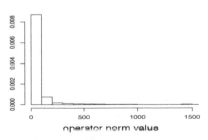

Example 3.7.2 $\beta = 1.1$, $n = 20$

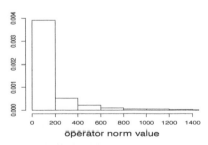

Example 3.7.2 $\beta = 1.1$, $n = 40$

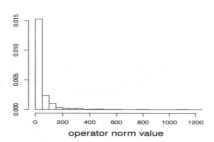

Example 3.7.2 $\beta = 1.2$, $n = 20$

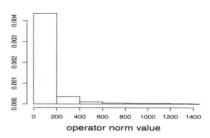

Example 3.7.2 $\beta = 1.2$, $n = 40$

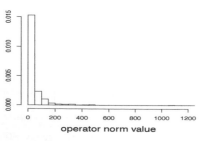

Example 3.7.2 $\beta = 1.5$, $n = 20$

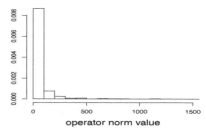

Example 3.7.2 $\beta = 1.5$, $n = 40$

Part II

Part II

Chapter 4

SPECTRAL DISTRIBUTION

In Chapter 3, we have encountered the infinite-dimensional MA(∞) process. A key quantity to analyze this model is the sequence of sample autocovariance matrices $\{\hat{\Gamma}_u\}$. There we used some regularization on $\{\hat{\Gamma}_u\}$ to obtain consistent estimators for their population counterpart. We now wish to explore further asymptotic properties of $\{\hat{\Gamma}_u\}$. These properties turn out to be quite interesting. Moreover, as we shall see later, they can also be used for statistical inference.

One natural way to study the large sample behaviour of a sample autocovariance matrix of any given order is through its limiting spectral distribution (LSD). Here we find common ground with high-dimensional random matrices. LSD of various random matrices occupy a central position in the literature of random matrix theory (RMT). This chapter collects the basic concepts and results in RMT that we shall need. We shall use them crucially in Chapters 6, 7, 8 and 10.

Of course the joint convergence of several sample autocovariance matrices together is also going to be important for statistical inference. The most natural way to do that is to consider the convergence as elements of a sequence of non-commutative *-probability spaces (NCP). The required notions on such spaces will be developed in the next chapter and we will consider the joint convergence of sample autocovariance matrices after that.

4.1 LSD

The following definition of a spectral distribution and its limit is valid for both random and non-random matrices.

Definition 4.1.1. *(ESD, EESD, and LSD) The empirical spectral distribution (ESD) of a $p \times p$ (random) matrix R_p is the (random) probability distribution with mass p^{-1} at each of its eigenvalues. If it converges weakly (almost surely) to a (non-degenerate) non-random probability distribution, then the latter will be called the limiting spectral distribution (LSD) of R_p. The expectation of ESD will be called the EESD. This is a non-random probability distribution function.*

For a non-random matrix, ESD and EESD are identical. There are other notions of convergence of the ESD for random matrices. However in this book we will stick to the almost sure convergence as described above.

Example 4.1.1. Let $A_p = 0.5I_p$, where I_p is as in (1.9). As all its eigenvalues are 0.5, its ESD is degenerate at 0.5 and the LSD is also so.

Example 4.1.2. Let $B_p = 0.5(I_p + J_p)$, where I_p and J_p are respectively as in (1.9) and (1.10). Note that B_p has $(p-1)$-many eigenvalues equal to 0.5 and one eigenvalue equals $1 + 0.5(p-1)$. Therefore, ESD of B_p, say F^{B_p}, can be written as

$$F^{B_p}(x) = \begin{cases} 0, & \text{if } -\infty < x < 0.5, \\ 1 - p^{-1}, & \text{if } 0.5 \leq x < 1 + 0.5(p-1), \\ 1, & \text{if } 1 + 0.5(1-p) \leq x < \infty. \end{cases} \tag{4.1}$$

Hence, LSD of B_p is degenerate at 0.5.

Example 4.1.3. Let

$$C_p = ((I(i = j, 1 \leq i \leq [p/2]) - I(i = j, [p/2] + 1 \leq i \leq p))),$$

where $[x]$ denotes the largest integer contained in x. Its $[p/2]$-many eigenvalues are equal to 1 and $(p - [p/2])$-many eigenvalues are equal to -1. Therefore, the LSD of C_p is the distribution $2Ber(0.5) - 1$, where $Ber(0.5)$ is the Bernoulli variable with success probability 0.5.

Example 4.1.4. Let $D_p = ((I(i + j = p + 1)))$. In this case also, LSD of D_p is the distribution $2Ber(0.5) - 1$.

Incidentally, the study of the limit spectrum of non-hermitian matrices is extremely difficult and very few results are known for general random non-hermitian sequences. Clearly, the sample autocovariances $\{\hat{\Gamma}_u : u \geq 1\}$ are not hermitian and there are no LSD results known for these matrices. We shall only consider certain symmetrized version of these matrices. This does entail some loss of information from the statistical viewpoint but we are limited by the current state of knowledge in RMT.

Two widely used approaches to establish the LSD of symmetric square random matrices are (i) the *moment method* and (ii) the *method of Stieltjes transform*. Below is a brief description of these methods. For more details see Bai and Silverstein [2009]. We shall primarily use the moment method to establish different LSD. We shall also use Stieltjes transforms, most frequently to establish link with many LSD results known in the literature.

4.1.1 Moment method

The h-th order moment of the ESD of any $p \times p$ real symmetric random matrix R_p equals

$$\beta_h(R_p) := \frac{1}{p}\text{Tr}(R_p^h).$$

This relation is known as the *trace-moment formula*.

We now state a lemma which ensures convergence of EESD.

Consider the following conditions.

(M1) For every $h \geq 1$, $E(\beta_h(R_p)) \to \beta_h$ and

(C) The sequence $\{\beta_h\}$ satisfies *Carleman's condition*, $\sum_{h=1}^{\infty} \beta_{2h}^{-\frac{1}{2h}} = \infty$.

Then we have the following lemma.

Lemma 4.1.1. *If (M1) and (C) holds, then the EESD of R_p converges to the distribution F determined by the moments $\{\beta_h\}$.*

The following lemma ensures convergence of ESD. Consider the additional condition.

(M4) $\sum_{n=1}^{\infty} E(\beta_h(R_p) - E(\beta_h(R_p)))^4 < \infty$, $\forall h \geq 1$.

We omit the proof of the following lemma. For more details, see for example Bai and Silverstein [2009] and Bose [2018].

Lemma 4.1.2. *If (M1), (M4) and (C) hold, then the ESD of R_p converges almost surely to the distribution F determined by the moments $\{\beta_h\}$.*

Clearly (M1) is the most crucial condition in this method since it identifies the moments of the LSD. Later in Chapter 5, we shall see that the (M1) condition for R_p is ensured by the convergence of an appropriate sequence of NCP generated by R_p.

The following lemma will be useful to us.

Lemma 4.1.3. *(a) Let $\{\mu_p\}$ be a sequence of probability measures on $\mathbb{R}$. Suppose for all $k \geq 1$ and for some $C > 0$,*

$$\lim_{p \to \infty} \int_{\mathbb{R}} x^k d\mu_p = m_k \quad and \quad |m_k| \leq C^k. \tag{4.2}$$

Then $\{m_k\}$ is a moment sequence, there is a unique probability measure μ on $\mathbb{R}$ such that

$$m_k = \int_{\mathbb{R}} x^k d\mu, \quad \forall k \geq 1 \quad and, \tag{4.3}$$

as $p \to \infty$, $\{\mu_p\}$ converges weakly to μ.

(b) Let R_p be a $p \times p$ real symmetric random matrix. Suppose for all $k \geq 1$ and for some $C > 0$,

$$\lim_{p \to \infty} \frac{1}{p} E \, Tr(R_p^k) = m_k \quad and \quad |m_k| \leq C^k. \tag{4.4}$$

Then there is a unique probability measure μ on $\mathbb{R}$ such that (4.3) holds. Moreover, the sequence of EESD converges in distribution to μ.

Proof. (a) By the convergence in (4.2), it follows easily that $\{m_k\}$ is a moment sequence. By the bound given in (4.2), $\sum_{k=1}^{\infty} m_{2k}^{-1/2k} = \infty$ and hence $\{m_k\}$ determines a unique probability distribution.

Again, by the convergence in (4.2), the sequence $\{\mu_p\}$ is tight. Fix any subsequence. Then there is a further subsequence that converges weakly. The corresponding sequence of moments of any given order also converges to the limit moment of the same order. Since the limit moments $\{m_k\}$ determine the distribution uniquely, all subsequential limits are same. Thus, $\{\mu_p\}$ converges weakly to the distribution with moments $\{m_k\}$.

(b) follows from (a) by observing that

$$\frac{1}{p} E \mathrm{Tr}(R_p^k) = \int_{\mathbb{R}} x^k d\mu_p, \ \forall k \geq 1, \tag{4.5}$$

where μ_p is the EESD of R_p. $\qquad\qquad\qquad\qquad\qquad\qquad\qquad\square$

4.1.2 Method of Stieltjes transform

Another widely used method to establish the LSD is the method of *Stieltjes transform*. We give a brief description of this method here. See Bai and Silverstein [2009] for further details on this method. Let $i = \sqrt{-1}$. The Stieltjes transform of any real random variable X or its probability measure μ on $\mathbb{R}$ equals

$$m_X(z) = m_\mu(z) = \int \frac{1}{x-z} \mu(dx), \quad z \in \mathbb{C}^+ := \{x+iy : x \in \mathbb{R}, y > 0\}. \tag{4.6}$$

Note that the integral above is always finite for $z \in \mathbb{C}^+$. Some basic properties of the Stieltjes transform are given in the following lemma. We omit its proof and leave it as an exercise. We shall need the following notation. Define for every $y > 0$, the function on $\mathbb{R}$ as ($\mathcal{I}$ denotes imaginary part)

$$f_{\mu,y}(x) = \frac{1}{\pi} \mathcal{I} m_\mu(x + iy).$$

Lemma 4.1.4. *The following properties hold for* $m_\mu(z)$.

(a) The function m_μ *is analytic and its range is contained in* $\mathbb{C}^+$.

(b) The support of μ *is a subset of* $\mathbb{R}^+$, *if and only if* $\mathcal{I}(z m_\mu(z)) \leq 0$.

(c) $\lim_{y\to\infty} \iota y m_\mu(iy) = -1$.

(d) The point masses of μ *are given by*

$$\mu\{t\} = \lim_{y \searrow 0} y \mathcal{I} m_\mu(t + iy), \ t \in \mathbb{R}.$$

(e) The function $f_{\mu,y}(\cdot) : \mathbb{R} \to \mathbb{R}^+$ *is a probability density function. As* $y \to 0$, *the corresponding sequence of probability measures converges to* μ *weakly.*

(f) For every bounded continuous function $f : \mathbb{R} \to \mathbb{R}$,

$$\int_{\mathbb{R}} f(t) d\mu(t) = \frac{1}{\pi} \lim_{y \searrow 0} \mathcal{I} \int_{\mathbb{R}} f(x) m_\mu(x + iy) dx.$$

(g) (Inversion formula) For all continuity points a, b of μ we have

$$\mu(a, \ b) = \frac{1}{\pi} \lim_{y \searrow 0} \mathcal{I} \int_a^b m_\mu(x + iy) dx.$$

(h) If $m_\mu(\cdot)$ has a continuous extension to $\mathbb{C}^+ \bigcup \mathbb{R}$, then μ has a density (with respect to the Lebesgue measure) given by

$$f_\mu(t) = \frac{1}{\pi} \lim_{y \searrow 0} \mathcal{I} m_\mu(t + iy), \ t \in \mathbb{R}.$$

(i) Suppose μ has moments $\{m_n\}$ and its support is contained in $[-C, \ C]$. Then

$$m_\mu(z) = -\sum_{n=0}^{\infty} \frac{m_n}{z^{n+1}}, \quad \forall z \in \mathbb{C}, \ |z| > C.$$

Moreover,

$$\lim_{z \in \mathbb{C}^+, |z| \to \infty} z m_\mu(z) = -1.$$

The following lemma provides a necessary and sufficient condition for the limit of Stieltjes transforms of a sequence of probability measures to be the Stieltjes transform of a probability measure.

Lemma 4.1.5. (Geronimo and Hill [2003]) Suppose that $\{\mathbb{P}_p\}$ is a sequence of probability measures on the real line with Stieltjes transforms $\{m_{\mathbb{P}_p}\}$. If $\lim_{p \to \infty} m_{\mathbb{P}_p}(z) = m(z)$ for all $z \in \mathbb{C}^+$, then there exists a probability measure $\mathbb{P}$ with Stieltjes transform m if and only if

$$\lim_{v \to \infty} iv m(iv) = -1 \tag{4.7}$$

in which case $\mathbb{P}_p$ converges to $\mathbb{P}$ in distribution.

Now consider a sequence of real symmetric (random) matrices $\{R_p\}$. Let F^{R_p} be the ESD of R_p. Then the Stieltjes transform of F^{R_p}, say m_{R_p}, is given by

$$m_{R_p}(z) - p^{-1} \mathrm{Tr}((R_p - zI_p)^{-1}).$$

Note that $(R_p - zI_p)^{-1}$ is the *resolvent* of the matrix R_p and its points of singularity are at the eigenvalues of R_p which are not in $\mathbb{C}^+$. To prove that F^{R_p} converges to a probability distribution F (say) almost surely, one needs to check whether $\{m_{R_p}\}$ satisfies the conditions of Lemma 4.1.5 almost surely. This is usually accomplished by the following steps.

1. Show that $E(m_{R_p}(z)) \to m(z)$ for $z \in \mathbb{C}^+$ where $m(z)$ satisfies (4.7). This ensures convergence of the EESD.

2. Show that $m_{R_p}(z) - E(m_{R_p}(z)) \to 0$ almost surely for each $z \in \mathbb{C}^+$. This is often achieved by using martingale convergence techniques.

Often, $m(z)$ appears as a solution of a functional equation which has multiple solutions. In that case, there is an additional step required to identify the correct solution.

Example 4.1.5. (Wigner matrix and the semi-circle law) A very well-known matrix in RMT is the scaled *Wigner matrix* $W_p = ((x_{i,j}/\sqrt{p}))_{p \times p}$ where $x_{i,j} = x_{j,i}$ are i.i.d. mean 0, variance 1. Assume that they are uniformly bounded. It is known that its LSD is the *semi-circle* law. We outline a proof of this which is based on the method of Stieltjes transform.

For each $1 \le k \le p$, let $w_{p,k}$ be the $(p-1) \times 1$ vector which is the k-th column of W_p with the k-th element removed. Let $W_{p,k}$ be the $(p-1) \times (p-1)$ matrix obtained from W_p after removing the k-th row and the k-th column. Let m_{W_p} denote the Stieltjes transform of the ESD of W_p. By using the rank one perturbation formula for inverses of matrices,

$$
\begin{aligned}
m_{W_p}(z) &= \frac{1}{p}\mathrm{Tr}((W_p - zI_p)^{-1}) \\
&= \frac{1}{p}\sum_{k=1}^{p}\left(p^{-1/2}x_{k,k} - z - a_{p,k}^*(W_{p,k} - zI_{p-1})^{-1}a_{p,k}\right)^{-1}. \quad (4.8)
\end{aligned}
$$

Note that for each k, $a_{p,k}$ and $W_{p,k}$ are independent and $a_{p,k}$ has i.i.d. entries with zero mean and variance p^{-1}. Hence, the quadratic form $a_{p,k}^*(W_{p,k} - zI_{p-1})^{-1}a_{p,k}$ concentrates around its (conditional) mean $p^{-1}(W_{p,k} - zI_{p-1})^{-1}$. Now note that for large p, the ESD of $W_{p,k}$ should be close to the ESD of W_p. Therefore for each k, $p^{-1}\mathrm{Tr}(W_{p,k} - zI_{p-1})^{-1}$ can be approximated by $m_{W_{p,k}}(z)$ and subsequently by $m_{W_p}(z)$. Finally, note that the contribution from the terms $p^{-1/2}x_{k,k}$ can be neglected. Hence, we have the following approximate identity

$$
m_{W_p}(z) \approx \frac{1}{-z - m_{W_p}(z)} \quad (4.9)
$$

for large enough p. From these heuristics, it is expected that for each $z \in \mathbb{C}^+$, the limiting Stieltjes transform $m(z)$ satisfies the identity

$$
m(z)(1 + zm(z)) = -1.
$$

For every z, this equation has two solutions for $m(z)$ and it can be shown that there is only one valid Stieltjes transform solution, namely that of the semi-circle law. The details are provided later in Section 4.2.

Example 4.1.6. (Unadjusted sample covariance matrix and the Marčenko–Pastur law) Let $B_p = ((n^{-1}\sum_{k=1}^{n}x_{ki}x_{kj}))_{p \times p}$ where $\{x_{i,j}\}$ are i.i.d. mean 0, variance 1 and uniformly bounded random variables. This (and its mean adjusted version) is usually known as the S-matrix in RMT. Its LSD is known as the *Marčenko–Pastur law*. We outline a proof of this based on the method of Stieltjes transform.

Define $X_k = (x_{k1}, x_{k2}, \ldots, x_{kp})$ for all $1 \le k \le n$. Suppose $p/n \to y \in (0, \infty)$. An important ingredient here is the following representation of the resolvent $U_p(z) = (B_p - zI_p)^{-1}$.

$$zU_p(z) + I_p = U_p(z)B_p = \frac{1}{n}\sum_{k=1}^{n} U_p(z)X_k X_k^*. \qquad (4.10)$$

Define $U_{-k,p}(z) = (B_p - n^{-1}X_k X_k^* - zI_p)^{-1}$, and use the *rank one perturbation formula* for inverses to write

$$U_p(z) = U_{-k,p}(z) - \frac{\frac{1}{n}U_{-k,p}(z)X_k X_k^* U_{-k,p}(z)}{1 - \frac{1}{n}X_k^* U_{-k,p}(z)X_k}, \quad \forall k \ge 1. \qquad (4.11)$$

Substituting this in (4.10), after simple algebra, one has

$$\frac{z}{p}\mathrm{Tr}(U_p(z)) + 1 = \frac{n}{p} - \frac{1}{p}\sum_{k=1}^{p}\frac{1}{1 - \frac{1}{n}X_k^* U_{-k,p}(z)X_k}. \qquad (4.12)$$

In addition, for each k, by the structure of X_k and the fact that it is independent of $(B_p - n^{-1}X_k X_k^*)$, we have the approximation

$$X_k^* U_{-k,p}X_k \approx E(X_k^* U_{-k,p}X_k | B_p - n^{-1}X_k X_k^*) = \mathrm{Tr}(U_{-k,p}(z)).$$

Further, we have the approximation

$$\mathrm{Tr}(U_p(z)) \approx \mathrm{Tr}(U_{-k,p}(z)).$$

Now let

$$m_{B_p}(z) = p^{-1}\mathrm{Tr}(U_p(z))$$

be the Stieltjes transform of the ESD of B_p.

Using the above approximations and replacing p/n by its limiting value y, we have the approximate equation

$$yzm_{B_p}(z) + y \approx 1 - (1 + ym_{B_p}(z))^{-1} \quad \text{or}$$

$$m_{B_p}(z) \approx \frac{1}{1 - y - yzm_{B_p}(z) - z}, \quad \forall z \in \mathbb{C}^+. \qquad (4.13)$$

Therefore, for each $z \in \mathbb{C}^+$, the limiting Stieltjes transform $m(z)$ satisfies the identity

$$m(z) \approx \frac{1}{1 - y - yzm(z) - z}, \quad \forall z \in \mathbb{C}^+. \qquad (4.14)$$

Again, this is a quadratic equation in $m(z)$ for every z but the only solution that is a Stieltjes transform is given by

$$m(z) = \frac{1 - y - z + i\sqrt{((1 + \sqrt{y})^2 - z)(z - (1 - \sqrt{y})^2)}}{2yz}, \quad z \in \mathbb{C}^+. \qquad (4.15)$$

This is the Stieltjes transform of the Marčenko–Pastur law. Details are provided in Section 4.3.1.

Let μ be a compactly supported probability measure on $\mathbb{R}$ with $\mu(-K, K) = 1$ for some $K > 0$. Then by Lemma 4.1.4(i), we have the following formal power series expansion of the Stieltjes transform $m(z)$ for $|z| > K$ and $z \in \mathbb{C}^+$,

$$m_\mu(z) = -\frac{1}{z} E_\mu \left(\frac{1}{1 - \frac{X}{z}} \right) = -\frac{1}{z} - \frac{E_\mu(X)}{z^2} - \frac{E_\mu(X^2)}{z^3} - \cdots . \qquad (4.16)$$

This relation is crucial in linking the moment approach and the Stieltjes transform approach. Since $m_\mu(z)$ is analytic for $z \in \mathbb{C}^+$, in principle it suffices to identify it only for large enough $z \in \mathbb{C}^+$.

The following observation on scale change is useful. We omit its proof.

Lemma 4.1.6. *Suppose X is a random variable with Stieltjes transform $m_X(z)$. Then for any $\sigma > 0$, the Stieltjes transform of σX is given by*

$$m_{\sigma X}(z) = \sigma^{-1} m_X(z\sigma^{-1}), \quad \forall z \in \mathbb{C}^+. \qquad (4.17)$$

We have already seen two examples of random matrices and how their LSD can be obtained by the method of Stieltjes transform. In the next sections we collect a few standard random matrix models that will be relevant to us and list the results on their LSD. Though all the following results hold for appropriate triangular sequence of entries of random matrices, for simplicity of notation here we stick to non-triangular sequence of entries.

4.2 Wigner matrix: Semi-circle law

Definition 4.2.1. *(Wigner matrix) A Wigner matrix W_p of order p is a square symmetric random matrix with independent mean 0 variance 1 entries on and above the diagonal.*

We shall often write W for W_p if there is no confusion about the dimension of the matrix. As we proceed, further restrictions will be imposed on the entries of these matrices as required.

Definition 4.2.2. *(Semi-circle law) The standard semi-circle law is defined by the probability density*

$$f(x) = \begin{cases} \frac{1}{4\pi} \sqrt{4 - x^2}, & \text{if } -2 < x < 2, \\ 0, & \text{otherwise.} \end{cases} \qquad (4.18)$$

Its moment sequence is given by

$$\beta_h = \begin{cases} \frac{h!}{(h/2)!(1+h/2)!}, & \text{if } h \text{ is even,} \\ 0, & \text{if } h \text{ is odd.} \end{cases} \qquad (4.19)$$

Its Stieltjes transform $m(z)$, satisfies the quadratic equation

$$m^2(z) + zm(z) + 1 = 0, \quad \forall z \in \mathbb{C}^+. \tag{4.20}$$

Only one solution of the above equation yields a valid Stieltjes transform and that is given by

$$m(z) = \frac{-z - \sqrt{z^2 - 4z}}{2}. \tag{4.21}$$

It can be shown easily that if $\{\beta_h\}$ satisfies (4.19), then

$$\sum_{h=1}^{\infty} \beta_{2h}^{-1/2h} = \infty. \tag{4.22}$$

We shall need the above facts later in Chapters 6 and 7.

Assuming that the entries are i.i.d. Gaussian, Wigner [1958] showed that the EESD of $p^{-1/2}W_p$ converges to the semi-circle law. When the entries are i.i.d. with finite 4-th moment, Arnold [1967] and Arnold [1971] showed that the ESD converges almost surely to the same limit. There has been much subsequent development on the necessary and sufficient conditions for the convergence of the ESD of $p^{-1/2}W_p$. We quote a recent result in this direction that will be relevant to us (see for example Anderson et al. [2009]).

Consider the following classes of independent random variables.

$$\begin{aligned}
\mathcal{L}_r \;=\; & \text{collections of independent random variables} && (4.23) \\
& \{\epsilon_{i,j}\} \text{ such that } \sup_{i,j} E|\epsilon_{i,j}|^r < \infty,
\end{aligned}$$

$$\mathcal{L} \;=\; \bigcap_{r=1}^{\infty} \mathcal{L}_r, \tag{4.24}$$

$$\begin{aligned}
C(\delta, p) \;=\; & \text{collections of random variables } \{\varepsilon_{i,j}\} \text{ such that} \\
& |\varepsilon_{i,j}| \le \eta_p p^{\frac{1}{2+\delta}} \; \forall i,j \text{ and some } \eta_p \downarrow 0 \text{ as } p \to \infty. && (4.25)
\end{aligned}$$

Theorem 4.2.1. *Let $W_p = ((\omega_{i,j}))$ be a Wigner matrix of order p. Suppose $\{\omega_{i,j} : 1 \le i, j \le p\} \in \mathcal{L} \cup C(0, p) \; \forall p \ge 1$. Then, as $p \to \infty$, the LSD of $p^{-1/2}W_p$ is the standard semi-circle law.*

Next we consider a very specific polynomial in a Wigner and a deterministic matrix considered by Bai and Zhang [2010]. To state their theorem, we need the following class of independent random variables. Let

$$\begin{aligned}
U(\delta) \;=\; & \text{collections of independent } \{\varepsilon_{i,j}\} \text{ such that for all } \eta > 0 \\
& \lim \frac{\eta^{-(2+\delta)}}{np} \sum_{i=1}^{p} \sum_{j=1}^{n} E(|\varepsilon_{i,j}|^{2+\delta} I(|\varepsilon_{i,j}| > \eta p^{\frac{1}{2+\delta}})) = 0 && (4.26)
\end{aligned}$$

By a sequence of nested matrices $\{B_r\}$, we mean that for each $r \ge 1$, the submatrix constructed by the first r rows and columns of B_{r+1} is B_r. Consider the following class of matrices:

$$\mathcal{NND} \quad = \quad \text{set of all sequences of non-negative definite} \qquad (4.27)$$
$$\text{symmetric nested matrices } \{B_r\} \text{ whose LSD exists.}$$

Theorem 4.2.2. *(Bai and Zhang [2010]) Let $W_p = ((\omega_{i,j}))$ be a Wigner matrix of order p and A_p be a non-random square matrix of order p. Suppose $\{\omega_{i,j} : 1 \leq i, j \leq p\} \in \mathcal{L} \cup C(0,p) \; \forall p$ or $\{\omega_{i,j} : i, j \geq 1\} \in U(0)$ and $\{A_p\} \in \mathcal{NND}$. Let F^A denote the LSD of A_p. Then, as $p \to \infty$, the ESD of $p^{-1/2} A_p^{1/2} W_p A_p^{1/2}$ converges weakly (almost surely) to a non-random probability distribution whose Stieltjes transform $m(z)$ uniquely solves the following system of equations*

$$m(z) \quad = \quad -z^{-1} - z^{-1} g^2(z), \qquad (4.28)$$

$$g(z) \quad = \quad \int \frac{t}{-z - tg(z)} dF^A(t), \quad \forall z \in \mathbb{C}^+. \qquad (4.29)$$

Note that the assumptions on W in Theorem 4.2.2 are weaker than those in Theorem 4.2.1. (4.20) can be derived from (4.28) and (4.29) as follows: let $A_p = I_p$, where I_p is as in (1.9). Therefore, by Example 4.1.1, F^A is degenerate at 1. Hence, by (4.29)

$$g^2(z) = -(zg(z) + 1). \qquad (4.30)$$

Therefore, substituting (4.30) in (4.28), we have

$$zm(z) = -1 + (zg(z) + 1), \quad \text{that is,} \quad m(z) = g(z). \qquad (4.31)$$

Hence, by (4.30), $m(z)$ satisfies (4.20).

4.3   Independent matrix: Marčenko–Pastur law

Definition 4.3.1. *(Independent matrix) An independent matrix is a rectangular matrix with all independent mean 0 and variance 1 entries. We denote an independent matrix of order $p \times n$ by $Z_{p \times n}$.*

We shall often write Z for $Z_{p \times n}$, if there is no confusion about the dimension of the matrices. As we proceed, further restrictions will be imposed on the entries of these matrices as required. We shall consider various symmetrised version of Z, for example ZZ^* or ZAZ^* for an appropriate matrix A. We may even have two independent Z matrices and some symmetrised polynomials involving them. At first, we shall consider simple polynomials and state the LSD results for them via Stieltjes transform. Later we shall consider more general polynomials in several independent matrices and appropriate deterministic matrices after we develop the necessary notions on NCP.

Note that now we have two indices n and p. The classical RMT model assumes $n = n(p) \to \infty$ as $p \to \infty$ and

$$\frac{p}{n} \to y \in [0, \infty). \qquad (4.32)$$

The LSD results for the cases $y > 0$ and $y = 0$ are significantly different. Hence, we discuss them separately. For the $y > 0$ case, as p and n are comparable, it does not really matter whether we are assuming '$p = p(n) \to \infty$ as $n \to \infty$' or '$n = n(p) \to \infty$ as $p \to \infty$'. But for the case $y = 0$, to be technically consistent we shall assume the latter.

Just like the semi-circle law is tied to the Wigner matrix, the *Marčenko–Pastur law* is tied to the independent matrix.

Definition 4.3.2. *(Marčenko–Pastur law) The Marčenko–Pastur law say, MP_y, is parameterized by $y \in (0, \infty)$.*
For $y \in (0, 1]$, it has the probability density function

$$f_y(x) = \begin{cases} \frac{\sqrt{(b_+(y)-x)(x-b_-(y))}}{2\pi y x}, & \text{if } b_-(y) < x < b_+(y) \\ 0, & \text{otherwise.} \end{cases} \tag{4.33}$$

where $b_\pm(y) = (1 \pm \sqrt{y})^2$.
For $y \in (1, \infty)$, it is a mixture of a point mass at 0 and the probability density function $f_{1/y}$ with weights $1 - y^{-1}$ and y^{-1}, respectively.

This law has the moment sequence

$$\beta_h = \sum_{k=1}^{h} \frac{1}{k} \binom{h-1}{k-1} \binom{h}{k-1} y^{k-1}, \quad \forall h \geq 1. \tag{4.34}$$

Its Stieltjes transform $m(z)$ satisfies the quadratic equation

$$yz(m(z))^2 + (y + z - 1)m(z) + 1 = 0, \quad \forall z \in \mathbb{C}^+. \tag{4.35}$$

Only one solution of the above equation yields a valid Stieltjes transform and that is given by

$$m(z) = \frac{1 - y - z + i\sqrt{((1+\sqrt{y})^2 - z)(z - (1-\sqrt{y})^2)}}{2yz}, \quad z \in \mathbb{C}^+. \tag{4.36}$$

4.3.1 Results on Z: $p/n \to y > 0$

The well-known $p \times p$ *Wishart matrix* (without centering) can be written as $S_p = n^{-1}ZZ^*$. Marčenko and Pastur [1967] derived the LSD of S_p when the entries of Z are i.i.d. with finite fourth moment. Over the years several researchers reduced the moment assumptions. For examples, one can consult Wachter [1978] and Yin [1986]. The version of the Marčenko–Pastur law with minimal moment conditions appears to be the following. Recall the class $U(\delta)$ given in (4.26).

Theorem 4.3.1. *(Bai and Silverstein [2009]) Let $Z_{p \times n} = ((z_{i,j}))$ be a $p \times n$ independent matrix and let $S_p = n^{-1}ZZ^*$ be the Wishart matrix of order p. Suppose $\{z_{i,j} : i, j \geq 1\} \in U(0)$. Then, as $p, n(p) \to \infty$ and $p/n \to y > 0$, the ESD of S_p almost surely converges in distribution to MP_y.*

The following theorem discusses the convergence of the ESD of $A_p^{1/2} S_p A^{1/2}$, where A_p is as in Theorem 4.2.2.

Theorem 4.3.2. *(Bai and Silverstein [2009]) Let $Z_{p \times n} = ((z_{i,j}))$ be an independent matrix of order $p \times n$ and $S_p = n^{-1} Z Z^*$ be the Wishart matrix of order p. Suppose $\{z_{i,j} : i, j \geq 1\} \in U(0)$. Let $\{A_p\} \in \mathcal{NND}$ with LSD F^A. Then, as $p, n(p) \to \infty$ and $p/n \to y > 0$, the ESD of $A_p^{1/2} S_p A_p^{1/2}$ almost surely converges in distribution to the probability distribution with Stieltjes transform $m(z)$ which uniquely solves*

$$m(z) = \int \frac{dF^A(t)}{t(1 - y - yzm(z)) - z}. \tag{4.37}$$

Note that Theorem 4.3.1 is a particular case of Theorem 4.3.2. (4.35) can be derived from (4.37) when we susbtitute $A_p = I_p$. To see this, first note that by Example 4.1.1, F^A is degenerate at 1. Now, by (4.37) we have

$$m(z) = \frac{1}{1 - y - yzm(z) - z} \quad \text{or}$$

$$0 = yzm^2(z) + (y - 1 + z)m(z) + 1. \tag{4.38}$$

Hence, (4.35) is established from (4.37).

The following remark can be found in Couillet and Debbah [2011] and is easy to establish by the use of Theorem 4.3.2.

Remark 4.3.1. *Under the assumptions of Theorem 4.3.2, the ESD of $n^{-1} Z A_n Z^*$ almost surely converges in distribution to the probability distribution with Stieltjes transform $m(z)$ which uniquely solves*

$$m(z) = \int \frac{dF^A(t)}{ty(y - 1 - zym(z)) - zy} \quad \forall z \in \mathbb{C}^+. \tag{4.39}$$

Proof. Note that $\{A_p\} \in \mathcal{NND}$ is a nested sequence of matrices. By A_p and A_n, we respectively mean the p-th and the n-th matrix of the sequence. Let $R_{1,p} = A_p^{1/2} S_p A_p^{1/2}$, $R_{2,p} = n^{-1} Z A_n Z^*$, $R_{3,p} = n^{-1} Z^* A_p Z$ and $R_{4,p} = p^{-1} Z^* A_p Z$. Note that by Theorem 4.3.2, the LSD of $R_{1,p}$ exists. Therefore, it is easy to see that the LSD of $R_{i,p}$ for $i = 2, 3, 4$ also exists. Suppose for $1 \leq i \leq 4$, the Stieltjes transform of ESD and LSD of $R_{i,p}$ are respectively denoted by $m_{i,p}(z)$ and $m_i(z)$. Therefore,

$$m_{i,p}(z) \to m_i(z) \quad \forall z \in \mathbb{C}^+ \quad \text{almost surely.} \tag{4.40}$$

Moreover, by Theorem 4.3.2

$$m_1(z) = \int \frac{dF^A(t)}{t(1 - y - yzm_1(z)) - z}, \quad \forall z \in \mathbb{C}^+. \tag{4.41}$$

It is easy to see that $m_2(z)$ is point-wise equal to $m_4(z)$ once we replace y by y^{-1}. Therefore, to find $m_2(z)$ it is enough to show that

$$m_4(z) = \int \frac{y^2 dF^A(t)}{t(1 - y - zm_4(z)) - zy}, \quad \forall z \in \mathbb{C}^+. \tag{4.42}$$

To prove (4.42), first note that

$$
\begin{aligned}
m_{R_{3,p}}(z) &= \frac{1}{n}\mathrm{Tr}((n^{-1}Z^*AZ - zI_n)^{-1}) \\
&= \frac{p}{n}\frac{1}{p}\mathrm{Tr}((n^{-1}A_p^{1/2}ZZ^*A_p^{1/2} - zI_p)^{-1}) \\
&= \frac{p}{n}m_{R_{1,p}}(z).
\end{aligned}
$$

Hence,

$$
\begin{aligned}
m_3(z) = ym_1(z) &= \int \frac{ydF^A(t)}{t(1 - y - yzm_1(z)) - z} \\
&= \int \frac{ydF^A(t)}{t(1 - y - zm_3(z)) - z}, \quad \forall z \in \mathbb{C}^+. \tag{4.43}
\end{aligned}
$$

Moreover, by Lemma 4.1.6

$$
\begin{aligned}
m_4(z) = ym_3(zy) &= \int \frac{y^2 dF^A(t)}{t(1 - y - yzm_3(yz)) - zy} \\
&= \int \frac{y^2 dF^A(t)}{t(1 - y - zm_4(z)) - zy}, \quad \forall z \in \mathbb{C}^+. \tag{4.44}
\end{aligned}
$$

This proves (4.42). Thus,

$$m_2(z) = \int \frac{dF^A(t)}{ty(y - 1 - zym_2(z)) - zy}, \quad \forall z \in \mathbb{C}^+. \tag{4.45}$$

This completes the proof of Remark 4.3.1. □

4.3.2 Results on $Z: p/n \to 0$

The case $y = 0$ is quite different from the case $y > 0$. If we put $y = 0$ in the results obtained for $y > 0$, we obtain degenerate distributions. For example, if we put $y = 0$ in (4.34), β_h will be 1 for all $h \geq 1$. Therefore, LSD of S_p would be degenerate at 1. Hence, we need appropriate centering and scaling on S_p. Some of the known results in this regime are given below. Moreover, as mentioned just before Definition 4.3.2, to be technically consistent, all the results below assume $p \to \infty$, $n = n(p) \to \infty$ as $p \to \infty$.

Theorem 4.3.3. *(Bai and Yin [1988]) Let $Z_{p \times n}$ be an independent matrix whose entries are i.i.d. and have finite fourth order moment. Then, as $p/n \to 0$, the almost sure LSD of $\sqrt{np^{-1}}(n^{-1}ZZ^* - I_p)$ exists and it is distributed as the standard semi-circle variable with pdf (4.18).*

Theorem 4.3.4. *(Bao [2012]) Let $Z_{p \times n}$ be an independent matrix whose entries are i.i.d. and have finite fourth moment. Suppose $\{A_p\} \in \mathcal{NND}$ with LSD F^A. Then as $p/n \to 0$, the almost sure LSD of $\sqrt{np^{-1}}(n^{-1}A^{1/2}ZZ^*A^{1/2} - A)$ exists and its Stieltjes transform $m(z)$ uniquely solves the system of equations (4.28) and (4.29).*

Remark 4.3.2. *Consider all the assumptions in Theorems 4.2.2 and 4.3.4. Then the a.s. LSD of $p^{-1/2}A^{1/2}WA^{1/2}$ and $\sqrt{np^{-1}}(n^{-1}A^{1/2}ZZ^*A^{1/2} - A)$ are identical.*

Recall the notation (1.8). Consider the following class of $r \times r$ deterministic matrices:

$$\mathcal{N} = \text{all symmetric non-negative definite nested matrix sequences} \quad (4.46)$$
$$\{B_r\} \text{ such that } \sup_r ||B_r||_2 < \infty \text{ and } \lim r^{-1}\text{Tr}(B_r^i) \text{ exists for } i = 1, 2.$$

Theorem 4.3.5. *(Wang and Paul [2014]) Let $Z_{p \times n}$ be an independent matrix with i.i.d. entries having finite fourth moment, and A_p, B_n be two deterministic matrices. Suppose $\{A_p\} \in \mathcal{NND}$ with LSD F^A. Suppose $\{B_n\} \in \mathcal{N}$ and $\lim n^{-1}\text{Tr}(B_n^2) = d_2$. Then as $p/n \to 0$, the almost sure LSD of $\sqrt{np^{-1}}(n^{-1}A_p^{1/2}ZB_nZ^*A_p^{1/2} - A_p n^{-1}\text{Tr}(B_n))$ exists and its Stieltjes transform $m(z)$ uniquely solves the system of equations*

$$m(z) = -\int \frac{dF^A(t)}{z + d_2 tg(z)}, \tag{4.47}$$

$$g(z) = -\int \frac{tdF^A(t)}{z + d_2 tg(z)}, \quad \forall z \in \mathbb{C}^+. \tag{4.48}$$

Note that,

$$zm(z) = -\int \frac{zdF^A(t)}{z + d_2 tg(z)} = -1 + d_2 g(z) \int \frac{tdF^A(t)}{z + d_2 tg(z)} = -1 - d_2 g^2(z).$$

Therefore, an equivalent way to write (4.47) is,

$$m(z) = -z^{-1} - z^{-1}d_2 g^2(z), \quad \forall z \in \mathbb{C}^+. \tag{4.49}$$

Corollary 4.3.1. *Theorem 4.3.4 follows immediately from Theorem 4.3.5 by putting $B_n = I_n$ and $d_2 = 1$.*

Corollary 4.3.2. *Consider all the assumptions in Theorems 4.2.2 and 4.3.5. Then the almost sure LSD of $\sqrt{d_2}p^{-1/2}A^{1/2}WA^{1/2}$ and $\sqrt{np^{-1}}(n^{-1}A^{1/2}ZBZ^*A^{1/2} - An^{-1}\text{Tr}(B))$ are identical.*

Proof. Suppose LSD of $p^{-1/2}A^{1/2}WA^{1/2}$ is distributed as the random variable X. Therefore, LSD of $p^{-1/2}\sqrt{d_2}A^{1/2}WA^{1/2}$ is distributed as $\sqrt{d_2}X$. By Theorem 4.2.2, the Stieltjes transform of X is given by

$$m_X(z) = -z^{-1} - z^{-1}g_X^2(z), \tag{4.50}$$

$$g_X(z) = \int \frac{t}{-z - tg_X(z)}dF^A(t), \quad \forall z \in \mathbb{C}^+. \tag{4.51}$$

Therefore, by Lemma 4.1.6, the Stieltjes transform of $\sqrt{d_2}X$ is

$$
\begin{aligned}
m_{\sqrt{d_2}X}(z) &= d_2^{-1/2} m_X(d_2^{-1/2}z) = -z^{-1} - z^{-1} g_X^2(d_2^{-1/2}z), \text{ where} \\
g_X(d_2^{-1/2}z) &= \int \frac{\sqrt{d_2}\, t}{-z - t\sqrt{d_2}\, g_X(d_2^{-1/2}z)} dF^A(t).
\end{aligned}
\qquad (4.52)
$$

Let us define

$$
g_{\sqrt{d_2}X}(z) = d_2^{-1/2} g_X(d_2^{-1/2}z).
$$

Therefore,

$$
g_{\sqrt{d_2}X}(z) = \int \frac{t\, dF^A(t)}{-z - t\sqrt{d_2}\, g_X(d_2^{-1/2}z)} = \int \frac{t\, dF^A(t)}{-z - t d_2 g_{\sqrt{d_2}X}(z)}.
$$

Hence,

$$
\begin{aligned}
m_{\sqrt{d_2}X}(z) &= -z^{-1} - z^{-1} d_2 g_{\sqrt{d_2}X}^2(z), \text{ where} \\
g_{\sqrt{d_2}X}(z) &= \int \frac{t}{-z - t d_2 g_{\sqrt{d_2}X}(z)} dF^A(t).
\end{aligned}
$$

This agrees with (4.28) and (4.29). Hence, it completes the proof. □

Exercises

1. Verify the LSD claimed in Example 4.1.4.

2. Prove the moment method Lemma 4.1.2.

3. Show that the Stieltjes transform of a finite measure μ is finite for all $z \in \mathbb{C}^+$.

4. Learn the proof of the properties of the Stieltjes transform given in Lemma 4.1.4.

5. Learn the proof of the Stieltjes transform convergence Lemma 4.1.5.

6. Show that the quadratic equation of $m(z)$ in Example 4.1.5 has a unique Stieltjes transform solution and identify this solution.

7. Show that the quadratic equation of $m(z)$ in Example 4.1.6 has a unique Stieltjes transform solution and identify this solution.

8. Prove the change of variable Lemma 4.1.6.

9. Show that the standard semi-circle law moments are given by (4.19).

10. Using the moment sequence in (4.19), show that the Stieltjes transform of the standard semi-circle law is as in (4.20).

11. Show that the moments of the Marčenko–Pastur law are as in (4.34).

12. Using the moment sequence in (4.34), show that the Stieltjes transform of the Marčenko–Pastur law is as in (4.35).

13. Show that the moments of any compactly supported probability measure satisfies Carleman's condition.

14. Show that the Gaussian moments satisfies Carleman's condition.

15. Show that the semi-circle moments satisfies Carleman's condition.

16. Consider the Toeplitz matrix $T = ((t_{|i-j|}))$ where $t_0 = 1$, $t_1 = 1$ and $t_u = 0$ $\forall u \geq 2$. Find the LSD of T_n by the moment method.

17. Find LSD of a diagonal matrix with i.i.d. diagonal entries from Bernoulli(p).

18. Let $W_p = ((w_{ij}))_{p \times p}$ be a symmetric matrix with $w_{ij} \overset{\text{i.i.d.}}{\sim} \mathcal{N}(0,1)$ for $i \leq j$. Find the LSD of $p^{-1/2} W_p$ and $p^{-1} W_p^2$ using Lemma 4.1.2.

19. Let $\{A_p\}$ be a sequence of symmetric, nested and non-negative definite matrices whose LSD exist. Also suppose Z is a $p \times n$ independent matrix which satisfies assumptions in Theorem 4.3.2 and $p/n \to y > 0$. Then show that the LSD of $n^{-1} A_p^{1/2} Z Z^* A_p^{1/2}$ and $p^{-1} Z A_n Z^*$ exist. Establish the relationship between the Stieltjes transform of these two LSD.

Chapter 5

NON-COMMUTATIVE PROBABILITY

In the previous chapter we briefly discussed the convergence of a single sequence of matrices in terms of the convergence of its ESD. When we have
more than one sequence of matrices, how should their joint convergence be
viewed? Then the most natural object to consider is the non-commutative
∗-probability space (NCP) generated by polynomials of these matrices and
study the convergence of the elements of this space. Convergence of the spectral distribution of any given polynomial of matrices is closely related to the
above convergence.

As matrices are non-commutative objects, appearance of non-commutative
spaces is not surprising. As we know, commutative (classical) random variables
are attached to a probability space $(\mathcal{S}, E)$, which consists of a σ-field $\mathcal{S}$ and
an expectation operator E. Similarly, non-commutative variables are attached
to an NCP.

This chapter will serve as a brief introduction to *non-commutative probability* and related notions that will be used in later chapters, specially in
Chapters 6, 7, 8, and 10. In particular the crucial notion of *free independence*, a non-commutative analogue of classical independence, will be developed in this chapter. An excellent reference for a combinatorial approach to
non-commutative probability is Nica and Speicher [2006], from which the material of this chapter is mostly taken.

5.1 NCP and its convergence

Definition 5.1.1. *A non-commutative ∗-probability space (NCP),* $(\mathcal{A}, \varphi)$,
consists of a unital ∗-algebra $\mathcal{A}$ *over* $\mathbb{C}$ *and a (unital) linear functional*

$$\varphi : \mathcal{A} \to \mathbb{C} \text{ such that } \varphi(1_\mathcal{A}) = 1.$$

Thus, φ is the analogue of the (classical) expectation operator and is called
a *state* of the algebra $\mathcal{A}$. The elements $a \in \mathcal{A}$ are called non-commutative
random variables in $(\mathcal{A}, \varphi)$. If $a = a^*$, then a is called *self-adjoint*. The state

φ is said to be *tracial* and *positive* if

$$\varphi(ab) - \varphi(ba), \quad \forall a, b \in \mathcal{A} \text{ (tracial)}, \tag{5.1}$$
$$\varphi(a^*a) \geq 0, \quad \forall a \in \mathcal{A} \text{ (positive)}. \tag{5.2}$$

In this book, φ will always be tracial and positive.

The following are some examples of non-commutative $*$-probability spaces.

Example 5.1.1. Let $(\Omega, \mathcal{F}, P)$ be a probability space in the classical sense, i.e., Ω is a non-empty set, $\mathcal{F}$ is a σ-algebra of subsets of Ω and $P : \mathcal{F} \rightarrow [0, 1]$ is a probability measure. Let $\mathcal{A} = L^\infty(\Omega, P) = $ set of all measurable and essentially bounded (bounded up to a set of measure zero) functions $a : \Omega \rightarrow \mathbb{C}$ and let φ be defined by

$$\varphi(a) = \int_\Omega a(\omega) dP(\omega), \quad a \in \mathcal{A}.$$

Then $(\mathcal{A}, \varphi)$ is an NCP. *In this case, $\mathcal{A}$ is commutative and a^* is the complex conjugate of a.*

Example 5.1.2. Let d be a positive integer. Let $\mathcal{M}_d(\mathbb{C})$ be the algebra of $d \times d$ matrices with complex entries and usual matrix multiplication, and let $\text{tr} : \mathcal{M}_d(\mathbb{C}) \rightarrow \mathbb{C}$ be the normalized trace,

$$\text{tr}(a) = \frac{1}{d} \sum_{i=1}^d \alpha_{ii}, \quad \forall a = ((\alpha_{ij}))_{i,j=1}^d \in \mathcal{M}_d(\mathbb{C}).$$

Then $(\mathcal{M}_d(\mathbb{C}), \text{tr})$ is an NCP, where the $*$-operation is to take both the transpose of the matrix and the complex conjugate of the entries. Also let $\mathcal{M}_{(d)}(\mathbb{C})$ be the $*$-algebra of all $d \times d$ random matrices with usual matrix multiplication. Then $(\mathcal{M}_{(d)}(\mathbb{C}), E\text{tr})$ forms an NCP.

The following lemma provides two inequalities. Its proof is trivial and hence we omit it (see Nica and Speicher [2006] for part (a)).

Lemma 5.1.1. *Suppose $(\mathcal{A}, \varphi)$ is an NCP. Let $a, b, a_1, a_2, \ldots, a_k \in \mathcal{A}$ and φ be positive. Then the following results hold.*

(a) $|\varphi(ab)| \leq \sqrt{\varphi(a^*a)\varphi(b^*b)}$.

(b) Moreover, if φ is tracial, then there exists $h_1, h_2, \ldots, h_k \geq 1$ such that

$$|\varphi(a_1 a_2 \cdots a_k)| \leq \prod_{i=1}^k (\varphi(a_i^* a_i)^{h_i})^{1/2h_i}.$$

Since we shall always work with φ which is tracial and positive, Lemma 5.1.1(a,b) shall be available to us.

Often we deal with $*$-sub-algebras of $\mathcal{M}_d(\mathbb{C})$ and $\mathcal{M}_{(d)}(\mathbb{C})$.

Definition 5.1.2. *(∗-sub-algebra and span) Let $\mathcal{B}$ be a unital ∗-sub-algebra of $\mathcal{A}$. Then $(\mathcal{B}, \varphi)$ also forms an NCP. Let $1_\mathcal{A}$ be the identity element of $\mathcal{A}$. Consider $t \geq 1$. Let $\Pi(1_\mathcal{A}, a_i, a_i^* : 1 \leq i \leq t) \in \mathcal{A}$ be any polynomial in $\{1_\mathcal{A}, a_i, a_i^* : 1 \leq i \leq t\} \subset \mathcal{A}$. Let*

$$Span\{a_i, a_i^* : i \leq t\} = \{\Pi(1_\mathcal{A}, a_i, a_i^* : i \leq t) : \Pi \text{ is a polynomial}\}. \quad (5.3)$$

$Span\{a_i, a_i^* : 1 \leq i \leq t\}$ is called the ∗-algebra generated by $\{a_i, a_i^* : 1 \leq i \leq t\}$. Equipped with φ, it is a unital ∗-algebra and is called the NCP generated by $\{a_i, a_i^* : 1 \leq i \leq t\}$.

By span of a collection of infinitely many non-commutative variables $\{a_i, a_i^* : i \geq 1\}$, we mean

$$Span\{a_i, a_i^* : i \geq 1\} = \{\Pi(1_\mathcal{A}, a_{i_k}, a_{i_k}^* : k \leq t) : i_k, t \geq 1, \Pi \text{ is a polynomial}\}$$

and it is also an NCP.

For example, consider a class of $d \times d$ (random) matrices $\{M_i : 1 \leq i \leq r\}$. Then $(Span\{M_i, M_i^* : 1 \leq i \leq r\}, Etr)$ is an NCP.

The distribution and moments of non-commutative variables are defined as follows.

Definition 5.1.3. *(Distribution and moments)   Let $(\mathcal{A}, \varphi)$ be an NCP. Let $\Pi(a, a^*) \in \mathcal{A}$ be any polynomial in $a, a^* \in \mathcal{A}$. Then $\{\varphi(\Pi(a, a^*)) : \Pi \text{ is a polynomial}\}$ is called the ∗-distribution of a or a^*. In particular, if $a \in \mathcal{A}$ is self-adjoint, then $\{\varphi(a^k)\}_{k=1}^\infty$ is called the distribution of a.*

Consider $t \geq 1$. Then $\{\varphi(\Pi(a_i, a_i^* : 1 \leq i \leq t)) : \Pi \text{ is a polynomial}\}$ is called the joint distribution of $\{a_i : 1 \leq i \leq t\}$.

For a collection of infinitely many non-commutative variables $\{a_i : i \geq 1\}$, $\{\varphi(\Pi(a_{i_k}, a_{i_k}^* : 1 \leq k \leq t)) : i_k, t \geq 1, \Pi \text{ is a polynomial}\}$ is its joint distribution. Likewise we can also define the distribution in all the above cases. For simplicity we shall write "distribution" for "∗-distribution".

Now we shall define convergence of variables and of NCP.

Definition 5.1.4. *(Convergence of variables and of NCP)   Let $\mathcal{A}_N = Span\{a_i^{(N)}, a_i^{*(N)} : i \geq 1\}$, $\forall N \geq 1$ and $\mathcal{A} = Span\{a_i, a_i^* : i \geq 1\}$. We say that the sequence of NCP $\{(\mathcal{A}_N, \varphi_N)\}$ converges to $(\mathcal{A}, \varphi)$ if for any $t \geq 1$ and any polynomial Π*

$$\lim_{N \to \infty} \varphi_N\left(\Pi(a_i^{(N)}, a_i^{*(N)} : 1 \leq i \leq t)\right) = \varphi\left(\Pi(a_i, a_i^* : 1 \leq i \leq t)\right). \quad (5.4)$$

This is also described as the joint convergence of $\{a_i^{(N)}, a_i^{*(N)} : i \geq 1\}$ to $\{a_i, a_i^* : i \geq 1\}$. If $\{\varphi_N\}$ are tracial and positive, then φ is also so (see (5.1) and (5.2)). For a fixed $i \geq 1$, we say that $a_i^{(N)}$ converges in distribution to a_i if for any polynomial Π,

$$\lim \varphi_N(\Pi(a_i^{(N)}, a_i^{(N)*})) = \varphi(\Pi(a_i, a_i^*)). \quad (5.5)$$

Remark 5.1.1. *Suppose we are given a unital ∗-sub-algebra $\mathcal{A}_N$ as above and the left side of (5.4) exists for all polynomial Π. Then we can construct a polynomial algebra $\mathcal{A}$ of indeterminates $\{a_i, a_i^*\}$ which also includes an identity $1_{\mathcal{A}}$. We can define φ on $\mathcal{A}$ by Equation (5.4). Then $(\mathcal{A}, \varphi)$ is an NCP and $(\mathcal{A}_N, \varphi_N) \to (\mathcal{A}, \varphi)$ in the above sense.*

Let A_p be a square real symmetric random matrix of order p. Note that the convergence of $(\text{Span}\{A_p\}, p^{-1}E\text{Tr})$ guarantees the (M1) condition in the moment method (see Lemma 4.1.2). The following lemma connects LSD and NCP convergence. Proof of this lemma follows immediately from Lemma 4.1.2.

Lemma 5.1.2. *Let A_p be a real symmetric random matrix of order p. Suppose $(\text{Span}\{A_p\}, p^{-1}E\text{Tr}) \to (\text{Span}\{a\}, \varphi)$. Moreover suppose (M4) holds and $\{\varphi(a^k)\}$ satisfies Carleman's condition (C). Then as $p \to \infty$, the almost sure LSD of A_p exists and it is uniquely determined by the moment sequence $\{\varphi(a^k)\}$.*

Some examples are given below for better insight of Definition 5.1.4 and Lemma 5.1.2.

Example 5.1.3. $\{D_{in} : i \geq 1\}$ is the set of $n \times n$ diagonal matrices with

$$D_{in}(j, j) = \begin{cases} 1, & \text{if } j \neq i \\ X_i, & \text{if } j = i, \end{cases} \qquad X_i \overset{\text{i.i.d.}}{\sim} \text{Bernoulli}(0.5),$$

$\mathcal{A}_n = \text{Span}\{D_{in} : i \geq 1\}$ and $\varphi_n = \frac{1}{n}E\text{Tr}$. Then $(\mathcal{A}_n, \varphi_n)$ forms a sequence of NCP. It is easy to see that for any $i_1, \ldots i_k \geq 1$, $t_1, \ldots, t_k \in \mathbb{R}$ and $k \geq 1$,

$$\lim \varphi_n(D_{i_1 n} D_{i_2 n} \cdots D_{i_k n}) = 1 \quad \text{and}$$
$$\lim \varphi_n(t_1 D_{i_1 n} + \cdots + t_k D_{i_k n}) = t_1 + \cdots + t_k. \tag{5.6}$$

Therefore $(\mathcal{A}_n, \varphi_n)$ converges to the NCP generated only by the identity element. Moreover the LSD of any polynomial in $\{D_{in} : i \geq 1\}$ is degenerate.

Example 5.1.4. Consider the sequence of NCP $(\mathcal{A}_n, \varphi_n)$, given in Example 5.1.3, with

$$D_{in}(j, j) = \begin{cases} 1, & \text{if } j \leq i \\ X_i, & \text{if } j > i. \end{cases}$$

It is easy to see that for any $i_1, \ldots, i_k \geq 1$, $t_1, \ldots, t_k \in \mathbb{R}$ and $k \geq 1$,

$$\lim \varphi_n(D_{i_1 n} D_{i_2 n} \cdots D_{i_k n}) = E(X_{i_1} X_{i_2} \cdots X_{i_k}) \quad \text{and}$$
$$\lim \varphi_n(t_1 D_{i_1 n} + \cdots + t_k D_{i_k n}) = E(t_1 X_{i_1} + \cdots + t_k X_{i_k})$$

where $E(\cdot)$ is usual expectation of a random variable. Hence, $(\mathcal{A}_n, \varphi_n)$ converges to $(\text{Span}\{X_i : i \geq 1\}, E)$. Moreover for any polynomial Π, the LSD of $\Pi(D_{in} : i \geq 1)$ is identical to the probability distribution of $\Pi(X_i : i \geq 1)$.

Example 5.1.5. Let $P_{un} = ((I(i - j = u)))_{n \times n}$. Note that $P_{un} = P_{-un}^*$ and $P_{0n} = I_n$. Define $\mathcal{A}_n = \text{Span}\{P_{un} : u = 0, \pm 1, \pm 2, \ldots\}$ and take $\varphi_n = \frac{1}{n}\text{Tr}$. Using simple matrix algebra, one can show that for any $i_1, i_2, \ldots i_k = 0, \pm 1, \pm 2, \ldots, t_1, t_2, \ldots, t_k \in \mathbb{R}$ and $k \geq 1$,

$$\lim \varphi_n(P_{i_1 n} P_{i_2 n} \cdots P_{i_k n}) = I(i_1 + i_2 + \cdots + i_k = 0) \text{ and}$$

$$\lim \varphi_n(t_1 P_{i_1 n} + t_2 P_{i_2 n} + \cdots + t_k P_{i_k n}) = \sum_{j=1}^{k} t_k I(i_j = 0).$$

Let the NCP $(\mathcal{A} = \text{Span}(c_u : u = 0, \pm 1, \pm 2, \ldots), \varphi)$ be such that $\varphi(c_{i_1} c_{i_2} \cdots c_{i_k}) = I(i_1 + \cdots + i_k = 0)$, $c_u^* = c_{-u}$, $c_0 = 1_{\mathcal{A}}$ for all $u, i_1, i_2, \ldots, i_k = 0, \pm 1, \pm 2, \ldots$ and $k \geq 1$. Then $(\mathcal{A}_n, \varphi_n) \to (\mathcal{A}, \varphi)$. Also for any symmetric polynomial Π, the LSD of $\Pi(P_{un} : u = 0, \pm 1, \pm 2, \ldots)$ exists and it is uniquely determined by the moment sequence $\{\varphi(\Pi(c_u : u = 0, \pm 1, \pm 2, \ldots))^k\}$.

Example 5.1.6. Consider the $n \times n$ Hankel matrices $H_{kn} = I(i + j = n + k)$, $k = \pm 1, \pm 2, \ldots$. Let $\mathcal{A}_n = \text{Span}\{H_{kn} : |k| < K\}$ and $\varphi_n = \frac{1}{n}\text{Tr}$. Then $(\mathcal{A}_n, \varphi_n) \to (\text{Span}\{a_k : |k| < K\}, \varphi)$ where

$$\lim \varphi_n \left(\prod_{j=1}^{L} a_{i_j} \right) = \begin{cases} I\left(\sum_{j=1}^{m} a_{i_{2j-1}} = \sum_{j=1}^{m} a_{i_{2j}} \right), & \text{if } L = 2m \text{ for } m \geq 1 \\ 0, & \text{otherwise.} \end{cases}$$

In particular, a_k is marginally $2\text{Bernoulli}(0.5) - 1$ for all $|k| < K$. For detailed justification of this, see Bose and Gangopadhyay [2015].

Example 5.1.7. Let $F(\cdot)$ be a cumulative distribution function on $\mathbb{R}$ with $\int_{\mathbb{R}} |x|^k dF(x) \leq C^k$ for some $C > 0$ and all $k \geq 1$. Consider the $n \times n$ diagonal matrix D_n with $D_n(i, i) \overset{\text{i.i.d.}}{\sim} F$. Then $(\text{Span}\{D_n\}, \frac{1}{n}E\text{Tr})$ converges to $(\text{Span}\{X\}, E)$ where $X \sim F$ and the almost sure LSD of D_n has cumulative distribution function F.

The following theorem (see for example Anderson et al. [2009]) is relevant in this context and shall be used later in Chapter 6.

Theorem 5.1.1. *Let W_p be a Wigner matrix of order p. Then under the same assumptions as in Theorem 4.2.1,*

$$(Span\{p^{-1/2}W_p\}, E\text{tr}) \to (Span\{s\}, \varphi), \tag{5.7}$$

where s is the standard semi-circle variable with $\varphi(s^h) = \beta_h$, $\forall h$ and $\{\beta_h\}$ is as in (4.19).

Often the limit of random matrices (in the sense of Definition 5.1.4) can be expressed in terms of some *freely independent variables*. Free variables in the non-commutative world, is the analogue of independent random variables in the commutative world. In the next few sections, we develop the basics of free probability.

5.2   Essentials of partition theory

In the commutative case, random variables (say with bounded support) are independent if and only if all joint moments obey the product rule. This independence can also be formulated in terms of cumulants. Suppose $X_1, \ldots, X_k$ are bounded real random variables. Their joint *moment generating function* (m.g.f.) is defined as

$$M(t_1, \ldots, t_k) = E \exp\{\sum_{i=1}^{k} t_i X_i\}, \ t_i \in \mathbb{R} \quad \forall i.$$

The *cumulant generating function* (c.g.f.) is given by

$$
\begin{aligned}
\mathbb{K}(t_1, \ldots, t_k) &= \log M(t_1, \ldots, t_k) \\
&= \sum_{j_1, \ldots, j_k = 1}^{\infty} \mathbb{K}_{j_1, \ldots j_k} \frac{\prod_{i=1}^{k} t_i^{j_i}}{\prod_{i=1}^{k} j_i!}.
\end{aligned}
$$

The coefficients $\{\mathbb{K}_{j_1, \ldots j_k}\}$ are called the joint *cumulants* of $X_1, \ldots, X_k$.

Example 5.2.1. Suppose X has the Gaussian distribution with mean μ and variance σ^2. Then the m.g.f. and the c.g.f. of X are

$$M(t) = \exp\{\mu t + \frac{\sigma^2}{2}\}, \quad \mathbb{K}(t) = \mu t + \frac{\sigma^2}{2}$$

so that $\mathbb{K}_1(X) = \mu$, $\mathbb{K}_2(X) = \sigma^2$ and all other cumulants are 0. This is a characterizing property of the Gaussian distribution.

Example 5.2.2. Suppose X has the Poisson distribution with mean λ. Then the m.g.f. and the c.g.f. of X are

$$M(t) = \exp\{\lambda(e^t - 1)\}, \quad \mathbb{K}(t) = \lambda(e^t - 1)$$

so that all cumulants of X are $\mathbb{K}_n(X) = \lambda$. This is a characterizing property of the Poisson distribution.

It is well-known that the cumulants and moments are related via the *Möbius function* on the partially ordered set ($POSET$) of *all partitions*. Using this it can be shown that independence is also equivalent to the vanishing of all mixed cumulants.

In the non-commutative case, we also have the notion of joint cumulants, called *free cumulants*. These can be uniquely obtained from the moments and vice-versa via a different Möbius function and its inverse on the POSET of all *non-crossing partitions*. Non-commutative variables are said to be *free* (freely independent) if and only if all their mixed free cumulants vanish. Before we develop the concepts related to free independence, we need the notion of Möbius function and some notions from the theory of partitions, specially non-crossing partitions.

5.2.1   Möbius function

Let P be a finite partially ordered set (POSET) with the partial order $\leq$. We also assume that P is a *lattice*. Let

$$P^{(2)} = \{(\pi, \sigma) : \pi, \sigma \in P, \pi \leq \sigma\} \tag{5.8}$$

be the set of intervals of P.

For any two complex-valued functions $F, G : P^{(2)} \to \mathbb{C}$, their *convolution* $F * G : P^{(2)} \to \mathbb{C}$ is defined by:

$$(F * G)(\pi, \sigma) \quad = \quad \sum_{\substack{\rho \in P \\ \pi \leq \rho \leq \sigma}} F(\pi, \rho) G(\rho, \sigma). \tag{5.9}$$

F is said to be *invertible* if there exists (unique) G (called the inverse of F), for which

$$(F * G)(\pi, \sigma) = (G * F)(\pi, \sigma) = I(\pi = \sigma), \ \forall \pi \leq \sigma \in P. \tag{5.10}$$

It is easy to show that F is invertible if and only if $F(\pi, \pi) \neq 0$ for every $\pi \in P$ (see Proposition 10.4 in Nica and Speicher [2006])

Definition 5.2.1. *(Zeta function and Möbius function) The Zeta function ξ of P is defined by*

$$\xi(\pi, \sigma) = 1, \ \ \forall \ (\pi, \sigma) \in P^{(2)}. \tag{5.11}$$

The Möbius function μ of P is the inverse of ξ.

Therefore,

$$(\xi * \mu)(\pi, \sigma) = (\mu * \xi)(\pi, \sigma) = I(\pi = \sigma), \ \forall \pi \leq \sigma \in P. \tag{5.12}$$

5.2.2   Partition and non-crossing partition

Let S be a finite totally ordered set. We call $\pi = \{V_1, V_2, \ldots, V_r\}$ a *partition* of the set S if and only if V_i $(1 \leq i \leq r)$ are pairwise disjoint, non-void subsets of S such that $V_1 \cup V_2 \cup \cdots \cup V_r = S$. We call $V_1, V_2, \ldots, V_r$ the *blocks* of π. Given two elements $p, q \in S$, we write $p \sim_\pi q$ if p and q belong to the same block of π.

A partition π is called a *pair partition* if each block of π contains exactly two elements.

Let $\pi = \{V_1, V_2, \ldots, V_r\}$ and $\sigma = \{U_1, U_2, \ldots, U_k\}$ be two partitions of S. Then we call $\pi \leq \sigma$ if for every fixed $1 \leq i \leq r$ there exists a $1 \leq j \leq k$ such that $V_i \subset U_j$. Therefore, the set of all partitions of S forms a POSET.

Definition 5.2.2. *A partition π of S is called crossing if there exists $p_1 < q_1 < p_2 < q_2$ in S such that $p_1 \sim_\pi p_2$ and $q_1 \sim_\pi q_2$ but (p_1, p_2) and (q_1, q_2) are not in the same block. If π is not crossing, then it is called non-crossing.*

Consider the following sets:

$$NC(n) = \{\pi : \pi \text{ is a non-crossing partition of } \{1,2,3,\ldots,n\}\}, \quad (5.13)$$

$$NC_2(2n) = \{\pi : \pi \in NC(2n) \text{ and it is a pair partition}\}, \quad (5.14)$$

$$NCE(2n) = \{\pi \in NC(2n) : \text{all blocks of } \pi \text{ are of even size}\}. \quad (5.15)$$

All the above sets are POSET and are lattice. In $NC(n)$, the smallest element is $0_n = \{\{1\},\{2\},\ldots,\{n\}\}$ and the largest element is $1_n = \{1,2,\ldots,n\}$. Moreover, $NC_2(2n) \subset NCE(2n) \subset NC(2n)$.

The semi-circle law is intimately connected to non-crossing partitions. Indeed, it can be shown that $\{\beta_h\}$ given in (4.19) satisfies

$$\beta_{2h} = \#NC_2(2h) = \#NC(h), \ \forall h \geq 1. \quad (5.16)$$

We shall need the Möbius function μ on the POSET $NC(n)$ given below.

$$\mu(0_n, 1_n) = (-1)^{n-1}C_{n-1}, \text{ where } C_n = \frac{1}{n+1}\binom{2n}{n}, \ \forall n \geq 1. \quad (5.17)$$

For other pairs $\pi \leq \sigma \in NC(n)$, $\mu(\pi,\sigma)$ can be obtained as follows. It is known that the interval $[\pi,\sigma]$ has the *canonical factorization*

$$[\pi,\sigma] \cong NC(1)^{k_1} \times NC(2)^{k_2} \times NC(n)^{k_n}. \quad (5.18)$$

Let $s_n = \mu(0_n, 1_n), \forall n$. Then it can be proved that

$$\mu(\pi,\sigma) = s_1^{k_1} s_2^{k_2} \cdots s_n^{k_n}. \quad (5.19)$$

For details, see Chapter 10 in Nica and Speicher [2006]. Following example provides $\mu(\pi, 1_n)$ for $\pi \in NCE(4)$ and $NCE(6)$.

Example 5.2.3. By (5.17), note that $s_1 = 1, s_2 = -1$ and $s_3 = 2$. Simple calculations using (5.19), lead to the following table.

Partition (π)	$[\pi, 1_n]$	$\mu(\pi, 1_n)$
$\pi_1 = \{(1,2,3,4)\}$	$NC(1)^4$	$s_1^4 = 1$
$\pi_2 = \{(1,2),(3,4)\}$	$NC(1)^2 \times NC(2)$	$s_1^2 s_2 = -1$
$\pi_3 = \{(1,4),(2,3)\}$	$NC(1)^2 \times NC(2)$	$s_1^2 s_2 = -1$
$\pi_4 = \{(1,2,3,4,5,6)\}$	$NC(1)^6$	$s_1^6 = 1$
$\pi_5 = \{(1,6),(2,3),(4,5)\}$	$NC(1)^3 \times NC(3)$	$s_1^3 s_3 = 2$
$\pi_6 = \{(1,2),(3,4),(5,6)\}$	$NC(1)^3 \times NC(3)$	$s_1^3 s_3 = 2$
$\pi_7 = \{(1,2),(3,6),(4,5)\}$	$NC(1)^2 \times NC(2)^2$	$s_1^2 s_2^2 = 1$
$\pi_8 = \{(1,6),(2,5),(3,4)\}$	$NC(1)^2 \times NC(2)^2$	$s_1^2 s_2^2 = 1$
$\pi_9 = \{(1,6),(3,4),(2,5)\}$	$NC(1)^2 \times NC(2)^2$	$s_1^2 s_2^2 = 1$
$\pi_{10} = \{(1,2,3,4),(5,6)\}$	$NC(1)^4 \times NC(2)$	$s_1^4 s_2 = -1$
$\pi_{11} = \{(1,2,5,6),(3,4)\}$	$NC(1)^4 \times NC(2)$	$s_1^4 s_2 = -1$
$\pi_{12} = \{(3,4,5,6),(1,2)\}$	$NC(1)^4 \times NC(2)$	$s_1^4 s_2 = -1$
$\pi_{13} = \{(1,4,5,6),(2,3)\}$	$NC(1)^4 \times NC(2)$	$s_1^4 s_2 = -1$
$\pi_{14} = \{(2,3,4,5),(1,6)\}$	$NC(1)^4 \times NC(2)$	$s_1^4 s_2 = -1$
$\pi_{15} = \{(1,2,3,6),(4,5)\}$	$NC(1)^4 \times NC(2)$	$s_1^4 s_2 = -1$

5.2.3   Kreweras complement

The *complementation map* $K : NC(n) \to NC(n)$ is defined as follows. We consider additional numbers $\bar{1}, \bar{2}, \ldots, \bar{n}$ and interlace them with $1, 2, \ldots, n$ in the following way:

$$1, \bar{1}, 2, \bar{2}, \ldots, n, \bar{n}.$$

Let π be a non-crossing partition of $\{1, 2, \ldots, n\}$. Then its Kreweras complement $K(\pi) \in NC(\bar{1}, \bar{2}, \ldots, \bar{n}) \cong NC(n)$ is defined to be the largest element among those $\sigma \in NC(\bar{1}, \bar{2}, \ldots, \bar{n})$ which have the property that

$$\pi \cup \sigma \in NC(1, \bar{1}, 2, \bar{2}, \ldots, n, \bar{n}).$$

where $\pi \cup \sigma$ is the partition whose blocks are the blocks from π and σ. Let π be a partition of the set S and $A \subset S$. Then by $K_\pi | A$ we mean the restriction of K_π on A. The following properties of Kreweras complement are useful. See Chapter 9 of Nica and Speicher [2006] for details.

Lemma 5.2.1. *(a)* $K : NC(n) \to NC(n)$ *is a bijection.*

   (b) $K(NCE(2n))$ *is in bijection with the set of all such* π *in* $NC(2n)$ *such that every block of* π *is contained either in* $\{1, 3, \ldots, 2n - 1\}$ *or in* $\{2, 4, \ldots, 2n\}$.

   (c) $NC_2(2n) \ni \pi \to (K_\pi | \{1, 3, \ldots, 2n - 1\})$ *is a bijection between* $NC_2(2n)$ *and* $NC(1, 3, \ldots, 2n - 1)$.

   (d) Let $|\pi|$ *be the total number of blocks in any partition* π. *Then for any* $\pi \in NC(n)$, *we have* $|\pi| + |K(\pi)| = n + 1$.

5.3   Free cumulant; free independence

Conventionally, free independence of non-commutative variables is defined via certain restrictions on their moments. On the other hand, there is a one to one correspondence between moments and what are known as free cumulants which are to be defined shortly. We choose to define free independence via free cumulants since this is more convenient for our purposes.

   Let $(\mathcal{A}, \varphi)$ be an NCP. Define a sequence of *multilinear functionals* $(\varphi_n)_{n \in \mathbb{N}}$ on $\mathcal{A}^n$ via

$$\varphi_n(a_1, a_2, \ldots, a_n) := \varphi(a_1 a_2 \cdots a_n). \tag{5.20}$$

Extend $\{\varphi_n\}$ to $\{\varphi_\pi (\pi \in NC(n), n \geq 1)\}$ *multiplicatively* in a recursive way by the following formula. If $\pi = \{V_1, V_2, \ldots, V_r\} \in NC(n)$, then

$$\varphi_\pi[a_1, a_2, \ldots, a_n] := \varphi(V_1)[a_1, a_2, \ldots, a_n] \cdots \varphi(V_r)[a_1, a_2, \ldots, a_n], \tag{5.21}$$

where

$$\varphi(V)[a_1, a_2, \ldots, a_n] := \varphi_s(a_{i_1}, a_{i_2}, \ldots, a_{i_s}) = \varphi(a_{i_1} a_{i_2} \cdots a_{i_s})$$

for $V = (i_1 < i_2 < \cdots < i_s)$. Observe the different use of the two types of braces ( ) and [ ] in (5.20) and (5.21).

Define the *joint free cumulant* of order n of $(a_1, a_2, \ldots, a_n)$ as

$$\kappa_n(a_1, a_2, \ldots, a_n) = \sum_{\sigma \in NC(n)} \varphi_\sigma[a_1, a_2, \ldots, a_n] \mu(\sigma, 1_n), \qquad (5.22)$$

where μ is the Möbius function on $NC(n)$.

The quantity $\kappa_n(a_1, a_2, \ldots, a_n)$ is called a (free) *mixed cumulant* if at least one pair a_i, a_j are different and $a_i \neq a_j^*$.

For any $\epsilon_i = 1, *, \ \forall 1 \leq i \leq n$, $\kappa_n(a^{\epsilon_1}, a^{\epsilon_2}, \ldots, a^{\epsilon_n})$ is called a *marginal free cumulant* of order n of $\{a, a^*\}$. For a self-adjoint element a,

$$\kappa_n(a) = \kappa_n(a, a, \ldots, a)$$

is called the n-th free cumulant of a. Note that mixed/marginal free cumulants are all special cases of joint free cumulants.

Just as in (5.21), $\{\kappa_n(a_1, a_2, \ldots, a_n) : n \geq 1\}$ has a multiplicative extension $\{\kappa_\pi : \pi \in NC(n)\}$. It is known that (see Proposition 11.4 in Nica and Speicher [2006])

$$\kappa_\pi[a_1, a_2, \ldots, a_n] := \sum_{\substack{\sigma \in NC(n) \\ \sigma \leq \pi}} \varphi_\sigma[a_1, a_2, \ldots, a_n] \mu(\sigma, \pi), \qquad (5.23)$$

where μ is the Möbius function on $NC(n)$. Note that

$$\begin{aligned}
\varphi_{1_n}[a_1, a_2, \ldots, a_n] &= \varphi_n(a_1, a_2, \ldots, a_n) = \varphi(a_1 a_2 \cdots a_n), \\
\kappa_{1_n}[a_1, a_2, \ldots, a_n] &= \kappa_n(a_1, a_2, \ldots, a_n).
\end{aligned}$$

Moreover, using the Möbius function, it can be shown that

$$\varphi(a_1 a_2 \cdots a_n) = \sum_{\pi \leq 1_n} \kappa_\pi[a_1, a_2, \ldots, a_n]. \qquad (5.24)$$

Therefore, (5.21)–(5.24) establish a one-to-one correspondence between free cumulants and moments.

Let $\mathbb{K}_n(\cdot)$ be the cumulant of order n in usual commutative probability space. Following example provides explicit formulae for κ_4 and κ_6 and compares them with $\mathbb{K}_4$ and $\mathbb{K}_6$.

Example 5.3.1. Recall the notation and the table given in Example 5.2.3. By (5.22) and for any variable X with $\varphi(X^{2k+1}) = 0 \ \forall k \geq 0$, we have

$$\begin{aligned}
\kappa_4(X) &= \varphi(X^4)\mu(\pi_1, 1_4) - (\varphi(X^2))^2(\mu(\pi_2, 1_4) + \mu(\pi_3, 1_4)) \\
&= \varphi(X^4) - 2(\varphi(X^2))^2, \qquad (5.25) \\
\kappa_6(X) &= \varphi(X^6)\mu(\pi_4, 1_6) + (\varphi(X^2))^3(\mu(\pi_5, 1_6) + \cdots + \mu(\pi_9, 1_6)) \\
&\quad + \varphi(X^4)\varphi(X^2)(\mu(\pi_{10}, 1_6) + \cdots + \mu(\pi_{15}, 1_6)) \\
&= \varphi(X^6) + 7(\varphi(X^2))^3 - 6\varphi(X^4)\varphi(X^2). \qquad (5.26)
\end{aligned}$$

On the other hand

$$\begin{aligned}
\mathbb{K}_4(X) &= \varphi(X^4) - 3(\varphi(X^2))^2, \\
\mathbb{K}_6(X) &= \varphi(X^6) + 30(\varphi(X^2))^3 - 15\varphi(X^4)\varphi(X^2).
\end{aligned} \tag{5.27}$$

Also moments can be expressed in terms of cumulants as follows.

$$\begin{aligned}
\varphi(X^4) &= \kappa_4(X) + 2\kappa_2(X) = \mathbb{K}_4(X) + 3(\mathbb{K}_2(X))^2 \tag{5.28} \\
\varphi(X^6) &= \kappa_6(X) + 5(\kappa_2(X))^3 + 6\kappa_4(X)\kappa_2(X) \\
&= \mathbb{K}_6(X) + 15(\mathbb{K}_2(X))^3 + 15\mathbb{K}_4(X)\mathbb{K}_2(X).
\end{aligned}$$

Example 5.3.2. Let s and X be respectively standard semi-circle and standard normal variables. The first six moments of s and X are $(0, 1, 0, 2, 0, 5)$ and $(0, 1, 0, 3, 0, 15)$ respectively. Therefore, the following table is immediate from (5.25) and (5.27).

Cumulants	s	X
$(\kappa_1, \kappa_2, \ldots, \kappa_6)$	$(0, 1, 0, 0, 0, 0)$	$(0, 1, 0, 1, 0, 4)$
$(\mathbb{K}_1, \mathbb{K}_2, \ldots, \mathbb{K}_6)$	$(0, 1, 0, -1, 0, 5)$	$(0, 1, 0, 0, 0, 0)$

Moreover, the above table and (5.28) together recover the moments up to order six of s and X.

It is easy to prove the following lemma by utilizing the moment-cumulant relations given above in (5.22) and (5.24), as applied to one self-adjoint variable.

Lemma 5.3.1. *A self-adjoint variable s is standard semi-circle (that is, its moment sequence is the same as the semi-circle moment sequence) if and only if its free cumulants are*

$$\kappa_n(s) = \begin{cases} 1, & \text{if } n = 2, \\ 0, & \text{if } n \neq 2. \end{cases} \tag{5.29}$$

In this context note that the usual cumulants of the standard Gaussian law has the above property. This is why the standard semi-circle variable/law is also known as the *free Gaussian* variable/law.

The *free cumulant generating function* $C(z)$ of a self-adjoint random variable a is defined as

$$C(z) = 1 + \sum_{n=1}^{\infty} \kappa_n(a)z^n \quad \forall z \in \mathbb{C} \text{ for which the series is defined.} \tag{5.30}$$

It can be shown that $C(z)$ satisfies the relation

$$C(-m(z)) = -zm(z), \tag{5.31}$$

where $m(z)$ is the Stieltjes transform defined in (4.6). If there exists a $C > 0$ such that $|\kappa_n(a)| \leq C^n \ \forall n$, then (5.30) exists for all $|z| < C^{-1}$. In that case (5.31) makes sense $\forall z \in \mathbb{C}^+$, $|z|$ large, since $m(z) \to 0$ as $|z| \to \infty$.

The free cumulant generating function $C(z)$ and relation (5.31) are useful to derive the Stieltjes transform of a random variable from its free cumulant generating function. We shall use it later.

The following classes of non-commutative variables will be useful in Chapter 7.

Definition 5.3.1. $\{s_1, s_2, \ldots, s_k\}$ *is called a semi-circle family if for each i, s_i is self-adjoint and their joint free cumulants vanish except for order 2.*

Definition 5.3.2. $\{c_1, c_2, \ldots, c_k\}$ *is called a circular family if their joint free cumulants, except those of order 2 vanish, and $\kappa_2(c_i, c_j) = \kappa_2(c_i^*, c_j^*) = 0$ for all i, j.*

Now we are ready to define *free independence* of random variables and *free product* of NCP.

Definition 5.3.3. *(Free independence) Let $(\mathcal{A}, \varphi)$ be an NCP. Unital $*$-subalgebras $(\mathcal{A}_i)_{i \in I}$ of $\mathcal{A}$ are said to be freely independent (strictly speaking, $*$-free) if for all $n \geq 2$, and all $a_1, a_2, \ldots, a_n$ from $(\mathcal{A}_i)_{i \in I}$, $\kappa_n(a_1, a_2, \ldots, a_n) = 0$ whenever at least two of the a_i are from different $\mathcal{A}_i$.*

Suppose $(Span\{a_{ij}^{(n)}, a_{ij}^{(n)*} : i \geq 0, 1 \leq j \leq k\}, \varphi_n)$ converges to $(Span\{a_{ij}, a_{ij}^* : i \geq 0, 1 \leq j \leq k\}, \varphi)$. Then $\{a_{ij}^{(n)}, a_{ij}^{(n)*} : i \geq 0\}, 1 \leq j \leq k$, are said to be asymptotically free if $Span\{a_{ij}, a_{ij}^* : i \geq 0\}$ are free across $1 \leq j \leq k$.

Definition 5.3.4. *(Free product) Let $(\mathcal{A}_i, \varphi_i)_{i \in I}$ be a family of NCP. Then there exists an NCP $(\mathcal{A}, \varphi)$, called free product of $(\mathcal{A}_i, \varphi_i)_{i \in I}$, such that $\mathcal{A}_i \subset \mathcal{A}, i \in I$ are freely independent in $(\mathcal{A}, \varphi)$ and $\varphi|\mathcal{A}_i = \varphi_i$.*

A consequence of freeness is that all joint moments of free variables are computable in terms of the moments of the individual variables. Of course, the algorithm for computing moments under freeness is different from (and more complicated than) the product rule under usual independence. In the following section we shall discuss such an algorithm.

5.4   Moments of free variables

Let s be the standard semi-circle variable. Recall that κ_r denotes the r-th order free cumulant defined in (5.22). Let $\{w_i : 1 \leq i \leq k\}$ be a family of non-commutative variables *which is closed under $*$ operation* and satisfies

$$\kappa_r(w_{l_1}, w_{l_2}, \ldots, w_{l_r}) = 0, \ \forall r \neq 2, \ l_1, l_2, \ldots, l_r \geq 1. \tag{5.32}$$

Moreover, s, $\{w_1, w_2, \ldots, w_k\}$, $\{b_i, b_i^*, i \geq 1\}$ and $\{d_i, d_i^*, i \geq 1\}$ are free. Later in Chapters 6 and 7, we shall encounter moments of the form

$$\varphi(d_0 s b_1 s d_1 s b_2 s d_2 \cdots s b_n s d_n), \ \text{and} \ \varphi(w_{l_1} b_{l_1} w_{l_2} b_{l_2} w_{l_3} b_{l_3} \cdots w_{l_r} b_{l_r}), \tag{5.33}$$

for all $l_1, l_2, \ldots, l_r, r \geq 1$. In this section, we shall discuss an algorithm for computing the expressions in (5.33) in terms of the moments of s, $\{w_1, w_2, \ldots, w_k\}$,

$\{b_i, b_i^*\}$ and $\{d_i, d_i^*\}$. The following two lemma are useful for this purpose. See Nica and Speicher [2006] for the proof of Lemma 5.4.1.

Lemma 5.4.1. *Let $(\mathcal{A}, \varphi)$ be an NCP and consider random variables $a_1, a_2, \ldots a_n, b_1, b_2, \ldots, b_n \in \mathcal{A}$ such that $\mathrm{Span}\{a_i, a_i^* : 1 \leq i \leq n\}$ and $\mathrm{Span}\{b_i, b_i^* : 1 \leq i \leq n\}$ are freely independent. Then we have*

$$\varphi(a_1 b_1 a_2 b_2 \cdots a_n b_n) = \sum_{\pi \in NC(n)} \kappa_\pi[a_1, a_2, \ldots, a_n] \, \varphi_{K(\pi)}[b_1, b_2, \ldots, b_n],$$

where $K(\pi)$ is the Kreweras Complement of π defined in Section 5.2.

The next lemma is useful to compute the first factor in (5.33).

Lemma 5.4.2. *Suppose φ is tracial. Then under the freeness assumption mentioned above, the following holds.*

(a)

$$\varphi(d_0 s b_1 s d_1 s b_2 \cdots s d_n)$$

$$= \sum_{\pi \in NC_2(2n)} \varphi_{K(\pi)}[b_1, d_1, \ldots, b_n, d_n d_0] \tag{5.34}$$

$$= \sum_{\pi \in NC(n)} \varphi_\pi[b_1, \ldots, b_n] \, \varphi_{K(\pi)}[d_1, d_2, \ldots, d_n d_0] \tag{5.35}$$

$$= \sum_{\pi \in NC(n)} \varphi_\pi[d_1, d_2, \ldots, d_n d_0] \, \varphi_{K(\pi)}[b_1, \ldots, b_n]. \tag{5.36}$$

(b) Fix $1 = k_0 < k_1 < \cdots < k_t \leq n$ and let $\mathcal{S} \subset NC_2(2n)$ be given by

$$\mathcal{S} = \{\pi \in NC_2(2n) : \{2k_i, 2k_{i+1} - 1\} \in \pi, \ 0 \leq i \leq t, \ k_{t+1} = k_0\}.$$

Then

$$\sum_{\pi \in \mathcal{S}} \varphi_{K(\pi)}[b_1, d_1, b_2, d_2, \ldots, b_n, d_n d_0] \tag{5.37}$$

$$= \varphi\Big(\prod_{s=0}^{t} b_{k_s}\Big) \prod_{s=1}^{t+1} \varphi(d_{k_{s-1}} s b_{k_{s-1}+1} s d_{k_{s-1}+1} \cdots s d_{k_s - 1}),$$

where $k_0 = 1, d_{k_{t+1}-1} = d_n d_0$.

Proof. Relation (5.34) follows from Lemma 5.4.1. Just use the fact that if s is a standard semi-circle, then $k_2(s, s) = 1$ and all other cumulants are 0. By freeness of $\{b_i\}$ and $\{d_i\}$, and by Properties (a)–(c) of $K(\pi)$ in Section 5.2, (5.35) and (5.36) follow from (5.34).

We now prove (5.37). Consider the following subsets of $\{1, 2, \ldots, 2n\}$ as

$$\mathcal{S}_0 = \{2k_i - 1, 2k_i : 1 \leq i \leq t\},$$

$$\mathcal{S}_i = \{2k_{i-1} + 1, 2k_{i-1} + 2, \ldots, 2k_i - 2\}, \ 1 \leq i \leq t+1, \ k_{t+1} = n+1.$$

By $NC(\mathcal{S}_i)$ and $NC_2(\mathcal{S}_i)$, respectively, we mean the sets of all non-crossing partitions and non-crossing pair partitions of indices in $\mathcal{S}_i$. Let

$$\sigma_0 = \{\{2k_i, 2k_{i+1} - 1\} : 0 \leq i \leq t, \ k_{t+1} = 1\} \in NC_2(\mathcal{S}_0). \tag{5.38}$$

Note that, as $\mathcal{S}$ contains only non-crossing partitions, we have

$$\mathcal{S} = \{\sigma_0 \cup \sigma_1 \cup \cdots \cup \sigma_{t+1} : \sigma_i \in NC_2(\mathcal{S}_i), \ \forall 1 \leq i \leq (t+1)\}. \tag{5.39}$$

Now to understand the nature of the Kreweras complement $K(\pi)$ for $\pi \in \mathcal{S}$, consider the following subsets and partitions of $\{1, 2, \ldots, 2n\}$. For $1 \leq i \leq t+1$, let

$$
\begin{aligned}
W_i &= \{2k_{i-1}, 2k_{i-1} + 1, \ldots, 2k_i - 2\}, \\
W_i^- &= \{2k_{i-1} + 1, 2k_{i-1} + 2, \ldots, 2k_i - 3\}, \\
1_{W_i} &= \{2k_{i-1}, 2k_{i-1} + 1, \ldots, 2k_i - 2\}, \\
0_{W_i^-} &= \{2k_{i-1} + 1\}, \{2k_{i-1} + 2\}, \ldots, \{2k_i - 3\}.
\end{aligned}
$$

Since the Kreweras complement $K(\pi)$ is non-crossing, it must be of the following form,

$$K(\pi) = \tau_0 \cup \tau_1(\sigma_1) \cup \tau_2(\sigma_2) \cup \ldots \cup \tau_{t+1}(\sigma_{t+1}), \tag{5.40}$$

where the blocks $\{\tau_0, \tau_i(\sigma_i)\}$ satisfy

$$
\begin{aligned}
\tau_0 &= \{2k_i - 1 : 0 \leq i \leq t\} \\
\{\{2k_{i-1}, 2k_i - 2\}, 0_{W_i^-}\} &\leq \tau_i(\sigma_i) \leq 1_{W_i}, \ \forall 1 \leq i \leq t+1.
\end{aligned}
$$

Clearly, $\tau_i(\sigma_i)$ depends only on $\sigma_i \in NC_2(\mathcal{S}_i)$ but on no other σ_k, $k \neq i$. Hence, by multiplicative property (5.21) of φ,

$$\sum_{\pi \in \mathcal{S}} \varphi_{K(\pi)}[b_1, d_1, b_2, d_2, \ldots, b_n, d_n d_0] \tag{5.41}$$

$$= \varphi(\prod_{i=0}^{t} b_{k_s}) \prod_{i=1}^{t+1} \sum_{\sigma_i \in NC_2(\mathcal{S}_i)} \varphi_{\tau_i(\sigma_i)}[b_{k_{i-1}+1}, d_{k_{i-1}+1}, \ldots, b_{k_i-1}, d_{k_i-1} d_{k_{i-1}}].$$

Now, note that the set of blocks $\mathcal{G}_i = \{\tau_i(\sigma_i) : \sigma_i \in NC_2(\mathcal{S}_i)\}$ is in one-to-one correspondence with the set of Kreweras complements

$$\{K(\pi) : \ \pi \in NC_2(2k_i - 2k_{i-1} - 2)\}.$$

This one-to-one correspondence is obvious when one sets $2k_{i-1} + j \leftrightarrow j$, $\forall 1 \leq j \leq 2k_i - 2k_{i-1} - 3$, and $2k_i - 2 \leftrightarrow 2k_i - 2k_{i-1} - 2$.

Let $2k_i - 2k_{i-1} - 2 = \omega_i$ (say). Hence, by (5.41), we have

$$\sum_{\pi \in \mathcal{S}} \varphi_{K(\pi)}[b_1, d_1, b_2, d_2, \ldots, b_n, d_n d_0] \tag{5.42}$$

$$= \varphi(\prod_{s=0}^{t} b_{k_s})(\prod_{i=1}^{t+1} \sum_{\sigma_i \in NC_2(\omega_i)} \varphi_{K(\sigma_i)}[b_{k_{i-1}+1}, d_{k_{i-1}+1}, \ldots, b_{k_i-1}, d_{k_i-1} d_{k_{i-1}}])$$

$$= \varphi(\prod_{s=0}^{t} b_{k_s})(\prod_{i=1}^{t+1} \varphi(d_{k_{i-1}} s b_{k_{i-1}+1} s d_{k_{i-1}+1} \cdots s d_{k_i-1})).$$

Hence, (5.37) is justified. $\qquad\qquad\qquad\qquad\qquad\qquad\qquad\qquad\square$

Next we shall see how to compute the second term of (5.33). By Lemma 5.4.1, for every $m \geq 1$,

$$\varphi(w_{l_1} b_{l_1} \cdots w_{l_r} b_{l_r}) = 0 \quad \text{if} \quad r = 2m - 1$$

and if $r = 2m$,

$$\varphi(w_{l_1} b_{l_1} \cdots w_{l_{2m}} b_{l_{2m}}) = \sum_{\pi \in NC_2(2m)} \varphi_{K(\pi)}[b_{l_1}, \ldots, b_{l_{2m}}] \kappa_\pi[w_{l_1}, \ldots, w_{l_{2m}}].$$

$$\tag{5.43}$$

Any $\pi \in NC_2(2m)$ is of the form $\{(t_1, t_2), (t_3, t_4), \ldots, (t_{2m-1}, t_{2m})\}$, where $t_1 < t_2, t_3 < t_4, \ldots, t_{2m-1} < t_{2m}$. Hence, by (5.43), and the multilinear property of κ_π, the above expression for $r = 2m$ can be written as

$$\sum_{\pi = \{(t_1, t_2), \ldots, (t_{2m-1}, t_{2m})\} \in NC_2(2m)} \prod_{i=1}^{m} \kappa_2(w_{l_{t_{2i-1}}}, w_{l_{t_{2i}}}) \varphi_{K(\pi)}[b_{l_1}, \ldots, b_{l_k}].$$

$$\tag{5.44}$$

5.5   Joint convergence of random matrices

The following theorem describes the joint convergence of several Wigner and deterministic matrices. See for example Anderson et al. [2009].

Theorem 5.5.1. *Let* $W_p^{(1)}, W_p^{(2)}, \ldots, W_p^{(r)}$ *be* r *independent Wigner matrices of order* p *such that each matrix individually satisfies the assumptions of Theorem 4.2.1. Let* $D_p^{(1)}, D_p^{(2)}, \ldots, D_p^{(2q)}$ *be* $2q$ *constant matrices of order* p *with bounded norm such that, for* $\epsilon = 0, 1$, $(Span\{D_p^{(2i-\epsilon)}, D_p^{(2i-\epsilon)*} : 1 \leq i \leq q\}, p^{-1}Tr)$ *converges. Then the following statements hold. As* $p \to \infty$,

(a) $p^{-1/2}W_p^{(1)}, p^{-1/2}W_p^{(2)}, \ldots, p^{-1/2}W_p^{(r)}$ *are asymptotically free.*

(b) For $\epsilon = 0$ *or* 1, *the collections* $\{p^{-1/2}W_p^{(i)}\}$ *and* $\{D_p^{(2i-\epsilon)}, D_p^{(2i-\epsilon)*}\}$ *are asymptotically free.*

(c) The collections $\{p^{-1}W_p^{(i)}D_p^{(2j)}W_p^{(i)}, p^{-1}W_p^{(i)}D_p^{(2j)*}W_p^{(i)} : i, j \geq 1\}$ *and* $\{D_p^{(2i-1)}, D_p^{(2i-1)*}\}$ *are asymptotically free.*

(d) Let $\epsilon_i = 1, *, \forall 1 \leq i \leq 2k$. *To compute* $\lim p^{-1}E\mathrm{Tr}(p^{-k}\prod_{i=1}^{K}D_p^{(2i-1)\epsilon_{2i-1}}W_p^{(i)}D_p^{(2i)\epsilon_{2i}}W_p^{(i)})$ *one can assume that the collections* $\{W_p^{(i)}\}$, $\{D_p^{(2i-1)}, D_p^{(2i-1)*}\}$ *and* $\{D_p^{(2i)}, D_p^{(2i)*}\}$ *are asymptotically free.*

We omit the proof. For (a) and (b) see Anderson et al. [2009]. (c) follows from (a), (b) and Theorem 11.12, page 180 of Nica and Speicher [2006]. (d) is immediate from (a), (b), and (c).

5.5.1 Compound free Poisson

In Remark 4.3.1 we have described the convergence of ESD of $n^{-1}ZAZ^*$. We now take a second look at that result via free independence.

For any probability measure μ, let $m_n(\mu) = \int x^n d\mu(x)$, $n \geq 1$ be its moments (assumed to be finite). Let Y be a random variable with probability distribution μ and m.g.f. $M_Y(\cdot)$. Let X be a random variable with the m.g.f.

$$M_X(t) = \exp\{\lambda(M_Y(t) - 1)\}.$$

Then the distribution of X is called the (classical) compound Poisson distribution with rate λ and jump distribution μ. It is easy to see that the cumulants of X satisfy

$$\mathbb{K}_n(X) = \lambda m_n(\mu), \ n \geq 1.$$

We are now ready to define the free analogue of the classical compound Poisson distribution. For any probability measure μ, let $k_n(\mu)$ denote its nth free cumulant.

Definition 5.5.1. *(Compound free Poisson distribution) A probability measure* μ *on* $\mathbb{R}$ *is called a compound free Poisson distribution if for some* $\lambda > 0$ *and some compactly supported probability measure* ν *on* $\mathbb{R}$,

$$k_n(\mu) = \lambda m_n(\nu), \ \forall n \geq 1.$$

The quantities λ *and* ν *are respectively known as the rate and the jump distribution.*

This definition will be useful in Chapters 6 and 7 to describe the LSD of $\{\frac{1}{2}(\hat{\Gamma}_u + \hat{\Gamma}_u^*)\}_{u \geq 0}$ when the coefficient matrices $\psi_j = \lambda_j I_p$, for all $j \geq 0$ and I_p is as in (1.9). We give two simpler examples of compound free Poisson.

Example 5.5.1. Let $(\mathcal{A}, \varphi)$ be a non-commutative probability space. Let $s, a \in \mathcal{A}$ where s is a standard semi-circle variable with moment sequence (4.19), and moreover s and a are *-free. Then using the developments of this chapter, the free cumulants of sas can be easily computed and are given by

$$k_n(sas, sas, \dots, sas) = \varphi(a^n), \ \forall n \geq 1. \tag{5.45}$$

In particular, if a is self-adjoint with distribution ν, then sas is a compound free Poisson random variable with rate $\lambda = 1$ and jump distribution ν.

Example 5.5.2. (LSD of $n^{-1}ZAZ^*$). Recall $\mathcal{L}$ and $C(\delta, p)$ respectively from (4.23) and (4.25). Suppose $Z = ((z_{i,j}))_{p \times n}$ is an independent matrix such that $\{z_{i,j} : 1 \le i \le p, \ 1 \le j \le n\} \in \mathcal{L} \cup C(\delta, p)$, $\forall p \ge 1$ and for some $\delta > 0$ and where $n = n(p))$ is such that $pn^{-1} \to y \in (0, \infty)$. Suppose further that A is a self-adjoint matrix of order n with compactly supported LSD. We already know that the almost sure LSD of $n^{-1}ZAZ^*$ exists and that can be proved using the Stieltjes transform method. See Remark 4.3.1. We now identify the LSD.

It can be shown that the limiting free cumulants of $n^{-1}ZAZ^*$ are given by

$$\lim_n k_r (n^{-1}ZAZ^*, n^{-1}ZAZ^*, \ldots, n^{-1}ZAZ^*) = y^{r-1}\varphi(a^r), \ \forall r \ge 1. \quad (5.46)$$

Since all the free cumulants converge, all the moments converge too and the limit moments are determined by the limit free cumulants. Since the limit free cumulants are those of the compound free Poisson distribution with rate y^{-1} and jump distribution ya, LSD of $n^{-1}ZAZ^*$ is compound free Poisson with these parameters.

The LSD of $n^{-1}ZAZ^*$ can also be described in terms of polynomial of two free variables. Suppose $\{c, c^*\}$ is such that

$$k_n(c^{\epsilon_1}, c^{\epsilon_2}, \ldots, c^{\epsilon_n}) = \begin{cases} 1, & \text{if } n = 2, \ \epsilon_1 = 1 \text{ and } \epsilon_2 = * \\ y, & \text{if } n = 2, \ \epsilon_1 = * \text{ and } \epsilon_2 = 1 \\ 0, & \text{otherwise.} \end{cases} \quad (5.47)$$

Also suppose a has same distribution as the LSD of A. Moreover, $\{c, c^*\}$ and a are free. Then the LSD of $n^{-1}ZAZ^*$ is same as the law of cac^*.

The above result continues to hold if $\{z_{i,j}\}$ is a triangular sequence of independent random variables.

Exercises

1. Prove Lemma 5.1.1 on the bounds of φ.

2. Show that the state given in Example 5.1.2 is tracial and positive.

3. Provide an example of NCP where the state is positive but not tracial.

4. Consider the sequence of NCP $(\mathcal{A}_n, \varphi_n)$, given in Example 5.1.3, with

$$D_{in}(j, j) = \begin{cases} 1, & \text{if } j > i \\ X_i, & \text{if } j \le i. \end{cases}$$

Show that $(\mathcal{A}_n, \varphi_n)$ converges almost surely to the NCP generated only by the identity element.

5. Consider the sequence of NCP $(\mathcal{A}_n, \varphi_n)$, given in Example 5.1.4, with $X_i \overset{\text{i.i.d.}}{\sim} \mathcal{N}(0,1)$. Show that for all $i_1, i_2, \ldots, i_k, k \geq 1$, the almost sure LSD of $\sum_{j=1}^{k} D_{i_j n}^2$ is the chi-square distribution with k degrees of freedom.

6. Consider the set of matrices $\{P_{un} : u = 0, \pm 1, \pm 2, \ldots\}$ defined in Example 5.1.5. Show that the LSD of $P_{an} + P_{-an}$ and $i(P_{an} - P_{-an})$ are identical with the probability distribution of $2\cos(aU)$ where $U \sim U(0,1)$.

7. Learn the proof of Theorem 5.1.1 on the NCP convergence of the Wigner matrix.

8. Show that the convolution operation defined in (5.9) is associative.

9. Show that $F : P^{(2)} \to \mathbb{C}$ is invertible if and only if $F(\pi, \pi) \neq 0$ for every $\pi \in P$.

10. Provide an example of a crossing partition of $\{1, 2, 3, 4, 5, 6\}$.

11. Write down all the elements of $NC(4)$, $NC_2(4)$ and $NCE(4)$.

12. Show that the standard semi-circle moments satisfy

$$\beta_{2h} = \#NC(h) = \#NC_2(2h) \ \forall h \geq 1.$$

13. Show that the Möbius function μ on the POSET $NC(n)$ satisfies

$$\mu(0_n, 1_n) = (-1)^{n-1} C_{n-1}, \text{ where } C_n = \frac{1}{n+1}\binom{2n}{n}, \ \forall n \geq 1.$$

14. Learn the proof of Lemma 5.2.1 from Nica and Speicher [2006].

15. Find the Kreweras complements of $\sigma_1 = \{\{1,4\}, \{2,3\}, \{5,8\}, \{6,7\}\}$, $\sigma_2 = \{\{1,6\}, \{2,5\}, \{3,4\}, \{7,8\}, \{9,12\}, \{10,11\}\}$ and $\sigma_3 = \{\{1,10\}, \{2,7\}, \{8,9\}, \{3,6\}, \{4,5\}, \{11,14\}, \{12,13\}\}$ respectively in $NC(8)$, $NC(12)$ and $NC(14)$.

16. Verify that (5.23) holds.

17. Find $\mu(\sigma_1, 1_8)$, $\mu(\sigma_2, 1_{12})$ and $\mu(\sigma_3, 1_{14})$ where σ_1, σ_2 and σ_3 are in Exercise 15.

18. Prove Lemma 5.3.1.

19. Find marginal free cumulant of first six orders of a Poisson(1) and a Bernoulli(0.5) variables.

20. Learn the proof of Theorem 5.5.1.

21. Let s_1 and s_2 be free standard semi-circle variables. Also suppose a is a Bernoulli(0.4) variable, $b = 2a + 1$ and $d = 2a - 1$. Compute $\varphi(as_1bs_1as_1ds_1)$, $\varphi(as_1bs_2as_1ds_2)$ and $\varphi(as_1bs_1as_2ds_2)$.

22. Let s_1 and s_2 be free standard semi-circle variables. Find the distribution of $\frac{s_1 + s_2}{\sqrt{2}}$.

Chapter 6

GENERALIZED COVARIANCE MATRIX I

In the previous chapter we have briefly discussed a few results on the NCP convergence and LSD for some specific type of random matrices such as the Wigner matrix and matrices of the form $AZBZ^*A$. Now we broaden our scope significantly and tackle much more general matrices. Suppose we have matrices $Z_u = ((\varepsilon_{u,t,i}))_{p \times n}$, $1 \le u \le U$, where $\{\varepsilon_{u,t,i} : u,i,j \ge 0\}$ are independent with mean 0 and variance 1. Note that each Z_u is an independent matrix and moreover, they are independent among themselves.

Also suppose $\{B_{2i-1} : 1 \le i \le K\}$ and $\{B_{2i} : 1 \le i \le L\}$ are constant matrices of order $p \times p$ and $n \times n$ respectively. Without loss of generality, we assume that these collections are closed under the $*$ operation.

Consider all $p \times p$ matrices

$$\mathbb{P}_{l,(u_{l,1},u_{l,2},\dots,u_{l,k_l})} = \prod_{i=1}^{k_l} \left(n^{-1} A_{l,2i-1} Z_{u_{l,i}} A_{l,2i} Z^*_{u_{l,i}}\right) A_{l,2k_l+1}, \tag{6.1}$$

where $\{A_{l,2i-1}\}$, $\{A_{l,2i}\}$ and $\{Z_{u_{l,i}}\}$ are matrices from the collections $\{B_{2i-1}, B^*_{2i-1} : 1 \le i \le K\}$, $\{B_{2i}, B^*_{2i} : 1 \le i \le L\}$ and $\{Z_i : 1 \le i \le U\}$ respectively. As the sample covariance matrix (without centering) is a special case of the matrices given in (6.1), we call them *generalized covariance matrices*.

Consider the sequence of NCP $(\mathcal{U}_p, p^{-1}E\text{Tr})$, where

$$\mathcal{U}_p = \text{Span}\left(\mathbb{P}_{l,(u_{l,1},\dots,u_{l,k_l})} : l, k_l \ge 1\right). \tag{6.2}$$

Note that $\mathcal{U}_p$ forms a $*$-algebra. All the matrices discussed in the previous chapter that involved both Z and Z^* belong to this algebra. We shall put to use the machinery developed in Chapter 5 to show that the NCP $(\mathcal{U}_p, p^{-1}E\text{Tr})$ converges.

At the same time, in Lemma 5.1.2 we have seen how NCP convergence with some additional effort guarantees existence of the LSD. Using this idea we shall show that the LSD of any symmetric polynomial in $\{\mathbb{P}_{l,(u_{l,1},u_{l,2},\dots,u_{l,k_l})}\}$ exists and the limit can be expressed in terms of some freely independent variables.

However, most of the LSD results given in Chapter 4 are obtained using the method of Stieltjes transform. While this is relatively easier for the specific polynomials discussed there, it is not at all clear how this method could be used for any arbitrary symmetric polynomial. Hence, we fall back on the moment method to derive the existence of the LSD for any arbitrary symmetric polynomial. After this is established, a description of the LSD is obtained using free variables. Such a description is arguably more clean than a possible Stieltjes transform description. Nevertheless, we do derive the Stieltjes transform of the LSD for a large class of polynomials. Finally we provide a list of LSD results for specific matrices.

One of the major uses of the above results is to obtain the LSD of any symmetric polynomial in the sample autocovariance matrices $\{\hat{\Gamma}_u\}$, along with their joint convergence. This will be done in details in the later chapters, but let us briefly indicate this link. Recall the independent matrix Z in Definition 4.3.1 and the sequence of coefficient matrices $\{\psi_j\}$ in (3.3). Let $\{P_j : j = 0, \pm1, \pm2, \ldots\}$ be a sequence of $n \times n$ matrices where P_j has entries equal to one on the j-th upper diagonal and 0 otherwise. Note that $P_0 = I_n$ where I_n is the $n \times n$ identity matrix, and $P_j = P_{-j}^*$, $\forall j$. Define

$$\Delta_u = \frac{1}{n} \sum_{j,j'=0}^{q} \psi_j Z P_{j-j'+u} Z^* \psi_{j'}^*, \quad \forall u = 0, 1, 2, \ldots. \tag{6.3}$$

Clearly the matrices $\{\Delta_u\}$ fall within the setting of this chapter. In Chapter 8, we shall prove that $\{\hat{\Gamma}_u\}$ are well approximated by $\{\Delta_u\}$ and thereby establish their asymptotic properties.

6.1 Preliminaries

6.1.1 Assumptions

We first list all the assumptions that are required.

First, let us describe our assumption on $\{Z_u\}$. Recall the independent matrix in Definition 4.3.1. Let $Z_u = ((\varepsilon_{u,i,j}))_{p \times n}$, $1 \le u \le U$ be $p \times n$ independent matrices. Therefore, $\{\varepsilon_{u,i,j}\}$ are independently distributed with $E(\varepsilon_{u,i,j}) = 0$, $E|\varepsilon_{u,i,j}|^2 = 1$. Recall the classes $\mathcal{L}$ and $C(\delta, p)$ respectively in (4.23) and (4.25). We assume that

(A1) $\{\varepsilon_{u,i,j}\} \in \mathcal{L} \cup C(\delta, p)$ for some $\delta > 0$ and for all $1 \le u \le U$.

If there is only one u i.e., if $U = 1$, we will write $\varepsilon_{i,j}$ and Z respectively for $\varepsilon_{1,i,j}$ and Z_1. Assumption $(A1)$ will be weakened later for some corollaries and applications by means of truncation techniques.

Now we move to the assumptions on the deterministic matrices $\{B_i\}$.

(A2) $\{B_{2i-1} : 1 \le i \le K\}$ are norm bounded $p \times p$ matrices and $(\mathrm{Span}(B_{2i-1}, B_{2i-1}^* : 1 \le i \le K), p^{-1}\mathrm{Tr})$ converges.

(A3) $\{B_{2i} : 1 \leq i \leq L\}$ are norm bounded $n \times n$ matrices and $(\mathrm{Span}(B_{2i}, B_{2i}^* : 1 \leq i \leq L), n^{-1}\mathrm{Tr})$ converges.

Note that we do not assume the joint convergence of the entire collection $\{B_i : i \geq 1\}$.

6.1.2   Embedding

As $\{Z_u\}$, $\{B_{2i-1}\}$ and $\{B_{2i}\}$ are all of different orders, it is not possible to directly define an algebra of these matrices. Therefore, it becomes difficult to describe the limit of $(\mathcal{U}_p, p^{-1}E\mathrm{Tr})$ directly in terms of the limits of $\{Z_u\}$, $\{B_{2i-1}\}$ and $\{B_{2i}\}$. The solution is to embed all these matrices into matrices of order $(n + p)$. This idea is already available in the free probability literature. For example, see Benaych-Georges [2009], Benaych-Georges [2010] and Benaych-Georges and Nadakuditi [2012]. It works as follows:

Recall the Wigner matrix in Definition 4.2.1. We first embed Z_u into a Wigner matrix W_u of order $(n + p)$. Thus,

$$W_u = \begin{pmatrix} W_{p \times p}^{(1u)} & Z_u \\ Z_u^* & W_{n \times n}^{(2u)} \end{pmatrix}, \tag{6.4}$$

where $\{W^{(iu)} : i = 1, 2, \ u \geq 1\}$ are independent Wigner matrices and are independent of $\{Z_u\}$.

For any matrices B and D of order p and n respectively, let $\bar{B}$ and $\underline{D}$ of order $(n + p)$ be the matrices

$$\bar{B} = \begin{pmatrix} B & 0 \\ 0 & 0 \end{pmatrix}, \quad \underline{D} = \begin{pmatrix} 0 & 0 \\ 0 & D \end{pmatrix}. \tag{6.5}$$

It is easy to see that

$$\bar{\mathbb{P}}_{l,(u_{l,1}, u_{l,2}, \ldots, u_{l,k_l})} = \prod_{i=1}^{k_l} \left(n^{-1} \bar{A}_{l,2i-1} W_{u_{l,i}} \underline{A}_{l,2i} W_{u_{l,i}}^* \right) \bar{A}_{l,2k_l+1}. \tag{6.6}$$

Note that the right side of (6.6) is a polynomial in Wigner matrices and deterministic matrices, all of which have the same dimension $(n + p)$.

Consider the sequence of NCP $(\bar{\mathcal{U}}_p, (n + p)^{-1}E\mathrm{Tr})$, where

$$\bar{\mathcal{U}}_p = \mathrm{Span}\left(\bar{\mathbb{P}}_{l,(u_{l,1}, \ldots, u_{l,k_l})} : \ l, k_l \geq 1 \right). \tag{6.7}$$

Note that $\bar{\mathcal{U}}_p$ also forms a *-algebra. Clearly there is a relation between $(\mathcal{U}_p, p^{-1}E\mathrm{Tr})$ and $(\bar{\mathcal{U}}_p, (n + p)^{-1}E\mathrm{Tr})$. Moreover, convergence of $(\bar{\mathcal{U}}_p, (n + p)^{-1}E\mathrm{Tr})$ is easy to describe by using Theorem 5.5.1. The limiting NCP can be expressed in terms of some free variables. The idea behind the limit is explained in Section 6.2.1 and the statement of the main convergence result is presented in Theorem 6.2.1.

6.2   NCP convergence

6.2.1   Main idea

Before we rigorously state the result for the joint convergence of the general matrices (6.1) we show how freeness comes into the picture and hence how it motivates the limiting NCP of $(\mathcal{U}_p, p^{-1}E\text{Tr})$ by considering an example. For simplicity of illustration, let us consider a *self-adjoint* polynomial:

$$P = n^{-1}(B_1 Z_1 B_2 Z_1^* B_3 + B_5 Z_1 B_4 Z_1^* B_7 + B_9 Z_1 B_6 Z_1^* B_{11} + B_{13} Z_1 B_8 Z_1^* B_{15}).$$

Our primary goal is to show that for all $r \geq 1$, $\lim p^{-1}E\text{Tr}(P^r)$ exists.

Embed the matrices Z_1, B_i as described in (6.4) and (6.5). To deal with P, consider $\bar{P}$ which involves the enlarged matrices:

$$n\bar{P} = \bar{B}_1 W_1 \underline{B}_2 W_1 \bar{B}_3 + \bar{B}_5 W_1 \underline{B}_4 W_1 \bar{B}_7 + \bar{B}_9 W_1 \underline{B}_6 W_1 \bar{B}_{11} + \bar{B}_{13} W_1 \underline{B}_8 W_1 \bar{B}_{15}.$$

Note that for any integer r, whenever the limits below exist,

$$\lim(n+p)^{-1}\text{Tr}(\bar{B}_{2j+1}^r) = y(1+y)^{-1}\lim p^{-1}\text{Tr}(B_{2j+1}^r), \qquad (6.8)$$

$$\lim(n+p)^{-1}\text{Tr}(\underline{B}_{2j}^r) = (1+y)^{-1}\lim n^{-1}\text{Tr}(B_{2j}^r), \text{ and} \qquad (6.9)$$

$$\lim p^{-1}\text{Tr}(P^r) = y^{-1}(1+y)\lim(n+p)^{-1}\text{Tr}(\bar{P}^r). \qquad (6.10)$$

Now observe the following three facts:

(1) By (A2) and (A3), for any monomial m, the following limits exist as $p, n = n(p) \to \infty$:

$$(\text{Span}\{\bar{B}_{2i-1}, \bar{B}_{2i-1}^* : i \leq 8\}, (n+p)^{-1}E\text{Tr}) \to (\text{Span}\{\bar{b}_{2i-1}, \bar{b}_{2i-1}^* : i \leq 8\}, \varphi_1),$$

and

$$(\text{Span}\{\underline{B}_{2i}, \underline{B}_{2i}^* : i \leq 4\}, (n+p)^{-1}E\text{Tr}) \to (\text{Span}\{\underline{b}_{2i}, \underline{b}_{2i}^* : i \leq 4\}, \varphi_2), \quad \text{say.}$$

(2) Recall the classes $\mathcal{L}$ and $\mathcal{C}(\delta, p)$ defined in (4.23) and (4.25). By Theorem 5.1.1, if $\{\varepsilon_{1,i,j}\} \in \mathcal{L} \cap \mathcal{C}(0, p)$, then

$$\lim E(n+p)^{-1}\text{Tr}((n+p)^{-1/2}W_1)^r = E(s^r)$$

i.e.,

$$(\text{Span}\{(n+p)^{-1/2}W_1\}, (n+p)^{-1}E\text{Tr}) \to (\text{Span}\{s\}, \varphi_3), \quad \text{say}$$

where s is a standard semi-circle variable with the moment sequence (4.19).

(3) Finally, by Theorem 5.5.1 (d) and for the polynomial $\bar{P}$, *in the limit*, the matrices $(n+p)^{-1/2}W_1$, $\{\bar{B}_{2i-1}, \bar{B}_{2i-1}^* : i \leq 8\}$ and $\{\underline{B}_{2i}, \underline{B}_{2i}^* : i \leq 4\}$ are free variables.

Recall the free product of NCP in Definition 5.3.4. By (3), s, $\{\bar{b}_{2i-1}, \bar{b}_{2i-1}^* : i \leq 8\}$ and $\{\underline{b}_{2i}, \underline{b}_{2i}^* : i \leq 4\}$ are free in some NCP $(\mathcal{A}, \varphi)$.

Thus, using the above observations (1)–(3) in conjunction with equations (6.8)–(6.10), we can conclude that $L_1 = \lim(n+p)^{-1}\mathrm{Tr}(\bar{P}^r)$ and $L_2 = \lim p^{-1}\mathrm{Tr}(P^r)$ exist and

$$
\begin{aligned}
L_1 &= \lim \Big(\frac{n+p}{n}\Big)^r \frac{1}{n+p}\mathrm{Tr}\Big(\sum_{i=1}^{4} \bar{B}_{4i-3}\frac{W_1}{\sqrt{n+p}}B_{2i}\frac{W_1}{\sqrt{n+p}}\bar{B}_{4i-1}\Big)^r \\
&= \varphi\Big((1+y)\sum_{i=1}^{4} \bar{b}_{4i-3}s\underline{b}_{2i}s\bar{b}_{4i-1}\Big)^r \quad \text{and} \tag{6.11}
\end{aligned}
$$

$$
L_2 = y^{-1}(1+y)\varphi\Big((1+y)\sum_{i=1}^{4} \bar{b}_{4i-3}s\underline{b}_{2i}s\bar{b}_{4i-1}\Big)^r. \tag{6.12}
$$

The right side of (6.12), involving free variables, are then the limit moments of P and can be computed using Lemma 5.4.2. This is the idea we implement for the general matrices (6.1).

6.2.2 Main convergence

To describe the limits, consider $(\mathcal{S} = \mathrm{Span}\{s_u : u \geq 1\}, \varphi_s)$ to be an NCP of free standard semi-circular variables $\{s_u\}$. We use the following notation for relevant limits:

$$
(\mathrm{Span}\{B_{2i-1}, B_{2i-1}^* : i \leq K\}, \frac{1}{p}\mathrm{Tr}) \rightarrow
$$
$$
(\mathcal{A}_{\mathrm{odd}} = \mathrm{Span}\{b_{2i-1}, b_{2i-1}^* : i \leq K\}, \varphi_{\mathrm{odd}}), \tag{6.13}
$$
$$
(\mathrm{Span}\{\bar{B}_{2i-1}, \bar{B}_{2i-1}^* : i \leq K\}, \frac{1}{n+p}\mathrm{Tr}) \rightarrow
$$
$$
(\bar{\mathcal{A}}_{\mathrm{odd}} = \mathrm{Span}\{\bar{b}_{2i-1}, \bar{b}_{2i-1}^* : i \leq K\}, \bar{\varphi}_{\mathrm{odd}}), \tag{6.14}
$$

$$
(\mathrm{Span}\{B_{2i}, B_{2i}^* : i \leq L\}, \frac{1}{n}\mathrm{Tr}) \rightarrow
$$
$$
(\mathcal{A}_{\mathrm{even}} = \mathrm{Span}\{b_{2i}, b_{2i}^* : i \leq L\}, \varphi_{\mathrm{even}}). \tag{6.15}
$$

$$
(\mathrm{Span}\{\underline{B}_{2i}, \underline{B}_{2i}^* : i \leq L\}, (n+p)^{-1}\mathrm{Tr}) \rightarrow
$$
$$
(\bar{\mathcal{A}}_{\mathrm{even}} = \mathrm{Span}\{\underline{b}_{2i}, \underline{b}_{2i}^* : i \leq L\}, \bar{\varphi}_{\mathrm{even}}). \tag{6.16}
$$

Therefore, for any polynomial Π,

$$
\varphi_{\mathrm{odd}}(\Pi(b_{2i-1}, b_{2i-1}^* : i \leq K)) = \frac{1+y}{y}\bar{\varphi}_{\mathrm{odd}}(\Pi(\bar{b}_{2i-1}, \bar{b}_{2i-1}^* : i \leq K)),
$$

$$
\varphi_{\mathrm{even}}(\Pi(b_{2i}, b_{2i}^* : i \leq L)) = (1+y)\bar{\varphi}_{\mathrm{even}}(\Pi(\underline{b}_{2i}, \underline{b}_{2i}^* : i \leq L)).
$$

Recall the free product of NCP in Definition 5.3.4. Let

$$
(\mathcal{A}, \bar{\varphi}) = \text{free product of } (\mathcal{S}, \varphi_s), (\bar{\mathcal{A}}_{\mathrm{odd}}, \bar{\varphi}_{\mathrm{odd}}) \text{ and}(\bar{\mathcal{A}}_{\mathrm{even}}, \bar{\varphi}_{\mathrm{even}}). \tag{6.17}
$$

Consider the sub-algebra $\bar{\mathcal{U}}$ of $\mathcal{A}$ as

$$\text{Span}\left(\bar{p}_{l,(u_{l,1},...,u_{l,k_l})} := (1+y)^{k_l} \prod_{i=1}^{k_l} \left(\bar{\bar{a}}_{l,2i-1} s_{u_{l,i}} \underline{a}_{l,2i} s_{u_{l,i}}\right) \bar{a}_{l,2k_l+1} : l \geq 1\right)$$

(6.18)

where $\bar{a}_{l,2i-1} \in \{\bar{b}_{2i-1}, \bar{b}_{2i-1}^*\}$, $\underline{a}_{l,2i} \in \{\underline{b}_{2i}, \underline{b}_{2i}^*\}$ and $s_{u_{l,i}} \in \{s_u\}$. Note that $\bar{\mathcal{U}}$ forms a $*$-algebra.

Then we have the following theorem. For notational convenience, we write any collection of finite degree polynomials $\{\mathbb{P}_{l,(u_{l,1},u_{l,2},...,u_{l,k_l})} : l \geq 1\}$ simply as $\{\mathbb{P}_{l,(u_{l,1},...,u_{l,k_l})}\}$. Similarly, any collection of variables $\{\bar{p}_{l,(u_{l,1},...,u_{l,k_l})} : l \geq 1\}$ will be written as $\{\bar{p}_{l,(u_{l,1},...,u_{l,k_l})}\}$.

Theorem 6.2.1. *(Bhattacharjee and Bose [2017]) Suppose Assumptions $(A1) - (A3)$ hold and $p, n = n(p) \to \infty$, $p/n \to y > 0$. Then*

(a) $(\bar{\mathcal{U}}_p, (n+p)^{-1}E\text{Tr}) \to (\bar{\mathcal{U}}, \bar{\varphi})$, *and*

(b) for any polynomial Π

$$\lim \frac{1}{p} E\text{Tr}(\Pi\{\mathbb{P}_{l,(u_{l,1},...,u_{l,k_l})}\}) = \frac{1+y}{y} \bar{\varphi}(\Pi\{\bar{p}_{l,(u_{l,1},...,u_{l,k_l})}\}).$$

(6.19)

Hence, $(\mathcal{U}_p, p^{-1}E\text{Tr})$ *converges. The limit NCP may be denoted as* $(\mathcal{U} := \text{Span}(p_{l,(u_{l,1},...,u_{l,k_l})} : l \geq 1), \varphi)$, *say, where*

$$\varphi(\Pi\{p_{l,(u_{l,1},...,u_{l,k_l})}\}) = \frac{1+y}{y} \bar{\varphi}(\Pi\{\bar{p}_{l,(u_{l,1},...,u_{l,k_l})}\}).$$

Proof. (a) By Definition 5.1.4, we need to show that for any polynomial Π,

$$\lim \frac{1}{n+p} E\text{Tr}(\Pi\{\bar{\mathbb{P}}_{l,(u_{l,1},...,u_{l,k_l})}\}) = \bar{\varphi}(\Pi\{\bar{p}_{l,(u_{l,1},...,u_{l,k_l})}\}).$$

(6.20)

For this, we first embed $\{Z_u\}$ into the Wigner matrices $\{W_u\}$ of order $(n+p)$ as given in (6.4). Recall $\bar{B}$ and $\underline{D}$ respectively for the matrices B and D of orders p and n in (6.5). Therefore,

$$\Pi\{\bar{\mathbb{P}}_{l,(u_{l,1},...,u_{l,k_l})}\} = \Pi\{\prod_{i=1}^{k_l} \left(n^{-1}\bar{A}_{l,2i-1} W_{u_{l,i}} \underline{A}_{l,2i} W_{u_{l,i}}^*\right) \bar{A}_{l,2k_l+1}\}.$$

(6.21)

By using (6.14), (6.16) and Theorem 5.5.1(a, d), the NCP (Span($\{\bar{B}_{2i-1}, \bar{B}_{2i-1}^*$: $i \geq 1\}, (n+p)^{-1}\text{Tr}$), (Span$\{\underline{B}_{2i}, \underline{B}_{2i}^* : i \geq 1\}, (n+p)^{-1}\text{Tr}$) and (Span$\{(n+p)^{-1/2}W_u : 1 \leq u \leq U\}, (n+p)^{-1}E\text{Tr}$) respectively converge to $(\bar{A}_{\text{odd}}, \bar{\varphi}_{\text{odd}})$, $(\bar{A}_{\text{even}}, \bar{\varphi}_{\text{even}})$ and $(\mathcal{S}, \varphi_s)$ and they are only asymptotically free. Note that $\{\bar{B}_{2i-1}, \bar{B}_{2i-1}^*\}$ and $\{\underline{B}_{2i}, \underline{B}_{2i}^*\}$ are not in general asymptotically free. They are only asymptotically free in polynomials where $\{B_{2i-1}, B_{2i-1}^*\}$ and $\{B_{2i}, B_{2i}^*\}$ are respectively enclosed within (Z^*, Z) and (Z, Z^*). Therefore, (6.20) follows once we observe (6.17) holds.

(b) Note that for any polynomial Π,

$$\lim p^{-1} E\mathrm{Tr}(\Pi\{\mathbb{P}_{l,(u_{l,1},\ldots,u_{l,k_l})}\})$$

$$= \lim \frac{n+p}{p}(n+p)^{-1} E\mathrm{Tr}(\Pi\{\bar{\mathbb{P}}_{l,(u_{l,1},\ldots,u_{l,k_l})}\})$$

$$= y^{-1}(1+y)\lim(n+p)^{-1} E\mathrm{Tr}(\Pi\{\bar{\mathbb{P}}_{l,(u_{l,1},\ldots,u_{l,k_l})}\})$$

$$= y^{-1}(1+y)\bar{\varphi}(\Pi\{\bar{p}_{l,(u_{l,1},\ldots,u_{l,k_l})}\})$$

$$= \varphi(\Pi\{p_{l,(u_{l,1},\ldots,u_{l,k_l})}\}), \text{ (say)}.$$

This completes the proof of Theorem 6.2.1(b). $\square$

6.3   LSD of symmetric polynomials

Now suppose we wish to show the existence of the LSD of any symmetric polynomial in $\{\mathbb{P}_{l,(u_{l,1},u_{l,2},\ldots,u_{l,k_l})}\}$. We can do this by using the moment method. The previous theorem has already done the hard work by verifying the (M1) condition.

Theorem 6.3.1. *(Bhattacharjee and Bose [2017]) Suppose (A1)-(A3) hold and $p, n = n(p) \to \infty$, $p/n \to y > 0$. Then the LSD of any symmetric polynomial $\Pi\{\mathbb{P}_{l,(u_{l,1},u_{l,2},\ldots,u_{l,k_l})}\}$ exists almost surely and it is uniquely determined by the (usual) moment sequence*

$$\lim \frac{1}{p} E\mathrm{Tr}(\Pi\{\mathbb{P}_{l,(u_{l,1},\ldots,u_{l,k_l})}\})^k = \varphi(\Pi\{p_{l,(u_{l,1},\ldots,u_{l,k_l})}\})^k$$

$$= y^{-1}(1+y)\bar{\varphi}(\Pi\{\bar{p}_{l,(u_{l,1},\ldots,u_{l,k_l})}\})^k.$$

Proof. By Lemma 4.1.2, we need to establish (M1), (M4) and (C) as described in the moment method. The (M1) condition is nothing but (6.19) in Theorem 6.2.1(b). Now we shall establish (M4) and (C).

Proof of (M4). To establish (M4), we need the following lemma on traces of polynomials. Its proof is very technical and is deferred to Section A.1.

Lemma 6.3.1. *Suppose (A1)-(A3) hold and $p, n = n(p) \to \infty$, $p/n \to y > 0$. Let $\mathbb{P}_u \in \mathrm{Span}\{\mathbb{P}_{l,(u_{l,1},\ldots,u_{l,k_l})}\}$, $u \geq 0$. Let for $1 \leq i \leq T$, $m_i(\mathbb{P}_u, \mathbb{P}_u^* : u \geq 0)$ be polynomials. Let*

$$\mathcal{P}_i = \mathrm{Tr}(m_i(\mathbb{P}_u, \mathbb{P}_u^* : u \geq 0)) \text{ and } \mathcal{P}_i^0 = E\mathcal{P}_i$$

For $d \geq 1$, define

$$\mathcal{S}_d = \{\pi : \pi \text{ is a pair partition } \{(i_1, i_2), \ldots, (i_{2d-1}, i_{2d})\} \text{ of } \{1, 2, \ldots, 2d\}\}.$$

Then, for all $d \geq 1$,

$$\lim E\left[\Pi_{i=1}^T \left(\mathcal{P}_i - \mathcal{P}_i^0\right)\right] \tag{6.22}$$

$$= \begin{cases} 0 \text{ if } T = 2d - 1, \\ \sum_{\mathcal{S}_d} \Pi_{k=1}^d \lim E\left[\left(\mathcal{P}_{i_{2k-1}} - \mathcal{P}_{i_{2k-1}}^0\right)\left(\mathcal{P}_{i_{2k}} - \mathcal{P}_{i_{2k}}^0\right)\right], \text{ if } T = 2d. \end{cases}$$

In particular, fix any polynomial $\Pi(\mathbb{P}_u, \mathbb{P}_u^* : u \geq 0)$ and let $T = 4$ and $\mathcal{P}_i = \mathrm{Tr}\,(\Pi(\mathbb{P}_u, \mathbb{P}_u^* : u \geq 0))^h$ in Lemma 6.3.1. Then we have,

$$E\Big[\frac{1}{p}\mathrm{Tr}\,(\Pi(\mathbb{P}_u, \mathbb{P}_u^* : u \geq 0))^h - E\big(\frac{1}{p}\mathrm{Tr}\,(\Pi(\mathbb{P}_u, \mathbb{P}_u^* : u \geq 0))^h\big)\Big]^4 = O(p^{-4})$$

and hence (M4) is established.

Proof of Carleman's condition (C). We have to show that for any symmetric polynomial Π,

$$\sum_{k=1}^{\infty}\Big(y^{-1}(1+y)\bar{\varphi}(\Pi(\bar{p}_{l,(u_{l,1},\ldots,u_{l,k_l})} : l \geq 1))^{2k}\Big)^{-1/2k} = \infty. \qquad (6.23)$$

Now note that since

$$\Big(\frac{y}{1+y}\Big)^k \leq \frac{y}{1+y},$$

to prove (6.23), it is enough to show that

$$\bar{\varphi}(\Pi(\bar{p}_{l,(u_{l,1},\ldots,u_{l,k_l})} : l \geq 1))^{2k} \leq C^{2k}, \ \forall k \geq 1. \qquad (6.24)$$

The following lemma is useful in this proof.

Lemma 6.3.2. *Let s be a standard semi-circle variable. Then for all $\{\bar{a}_{2i-1}\} \in \{\bar{b}_{2i-1}, \bar{b}_{2i-1}^*\}$, $\{\bar{a}_{2i}\} \in \{\bar{b}_{2i}, \bar{b}_{2i}^*\}$, $h \geq 1$ and for some $C_1 > 0$, we have*

$$|\bar{\varphi}(\bar{a}_1 s \underline{a}_2 s \bar{a}_3 \cdots \underline{a}_{2h} s)| \leq C_1^{2h}.$$

Proof. Recall $\|\cdot\|_2$ defined in (1.8). By Assumptions (A2) and (A3), there exists $C > 0$ such that

$$\sup_{1 \leq i \leq K}\sup_p \|B_{2i-1}\|_2 = \sup_{1 \leq i \leq K}\sup_p \|\bar{B}_{2i-1}\|_2 \leq C, \ \text{ and} \qquad (6.25)$$

$$\sup_{1 \leq i \leq L}\sup_n \|B_{2i}\|_2 = \sup_{1 \leq i \leq L}\sup_n \|\bar{B}_{2i}\|_2 \leq C. \qquad (6.26)$$

Therefore, for all $h \geq 1$, $1 \leq i \leq K$,

$$\bar{\varphi}(\bar{b}_{2i-1}^* \bar{b}_{2i-1})^h = \lim \frac{1}{n+p}\mathrm{Tr}(\bar{B}_{2i-1}^* \bar{B}_{2i-1})^h$$

$$\leq \sup_p \|\bar{B}_{2i-1}^* \bar{B}_{2i-1}\|_2^h \leq C^{2h}. \qquad (6.27)$$

Similarly,

$$\bar{\varphi}(\underline{b}_{2i}^* \underline{b}_{2i})^h \leq C^{2h}, \ \ \forall h \geq 1, \ 1 \leq i \leq L. \qquad (6.28)$$

Also note that, for all $\bar{a}_{2i-1} \in \{\bar{b}_{2i-1}, \bar{b}_{2i-1}^* : i \leq K\}$, $\underline{a}_{2i} \in \{\underline{b}_{2i}, \underline{b}_{2i}^* : i \leq L\}$ and $h \geq 1$, by Lemma 5.1.1(b), there exists $\{h_i : i \leq 2h\}$ such that

$$|\bar{\varphi}(\bar{a}_1 \underline{a}_2 \cdots \bar{a}_{2h-1} \underline{a}_{2h})| \;\leq\; \prod_{i=1}^{h} \left[\bar{\varphi}((\bar{a}_{2i-1}^* \bar{a}_{2i-1})^{h_{2i-1}}) \right]^{1/h_{2i-1}}$$

$$\times \prod_{i=1}^{h} \left[\bar{\varphi}((\underline{a}_{2i}^* \underline{a}_{2i})^{h_{2i}}) \right]^{1/h_{2i}}.$$

Hence, by (6.27) and (6.28)

$$|\bar{\varphi}(\bar{a}_1 \underline{a}_2 \bar{a}_3 \cdots \bar{a}_{2h-1} \underline{a}_{2h})| \;\leq\; C^{2h}, \;\; \forall h \geq 1. \tag{6.29}$$

Therefore, applying (5.16) and (5.34) and using the trivial bound $\#NC_2(2h) \leq 2^{2h}$, $\forall h \geq 1$,

$$|\bar{\varphi}(\bar{a}_1 s \underline{a}_2 s \bar{a}_3 \cdots \underline{a}_{2h} s)| \leq C^{2h}(\#NC_2(2h)) \leq (2C)^{2h}.$$

Hence, the proof of Lemma 6.3.2 is complete.   □

Now by (6.18), note that we can write,

$$\Pi\{\bar{p}_{l,(u_{l,1},\ldots,u_{l,k_l})}\} = \sum_{i=1}^{T} g_i, \;\; \text{where}$$

$$g_i = \bar{a}_{1,i} s \underline{a}_{2,i} s \cdots \underline{a}_{2l_i,i} s, \;\; \forall i \geq 1, \tag{6.30}$$

$\bar{a}_{2j-1,i} \in \{\bar{b}_{2i-1}, \bar{b}_{2i-1}^* : i \geq 1\}$ and $\underline{a}_{2j,i} \in \{\underline{b}_{2i}, \underline{b}_{2i}^* : i \geq 1\}$. Now, by Lemma 5.1.1(b) and Lemma 6.3.2, there are $C_1, C_2 > 0$ such that

$$\bar{\varphi}(\Pi\{\bar{p}_{l,(u_{l,1},\ldots,u_{l,k_l})}\})^{2k} \;=\; \bar{\varphi}\left(\sum_{i=1}^{T} g_i\right)^{2k}$$

$$= \sum_{1 \leq i_1,\ldots,i_{2k} \leq T} \bar{\varphi}(g_{i_1} \cdots g_{i_{2k}})$$

$$\leq \sum_{1 \leq i_1,\ldots,i_{2k} \leq T} |\bar{\varphi}(g_{i_1} \cdots g_{i_{2k}})|$$

$$\leq C_1^{2 \sum_{j=1}^{2k} l_{i_j}} T^{2k} \leq C_2^{2k}. \tag{6.31}$$

Hence, (6.24) is proved and Carleman's condition (C) follows. This completes the proof of Theorem 6.3.1.   □

6.4   Stieltjes transform

It may now be observed that all the LSD results discussed in Chapter 4 are for matrices which are special cases of Δ of the form

$$\Delta = \frac{1}{n} \sum_{i=1}^{q} B_{4i-3} Z B_{2i} Z^* B_{4i-1}. \tag{6.32}$$

Moreover, the matrices $\{\Delta_u\}$, which are defined in (6.3) and which will approximate the sample autocovariance matrices, are also special cases of Δ. We assume appropriate conditions on $\{B_i\}$ so that Δ is symmetric. Our goal in this section is to describe the Stieltjes transform of the LSD of Δ through some recursive functional equations.

By Theorem 6.3.1, under (A1)-(A3), the almost sure LSD of Δ exists and it is characterized by the moment sequence

$$\lim \frac{1}{p} E\mathrm{Tr}(\Delta)^k = \frac{1+y}{y}\bar{\varphi}(\bar{\delta}^k), \ \forall k \geq 1, \tag{6.33}$$

where

$$\bar{\delta} = (1+y)\sum_{i=1}^{q} \bar{b}_{4i-3}s\underline{b}_{2i}s\bar{b}_{4i-1}. \tag{6.34}$$

Recall that $\{\bar{b}_{2i-1}\}$ and $\{\underline{b}_{2i}\}$ are respectively limits of $\{\bar{B}_{2i-1}\}$ and $\{\underline{B}_{2i}\}$. Moreover, s, $\{\bar{b}_{2i-1}\}$ and $\{\underline{b}_{2i}\}$ are free (by Theorem 6.3.1, as far as computing limits of polynomials of the form (6.32) is concerned).

Also note that $\bar{\delta}$ is self-adjoint and $\bar{\varphi}$ is positive. By (A2) and (A3), there is $C > 0$ such that $|\bar{\phi}(\bar{\delta}^k)| \leq C^k$, $\forall k$. Hence, by Lemma 4.1.3(b), there is a unique probability measure on $\mathbb{R}$, say $\bar{\mu}$, characterized by the moment sequence $\{\bar{\phi}(\bar{\delta}^k)\}$.

Let μ be the probability measure on $\mathbb{R}$ corresponding to the LSD of Δ. Note that by (6.33),

$$\int_{\mathbb{R}} x^k d\mu = \frac{1+y}{y} \int_{\mathbb{R}} x^k d\bar{\mu}, \ \forall k \geq 1. \tag{6.35}$$

Let δ_0 be the degenerate probability measure at 0. Then by (6.35), the following relation is immediate:

$$\bar{\mu} = \frac{y}{1+y}\mu + \frac{1}{1+y}\delta_0. \tag{6.36}$$

Let $m_{\bar{\mu}}(z)$ and $m_\mu(z)$ be respectively the Stieltjes transforms of $\bar{\mu}$ and μ. Then by (6.36), we have

$$m_{\bar{\mu}}(z) = \frac{y}{1+y}m_\mu(z) - \frac{1}{1+y}\frac{1}{z}, \ z \in \mathbb{C}^+. \tag{6.37}$$

We first describe $m_{\bar{\mu}}(z)$. Then it is easy to express $m_\mu(z)$ in terms of $m_{\bar{\mu}}(z)$.

We write infinite sums of the form $\displaystyle\sum_{1\leq i_1,\ldots,i_k<\infty} a_{i_1}\cdots a_{i_k}$ in the sense that

$$\varphi\Big(\sum_{1\leq i_1,\ldots,i_k<\infty} a_{i_1}\cdots a_{i_k}\Big) = \sum_{1\leq i_1,\ldots,i_k<\infty} \varphi(a_{i_1}\cdots a_{i_k}),$$

whenever

$$\sum_{1 \le i_1,\dots,i_k < \infty} |\varphi(a_{i_1} \cdots a_{i_k})| < \infty. \tag{6.38}$$

Moreover, we write $(1-a)^{-1} := \sum_{i=0}^{\infty} a^i$ in the above sense.

To simplify notation for our formula, let,

$$d_i = b_{4i-3}, \quad e_i = b_{4i-1}, \quad f_i = b_{2i}, \quad d_0 = e_0 = f_0 = 1, \tag{6.39}$$
$$d = \{d_0, d_1, \dots, d_q\}, \quad e = \{e_0, e_1, \dots, e_q\}, \quad f = \{f_0, f_1, \dots, f_q\},$$
$$\bar{d}_i = \bar{b}_{4i-3}, \quad \bar{e}_i = \bar{b}_{4i-1}, \quad \underline{f}_i = \underline{b}_{2i}, \quad \bar{d}_0 = \bar{e}_0 = \underline{f}_0 = 1,$$
$$\bar{d} = \{\bar{d}_0, \bar{d}_1, \dots, \bar{d}_q\}, \quad \bar{e} = \{\bar{e}_0, \bar{e}_1, \dots, \bar{e}_q\}, \quad \underline{f} = \{\underline{f}_0, \underline{f}_1, \dots, \underline{f}_q\}.$$

Let
$$\Pi = \Pi(d, e, z) \quad \text{any polynomial in } d, e \text{ where } z \in \mathbb{C}^+.$$

Suppose $\bar{\Pi} = \Pi(\bar{d}, \bar{e}, z)$ is its embedded version. Define

$$\bar{R}_{j,j_1,j_2,j_3}(\underline{f}, \bar{\Pi}) = (1+y)\bar{\varphi}(\bar{\Pi}\bar{d}_{j_1}\bar{e}_{j_3}\bar{\delta}^{j-1})\underline{f}_{j_2} \tag{6.40}$$

$$R_{j,j_1,j_2,j_3}(f, \Pi) = \frac{1+y}{y}\bar{\varphi}(\bar{\Pi}\bar{d}_{j_1}\bar{e}_{j_3}\bar{\delta}^{j-1})f_{j_2},$$

$$\bar{A}_{j_1,j_2,j_3}(z, \underline{f}, \bar{\Pi}) = \sum_{i=1}^{\infty} z^{-i}\bar{R}_{i,j_1,j_2,j_3}(\underline{f}, \bar{\Pi}),$$

$$A_{j_1,j_2,j_3}(z, f, \Pi) = \sum_{i=1}^{\infty} z^{-i}R_{i,j_1,j_2,j_3}(f, \Pi).$$

The next lemma guarantees existence of the above sums. Recall the states φ_{odd} and φ_{even} defined on the spaces $\mathcal{A}_{\text{odd}}$ and $\mathcal{A}_{\text{even}}$ respectively. Note that the variables $\{d_j, e_j\}$ are in $\mathcal{A}_{\text{odd}}$ and the variables $\{f_j\}$ are in $\mathcal{A}_{\text{even}}$.

Lemma 6.4.1. *Suppose for all sufficiently large* $|z|, z \in \mathbb{C}^+$, *and for some* $C > 0$, $|\varphi_{\text{odd}}((\Pi\Pi^*)^r)| \le C^r$ *for all* $r \ge 1$. *Then* $\bar{A}_{j_1,j_2,j_3}(z, \underline{f}, \bar{\Pi})$ *and* $A_{j_1,j_2,j_3}(z, f, \Pi)$ *exist for all sufficiently large* $|z|$, *in the sense of (6.38).*

Proof. Note that there is a $C > 0$ such that for any $\{a_{2i-1}\} \in \{b_{2i-1}, b_{2i-1}^*\}$, $\{a_{2i}\} \in \{b_{2i}, b_{2i}^*\}$ and $h \ge 1$, we have

$$|\bar{\varphi}(\bar{a}_1\bar{a}_3 \cdots \bar{a}_{2h-1})| \le C^h, \quad |\bar{\varphi}(\underline{a}_2\underline{a}_4 \cdots \underline{a}_{2h})| \le C^h, \quad |\bar{\varphi}(\bar{\delta}^h)| \le C^h. \tag{6.41}$$

Proof of (6.41) is along the same lines as the proof of Carleman's condition (C) in Theorem 6.3.1. Hence, we omit it.

Now, by Lemmas 5.1.1 and (6.41), for some $C_1 > 0$, we have

$$|\varphi_{\text{even}}((A_{j_1,j_2,j_3}(z,f,\Pi))^r)|$$

$$\leq C_1|\varphi_{\text{even}}(f_{j_2}^r)|\sum_{i_1,i_2,\ldots,i_r=1}^{\infty}\left(\prod_{k=1}^{r}|z|^{-i_k}|\bar{\varphi}(\bar{\Pi}\bar{d}_{j_1}\bar{e}_{j_3}\bar{\delta}^{i_k-1})|\right)$$

$$\leq C_1|\varphi_{\text{even}}(f_{j_2}^r)|(\bar{\varphi}(\bar{\Pi}\bar{\Pi}^*))^{r/2}(\bar{\varphi}((\bar{e}_{j_3}^*\bar{d}_{j_1}^*\bar{d}_{j_1}\bar{e}_{j_3})^2))^{r/4}$$

$$\times \sum_{i_1,i_2,\ldots,i_r=1}^{\infty}\left(\prod_{k=1}^{r}|z|^{-i_k}(\bar{\varphi}(\bar{\delta}^{4i_k-4}))^{1/4}\right)$$

$$\leq C_1C^{3r}\sum_{i_1,i_2,\ldots,i_r=1}^{\infty}\left(\prod_{k=1}^{r}|z|^{-i_k}C^{i_k}\right)$$

$$= C_1C^{4r}(|z|-C)^{-r}, \quad \text{for sufficiently large } |z|.$$

This completes the proof of Lemma 6.4.1. $\square$

The following Theorem provides $m_{\bar{\mu}}(z)$ and $m_{\mu}(z)$. The general formulae are quite cumbersome and involve recursive functional equations. However, as we shall see soon, these simplify to a great extent in our specific cases of interest. For ease of writing the formulae, let us use the notation $\displaystyle\sum_1 = \sum_{l_0,\ldots,l_{t-1}=1}^{q}$.

Theorem 6.4.1. *Assume (A1)-(A3) hold and $p,n \to \infty$, $p/n \to y > 0$.*

(a) The following recursive relation holds:

$$\bar{A}_{j_1,j_2,j_3}(z,\underline{f},\bar{\Pi}) = \frac{1+y}{z}\left[\bar{\varphi}(\bar{\Pi}\bar{d}_{j_1}\bar{e}_{j_3})\underline{f}_{j_2}\right. \tag{6.42}$$

$$\left. +\sum_{t=1}^{\infty}\sum_{1}\bar{\varphi}\left(\bar{A}_{l_0,l_{t-1},l_{t-1}}(z,\underline{f},\bar{\Pi}\bar{d}_{j_1}\bar{e}_{j_3})\prod_{u=0}^{t-2}\bar{A}_{l_{u+1},l_u,l_u}(z,\underline{f},1)\right)\underline{f}_{j_2}\right].$$

As a consequence, for $z \in \mathbb{C}^+$, $|z|$ large, $m_{\bar{\mu}}(z)$ is given by

$$m_{\bar{\mu}}(z) = -\frac{1}{z}\left[1+\sum_{t=1}^{\infty}\sum_{1}\bar{\varphi}\left(\prod_{u=1}^{t}\bar{A}_{l_u,l_{u-1},l_{u-1}}(z,\underline{f},1)\right)\right]. \tag{6.43}$$

(b) The following recursive relation holds:

$$A_{j_1,j_2,j_3}(z,f,\Pi) = \frac{1}{z}\left[\varphi_{odd}(\Pi d_{j_1}e_{j_3})f_{j_2}\right. \tag{6.44}$$

$$\left. +\sum_{t=1}^{\infty}\sum_{1}y^{t-1}\varphi_{even}\left(A_{l_0,l_{t-1},l_{t-1}}(z,f,\Pi d_{j_1}e_{j_3})\prod_{u=0}^{t-2}A_{l_{u+1},l_u,l_u}(z,f,1)\right)f_{j_2}\right].$$

As a consequence, for $z \in \mathbb{C}^+$, $|z|$ large, $m_\mu(z)$ is given by

$$m_\mu(z) = -\frac{1}{z}\left[1 + \sum_{t=1}^{\infty}\sum_{1} y^{t-1}\varphi_{even}\left(\prod_{u=1}^{t} A_{l_u,l_{u-1},l_{u-1}}(z,f,1)\right)\right]. \quad (6.45)$$

Proof. Proof of part (a) is quite technical and tedious and is deferred to Section A.2 of the Appendix. To prove part (b), by (6.40), note that for any polynomials $\Pi_1, \Pi_2, \ldots, \Pi_k$, of the same form as Π, we have

$$\bar{\varphi}\left(\prod_{k=1}^{r} \bar{R}_{j_k,j_{1k},j_{2k},j_{3k}}(\underline{f},\bar{\Pi}_k)\right)$$

$$= (1+y)^r \bar{\varphi}(\underline{f}_{j_{21}}\underline{f}_{j_{22}}\cdots\underline{f}_{j_{2r}})\prod_{k=1}^{r}\bar{\varphi}(\bar{\Pi}_k\bar{d}_{j_{1k}}\bar{e}_{j_{3k}}\bar{\delta}^{j_k-1})$$

$$= \frac{y^r}{1+y}\varphi_{even}(f_{j_{21}}f_{j_{22}}\cdots f_{j_{2r}})\prod_{k=1}^{r}\left(\frac{1+y}{y}\bar{\varphi}(\bar{\Pi}_k\bar{d}_{j_{1k}}\bar{e}_{j_{3k}}\bar{\delta}^{j_k-1})\right)$$

$$= \frac{y^r}{1+y}\varphi_{even}\left(\prod_{k=1}^{r}R_{j_k,j_{1k},j_{2k},j_{3k}}(f,\Pi_k)\right).$$

Thus,

$$\bar{\varphi}\left(\prod_{k=1}^{r}\bar{A}_{j_{1k},j_{2k}j_{3k}}(z,\underline{f},\bar{\Pi}_k)\right) = \frac{y^r}{1+y}\varphi_{even}\left(\prod_{k=1}^{r}A_{j_{1k},j_{2k}j_{3k}}(z,f,\Pi_k)\right). \quad (6.46)$$

Therefore, by (6.42) and (6.46), we have

$$\bar{A}_{j_1,j_2,j_3}(z,\underline{f},\bar{\Pi}) = \frac{1}{z}\Bigg[y\varphi_{odd}(\Pi d_{j_1}e_{j_3})\underline{f}_{j_2} \quad (6.47)$$

$$+ \sum_{t=1}^{\infty}\sum_{1}y^t\varphi_{even}\left(A_{l_0,l_{t-1},l_{t-1}}(z,f,\Pi d_{j_1}e_{j_3})\prod_{u=0}^{t-2}A_{l_{u+1},l_u,l_u}(z,f,1)\right)\underline{f}_{j_2}\Bigg].$$

Hence, (6.44) follows from the above equation and (6.40).

Now by (6.37), (6.43) and (6.46), we have

$$\frac{y}{1+y}m_\mu(z) - \frac{1}{1+y}\frac{1}{z} = -\frac{1}{z}\Bigg[1 + \sum_{t=1}^{\infty}\sum_{1}\frac{y^t}{1+y}\varphi_{even}$$

$$\times\left(\prod_{u=1}^{t}A_{l_u,l_{u-1},l_{u-1}}(z,f,1)\right)\Bigg].$$

Simplifying the above equation,

$$m_\mu(z) = -\frac{1}{z}\left[1 + \sum_{t=1}^{\infty}\sum_{1}y^{t-1}\varphi_{even}\left(\prod_{u=1}^{t}A_{l_u,l_u-1,l_{u-1}}(z,f,1)\right)\right].$$

This establishes (6.45) and hence completes the proof of Theorem 6.4.1. □

6.5   Corollaries

This section demonstrates application of Theorems 6.3.1 and 6.4.1. As we shall see, in specific cases, there is significant simplification of the general formulae. Corollaries 6.5.1–6.5.3 will be useful later when we deal with LSD of sample autocovariance matrices. Corollaries 6.5.4–6.5.6 show how some well-known LSD results can be quickly derived using Theorem 6.4.1.

Recall $\{\Delta_u\}$ defined in (6.3) and the coefficient matrices $\{\psi_j\}$ in (3.3). Suppose $\{\psi_j\} \subset \{B_{2i-1}, B_{2i-1}^*\}$ i.e., we assume:

(B) $\{\psi_j\}$ are norm bounded and they jointly converge.

Let us use the following notation to describe this convergence.

$$(\mathrm{Span}\{\psi_j, \psi_j^* : j \geq 0\}, \frac{1}{p}\mathrm{Tr}) \quad \to \quad (\mathrm{Span}\{\eta_j, \eta_j^* : j \geq 0\}, \varphi_{\mathrm{odd}}), \quad (6.48)$$

$$(\mathrm{Span}\{\bar{\psi}_j, \bar{\psi}_j^* : j \geq 0\}, \frac{1}{n+p}\mathrm{Tr}) \quad \to \quad (\mathrm{Span}\{\bar{\eta}_j, \bar{\eta}_j^* : j \geq 0\}, \bar{\varphi}_{\mathrm{odd}}). \quad (6.49)$$

Recall the NCPs $(\mathcal{A}_{\mathrm{odd}}, \varphi_{\mathrm{odd}})$ and $(\bar{\mathcal{A}}_{\mathrm{odd}}, \bar{\varphi}_{\mathrm{odd}})$ defined in (6.13) and (6.14). Clearly the NCP in the right side of (6.48) and (6.49) are $*$-sub-algebras of $(\mathcal{A}_{\mathrm{odd}}, \varphi_{\mathrm{odd}})$ and $(\bar{\mathcal{A}}_{\mathrm{odd}}, \bar{\varphi}_{\mathrm{odd}})$, respectively. Recall that $\bar{\varphi}$ is the state corresponding to the free product given in (6.17). Therefore, by Definition 5.3.4, the restriction of $\bar{\varphi}$ on $\bar{\mathcal{A}}_{\mathrm{odd}}$ is $\bar{\varphi}_{\mathrm{odd}}$. Further, for any polynomial Π

$$\bar{\varphi}(\Pi(\bar{\eta}_j, \bar{\eta}_j^* : j \geq 0)) = \bar{\varphi}_{\mathrm{odd}}(\Pi(\bar{\eta}_j, \bar{\eta}_j^* : j \geq 0))$$

$$= \frac{y}{1+y}\varphi_{\mathrm{odd}}(\Pi(\eta_j, \eta_j^* : j \geq 0)). \quad (6.50)$$

The following corollary is relevant since later in Chapter 8, we will deal with the LSD of $\hat{\Gamma}_u + \hat{\Gamma}_u^*$.

Corollary 6.5.1. *Suppose (A1), (B) hold and $p, n = n(p) \to \infty$, $p/n \to y > 0$. Then the almost sure LSD of $\frac{1}{2}(\Delta_u + \Delta_u^*)$ exists and its Stieltjes transform $m_u(z)$, for $z \in \mathbb{C}^+$ and $|z|$ large, is given by (θ denotes a $U(0, 2\pi)$ random variable),*

$$m_u(z) = \varphi_{\mathrm{odd}}((B_u(\lambda, z) - z)^{-1}), \quad where \quad (6.51)$$

$$K_u(z, \theta) = \varphi_{\mathrm{odd}}(h(\lambda, \theta)(B_u(\lambda, z) - z)^{-1}|\theta), \quad (6.52)$$

$$:= \sum_{j,k=0}^{q} \varphi_{\mathrm{odd}}(\eta_j \eta_k^*(B_u(\lambda, z) - z)^{-1})e^{i(j-k)\theta}$$

$$h(\lambda, \theta) = (\sum_{j=0}^{q} e^{ij\theta}\eta_j)(\sum_{j=0}^{q} e^{-ij\theta}\eta_j^*), \quad \lambda = \{\eta_j, \eta_j^* : j \geq 0\}, \quad (6.53)$$

$$B_u(\lambda, z) = E_\theta \left(\cos(u\theta)h(\lambda, \theta)(1 + y\cos(u\theta)K(z, \theta))^{-1}|\lambda \right), \quad (6.54)$$

$$:= \sum_{j,k=0}^{q} \eta_j \eta_k^* E_\theta \left(\cos(u\theta)e^{i(j-k)\theta}(1 + y\cos(u\theta)K(z, \theta))^{-1} \right).$$

Proof. Let
$$\mathbb{Z} = \{0, \pm 1, \pm 2, \ldots\}.$$
First note that $\{\Delta_u\}$ satisfy the form (6.1). Moreover, under (B) and (6.48), $\{\psi_j\}$ satisfy (A2) and (6.13). Also note that the matrices $\{P_u : u \in \mathbb{Z}\}$ satisfy (A3). We use the following notation for the convergence of these matrices.

$$(\text{Span}\{\underline{P}_u : u \in \mathbb{Z}\}, (n+p)^{-1}\text{Tr}) \;\to\; (\text{Span}\{\underline{c}_u : u \in \mathbb{Z}\}, \bar{\varphi}), \qquad (6.55)$$
$$(\text{Span}\{P_u : u \in \mathbb{Z}\}, n^{-1}\text{Tr}) \;\to\; (\text{Span}\{c_u : u \in \mathbb{Z}\}, \varphi_{\text{even}}).$$

Then for any polynomial Π

$$
\begin{aligned}
\bar{\varphi}(\Pi(\{\underline{c}_u : u \in \mathbb{Z}\})) &= \lim \frac{1}{n+p} \text{Tr}(\Pi(\{\underline{P}_u : u \in \mathbb{Z}\})) \\
&= \frac{1}{1+y} E_\theta(\Pi(\{e^{iu\theta} : u \in \mathbb{Z}\})) \\
&= \frac{1}{1+y} \varphi_{\text{even}}(\Pi(\{c_u : u \in \mathbb{Z}\})), \qquad (6.56)
\end{aligned}
$$

where $\theta \sim U(0, 2\pi)$.

Now applying Theorem 6.4.1, (6.34), (6.39), and (6.40) reduce to

$$
\bar{\delta} = 0.5(1+y)\left[\sum_{j,j'=0}^{q} \bar{\eta}_j s \underline{c}_{j-j'+u} s \bar{\eta}_{j'}^* + \sum_{j,j'=0}^{q} \bar{\eta}_j s \underline{c}_{j-j'-u} s \bar{\eta}_{j'}^* \right],
$$
$$d = \{\eta_0, \eta_1, \ldots, \eta_q\}, \quad e = \{\eta_0^*, \eta_1^*, \ldots, \eta_q^*\},$$
$$f = \{0.5 c_{j_1-j_2+a} : \; 0 \le j_1, j_2 \le q, a = -u, 0, u\},$$
$$R_{j,j_1,j_2,j_1-j_2+a}(f, \Pi) = \frac{1+y}{2y} \bar{\varphi}(\bar{\Pi}\bar{\eta}_{j_1}\bar{\eta}_{j_2}^* \bar{\delta}^{j-1}) c_{j_1-j_2+a},$$
$$A_{j_1,j_2,j_1-j_2+a}(z, f, \Pi) = \sum_{i=1}^{\infty} z^{-i} R_{j,j_1,j_2,j_1-j_2+a}(f, \Pi).$$

Also define,

$$
\tilde{R}_{j,j_1,j_2,j_1-j_2+a}(\theta, \Pi) = \frac{1+y}{2y} \bar{\varphi}(\bar{\Pi}\bar{\eta}_{j_1}\bar{\eta}_{j_2}^* \bar{\delta}^{j-1}) e^{i(j_1-j_2+a)\theta},
$$
$$
\tilde{A}_{j_1,j_2,j_1-j_2+a}(z, 0, \Pi) = \sum_{i=1}^{\infty} z^{-i} \tilde{R}_{j,j_1,j_2,j_1-j_2+a}(\theta, \Pi).
$$

Therefore, by (6.56) and for any polynomials $\Pi_1, \Pi_2, \ldots, \Pi_r$, we have

$$
\varphi_{\text{even}}\left(\prod_{k=1}^{r} R_{j_k,j_{1k},j_{2k},j_{1k}-j_{2k}+a_k}(f, \Pi_k)\right) = E_\theta\left(\prod_{k=1}^{r} \tilde{R}_{j_k,j_{1k},j_{2k},j_{1k}-j_{2k}+a_k}(\theta, \Pi_k)\right),
$$
$$
\varphi_{\text{even}}\left(\prod_{k=1}^{r} A_{j_{1k},j_{2k},j_{1k}-j_{2k}+a_k}(z, f, \Pi_k)\right) = E_\theta\left(\prod_{k=1}^{r} \tilde{A}_{j_{1k},j_{2k},j_{1k}-j_{2k}+a_k}(z, \theta, \Pi_k)\right).
$$

Note that in the context of Corollary 6.5.1,

$$\sum_1 = \sum_{l_0,l_1,\ldots,l_{t-1}=0}^{q} \sum_{l'_0,l'_1,\ldots,l'_{t-1}=0}^{q} \sum_{a_0,a_1,\ldots,a_{t-1}\in\{-u,u\}} .$$

Define,

$$G_{t,j_1,j_2}(z,f,\Pi) = \sum_1 \left(A_{l_0,l'_{t-1},l_{t-1}-l'_{t-1}+a_{t-1}}(z,f,\Pi\eta_{j_1}\eta^*_{j_2}) \right.$$
$$\left. \prod_{k=0}^{t-2} A_{l_{k+1},l'_k,l_k-l'_k+a_k}(z,f,1) \right),$$

$$\tilde{G}_{t,j_1,j_2}(z,\theta,\Pi) = \sum_1 \left(\tilde{A}_{l_0,l'_{t-1},l_{t-1}-l'_{t-1}+a_{t-1}}(z,\theta,\Pi\eta_{j_1}\eta^*_{j_2}) \right.$$
$$\left. \prod_{k=0}^{t-2} \tilde{A}_{l_{k+1},l'_k,l_k-l'_k+a_k}(z,\theta,1) \right).$$

Therefore from the above equations,

$$\varphi_{\text{even}}(G_{t,j_1,j_2}(z,f,\Pi)) = E_\theta(\tilde{G}_{t,j_1,j_2}(z,\theta,\Pi)). \qquad (6.57)$$

Next, for any polynomial $\Pi(\theta,z)$, define

$$K_u(z,\theta) = -2 \sum_{j_1,j_2=0}^{q} \tilde{A}_{j_1,j_2,j_1-j_2}(z,\theta,1), \quad \sum_2 = \sum_{l,l'=0}^{q} \sum_{a\in\{u,-u\}}, \qquad (6.58)$$

$$E_\theta(h(\lambda,\theta)\Pi(z,\theta)|\lambda) := \sum_{l,l'=0}^{q} \eta_l\eta^*_{l'} E_\theta(e^{i(l-l')\theta}\Pi(z,\theta)), \quad D = \sum_{j=1}^{\infty} z^{-j}\bar{\delta}^{j-1}.$$

Recall $B_u(\lambda,z)$ from (6.54). Therefore, note that

$$h(\bar{\lambda},\theta) = \left(\sum_{j=0}^{q} e^{ij\theta}\bar{\eta}_j \right)\left(\sum_{j=0}^{q} e^{-ij\theta}\bar{\eta}^*_j \right),$$

$$E_\theta(h(\bar{\lambda},\theta)\Pi(z,\theta)|\bar{\lambda}) = \sum_{l,l'=0}^{q} \bar{\eta}_l\bar{\eta}^*_{l'} E_\theta(e^{i(l-l')\theta}\Pi(z,\theta)),$$

$$B_u(\bar{\lambda},z) = E_\theta\left(\cos(u\theta)h(\bar{\lambda},\theta)(1+y\cos(u\theta)K(z,\theta))^{-1}|\bar{\lambda} \right),$$

$$= \sum_{j,k=0}^{q} \bar{\eta}_j\bar{\eta}^*_k E_\theta\left(\cos(u\theta)e^{i(j-k)\theta}(1+y\cos(u\theta)K(z,\theta))^{-1} \right).$$

Now, as $\{e^{ij\theta} : j \in \mathbb{Z}\}$ are commutative, we have

$$E_\theta(\tilde{G}_{t,j_1,j_2}(z,\theta,\Pi)) \tag{6.59}$$

$$= \sum_2 E_\theta\left(\tilde{A}_{l,l',l-l'+a}(z,\theta,\Pi\eta_{j_1}\eta_{j_2}^*)(-\cos(u\theta)K_u(z,\theta))^{t-1}\right)$$

$$= \sum_{l,l'=0}^{q}\sum_{j=1}^{\infty}\left[z^{-j}\frac{1+y}{y}\bar{\varphi}(\bar{\Pi}\bar{\eta}_{j_1}\bar{\eta}_{j_2}^*\bar{\eta}_l\bar{\eta}_{l'}^*\bar{\delta}^{j-1})\right.$$

$$E_\theta(e^{i(l-l')\theta}\cos(u\theta)(-\cos(u\theta)K_u(z,\theta))^{t-1})\bigg]$$

$$= \frac{1+y}{y}\bar{\varphi}(\bar{\Pi}\bar{\eta}_{j_1}\bar{\eta}_{j_2}^* E_\theta(h(\bar{\lambda},\theta)\cos(u\theta)(-\cos(u\theta)K_u(z,\theta))^{t-1}|\bar{\lambda})D).$$

Thus, (6.44) reduces to

$$A_{j_1,j_2,j_1-j_2+a}(z,f,\Pi) \tag{6.60}$$

$$= \frac{1}{2z}[\varphi_{\mathrm{odd}}(\Pi\eta_{j_1}\eta_{j_2}^*) + \sum_{t=1}^{\infty}y^{t-1}\varphi_{\mathrm{even}}(G_{t,j_1,j_2}(z,f,\Pi)]c_{j_1-j_2+a}$$

$$= \frac{1}{2z}[\varphi_{\mathrm{odd}}(\Pi\eta_{j_1}\eta_{j_2}^*) + \sum_{t=1}^{\infty}y^{t-1}E_\theta(\tilde{G}_{t,j_1,j_2}(z,\theta,\Pi))]c_{j_1-j_2+a}$$

$$= \frac{1}{2z}[\varphi_{\mathrm{odd}}(\Pi\eta_{j_1}\eta_{j_2}^*) + \frac{1+y}{y}\bar{\varphi}(\bar{\Pi}\bar{\eta}_{j_1}\bar{\eta}_{j_2}^* B_u(\bar{\lambda},z)D)]c_{j_1-j_2+a}.$$

Similarly, (6.45) reduces to

$$m_u(z) = -\frac{1}{z}[1 + \frac{1+y}{y}\bar{\varphi}(B_u(\bar{\lambda},z)D)]. \tag{6.61}$$

Note that for any polynomial Π,

$$\frac{1+y}{y}\bar{\varphi}(\bar{\Pi}\bar{D}) = \sum_{j=1}^{\infty}z^{-j}\frac{1+y}{y}\bar{\varphi}(\bar{\Pi}\bar{\delta}^{j-1}) = 2A_{0,0,0}(z,f,\Pi) \tag{6.62}$$

and by repeated use of (6.60), we have

$$2A_{0,0,0}(z,f,\Pi) = \frac{1}{z}[\varphi_{\mathrm{odd}}(\Pi) + \frac{1+y}{y}\bar{\varphi}(\bar{\Pi}B_u(\bar{\lambda},z)D)]$$

$$= \frac{1}{z}[\varphi_{\mathrm{odd}}(\Pi) + \frac{1}{z}\{\varphi_{\mathrm{odd}}(\Pi B_u(\lambda,z)) + \frac{1+y}{y}\bar{\varphi}(\bar{\Pi}(B_u(\bar{\lambda},z))^2 D)\}]$$

$$= \frac{1}{z}\sum_{j=0}^{\infty}z^{-j}\varphi_{\mathrm{odd}}(\Pi(B_u(\lambda,z))^j)$$

$$= \varphi_{\mathrm{odd}}(\Pi(z - B_u(\lambda,z))^{-1}). \tag{6.63}$$

Thus, by (6.61)–(6.63), we have

$$
\begin{aligned}
m_u(z) &= -\frac{1}{z}[1 + \varphi_{\text{odd}}(B_u(\lambda, z)(z \quad B_u(\lambda, z))^{-1})] \\
&= \varphi_{\text{odd}}((B_u(\lambda, z) - z)^{-1}).
\end{aligned}
$$

This proves (6.51).

Now we shall establish (6.52). For any polynomial $\Pi(z, \lambda)$, define

$$
\varphi_{\text{odd}}(h(\lambda, \theta)\Pi(z, \lambda)|\theta) = \sum_{j_1, j_2 = 0}^{q} \varphi_{\text{odd}}(\eta_{j_1}\eta_{j_2}^* \Pi(z, \lambda))e^{i(j_1 - j_2)\theta},
$$

$$
\bar{\varphi}(h(\bar{\lambda}, \theta)\Pi(z, \bar{\lambda})|\theta) = \sum_{j_1, j_2 = 0}^{q} \bar{\varphi}(\bar{\eta}_{j_1}\bar{\eta}_{j_2}^* \Pi(z, \bar{\lambda}))e^{i(j_1 - j_2)\theta}.
$$

Note that, by (6.54), (6.57)–(6.60), (6.62), (6.63), we have

$$
\begin{aligned}
K_u(z, \theta) &= -2 \sum_{j_1, j_2 = 0}^{q} \tilde{A}_{j_1, j_2, j_1 - j_2}(z, \theta, 1) \\
&= -\sum_{j_1, j_2 = 0}^{q} \frac{1}{z}[\varphi_{\text{odd}}(\eta_{j_1}\eta_{j_2}^*) + \sum_{t=1}^{\infty} y^{t-1} E_\theta(\tilde{G}_{t, j_1, j_2}(z, \theta, 1)]e^{i(j_1 - j_2)\theta} \\
&= -\frac{1}{z}[\varphi_{\text{odd}}(h(\lambda, \theta)|\theta) + \frac{1 + y}{y}\bar{\varphi}(h(\bar{\lambda}, \theta)B_u(\bar{\lambda}, z)D|\theta))] \\
&= -\frac{1}{z}[\varphi_{\text{odd}}(h(\lambda, \theta)|\theta) + \varphi_{\text{odd}}(h(\lambda, \theta)B_u(\lambda, z)(z - B_u(\lambda, z))^{-1}|\theta)] \\
&= \varphi_{\text{odd}}(h(\lambda, \theta)(B_u(\lambda, z) - z)^{-1}|\theta).
\end{aligned}
$$

This establishes (6.52) and the proof of Corollary 6.5.1 is complete. $\square$

Note the cumbersome expressions (6.51)–(6.54). However, we have a better description of the LSD of $\Delta_u + \Delta_u^*$ in the special case $\psi_0 = I_p$, $\psi_j = \lambda_j I_p$, $\lambda_j \in \mathbb{R}$, for all $j \geq 1$. The following corollary will be useful later in Chapter 8 and there we shall also obtain the limit Stieltjes transform. Recall the compound free Poisson distribution from Definition 5.5.1.

Corollary 6.5.2. *Suppose (A1) holds and $p, n = n(p) \to \infty$, $p/n \to y > 0$. Let $\psi_0 = I_p$, $\psi_j = \lambda_j I_p$, $1 \leq j \leq q$. Then the LSD of $\frac{1}{2}(\Delta_u + \Delta_u^*)$ is the compound free Poisson whose r-th order free cumulant equals*

$$
\kappa_{ur} = y^{r-1} E_\theta(\cos(u\theta)\tilde{h}(\lambda, \theta))^r, \quad \forall i \geq 0, \tag{6.64}
$$

where

$$
\tilde{h}(\lambda, \theta) = |\sum_{j=0}^{q} e^{ij\theta}\lambda_j|^2, \ \lambda_0 = 1, \ \lambda = (\lambda_1, \ldots, \lambda_q) \ \text{and} \ \theta \sim U(0, 2\pi). \tag{6.65}
$$

Proof. Note that we can write

$$n\Delta_u = Z\Big(\sum_{j,j'=0}^{q} \lambda_j \lambda_{j'} P_{j-j'+u} \Big) Z^*, \quad n\Delta_u^* = Z\Big(\sum_{j=0,j'=0}^{q} \lambda_j \lambda_{j'} P_{j-j'+u}^* \Big) Z^*$$

and hence

$$\frac{1}{2}(\Delta_u + \Delta_u^*) = n^{-1} Z\Big(\frac{1}{2} \sum_{j,j'=0}^{q} \lambda_j \lambda_{j'} (P_{j-j'+u} + P_{j-j'+u}^*) \Big) Z^*.$$

Note that by (6.56), for all $r \geq 1$,

$$\lim n^{-1} \mathrm{Tr}\Big(\frac{1}{2} \sum_{j,j'=0}^{q} \lambda_j \lambda_{j'} (P_{j-j'+u} + P_{j-j'+u}^*) \Big)^r$$

$$= E_\theta \Big(\frac{1}{2} \sum_{j,j'=0}^{q} \lambda_j \lambda_{j'} (e^{(j-j'+u)\theta} + e^{-(j-j'+u)\theta}) \Big)^r$$

$$= E_\theta (\cos(u\theta) \tilde{h}(\lambda, \theta))^r.$$

Therefore, invoking the discussion around (5.46), the LSD of $\frac{1}{2}(\Delta_u + \Delta_u^*)$ is a compound free Poisson with the r-th order free cumulant

$$y^{r-1} \lim \frac{1}{n} \mathrm{Tr}\Big(\frac{1}{2} \sum_{j,j'=0}^{q} \lambda_j \lambda_{j'} (P_{j-j'+u} + P_{j-j'+u}^*) \Big)^r = y^{r-1} E_\theta (\cos(u\theta) \tilde{h}(\lambda, \theta))^r,$$

where $\tilde{h}$ is as given in (6.65) and $\theta \sim U(0, 2\pi)$. Hence, the proof of Corollary 6.5.2 is complete. $\qquad \square$

The following corollary will be invoked later in Chapter 8, when we deal with the Stieltjes transform of the LSD of $\hat{\Gamma}_u + \hat{\Gamma}_u^*$ for the MA(0) process. See pages 1208–1209 of Jin et al. [2014] for the detailed expression of $m(z)$.

Corollary 6.5.3. *Suppose (A1) holds and $p, n = n(p) \to \infty$, $p/n \to y > 0$. Then for each $u \geq 1$, LSD of $(2n)^{-1} Z(P_u + P_u^*) Z^*$ exists almost surely and its Stieltjes transform $m(z)$ is given by the solution of*

$$(1 - y^2 m^2(z))(yzm(z) + y - 1)^2 = 1, \quad z \in \mathbb{C}^+. \tag{6.66}$$

Only one solution of (6.66) is a valid Stieltjes transform.

Proof. By Theorem 6.3.1, LSD of $(2n)^{-1} Z(P_u + P_u^*) Z^*$ exists almost surely. To obtain the Stieltjes transform, we now use Corollary 6.5.1. So assume $z \in \mathbb{C}^+$ and $|z|$ large. Note that $(2n)^{-1} Z(P_u + P_u^*) Z^* = \Delta_u$, $\forall u \geq 1$ with

$$\psi_0 = I_p, \quad \psi_j = 0, \ \forall j \geq 1. \tag{6.67}$$

By (6.67) and (6.48), $\eta_0 = 1_{\mathcal{A}_{\mathrm{odd}}}$ (the identity element of $\mathcal{A}_{\mathrm{odd}}$) and $\eta_j = 0$, $\forall j \geq 1$. Therefore, (6.53) reduces to

$$\lambda = 1_{\mathcal{A}_{\mathrm{odd}}}, \quad h(\lambda, \theta) = 1_{\mathcal{A}_{\mathrm{odd}}}. \tag{6.68}$$

By (6.51) and (6.52), we have ($z \in \mathbb{C}^+$, $|z|$ large)

$$\begin{aligned}
K_u(z, \theta) &= \varphi_{\mathrm{odd}}(1_{\mathcal{A}_{\mathrm{odd}}}(B_u(1_{\mathcal{A}_{\mathrm{odd}}}, z) - z)^{-1}) \\
&= \varphi_{\mathrm{odd}}(B_u(1_{\mathcal{A}_{\mathrm{odd}}}, z) - z)^{-1}), \tag{6.69} \\
m_u(z) &= \varphi_{\mathrm{odd}}((B_u(1_{\mathcal{A}_{\mathrm{odd}}}, z) - z)^{-1}). \tag{6.70}
\end{aligned}$$

Therefore, by (6.69) and (6.70), we have ($z \in \mathbb{C}^+$, $|z|$ large)

$$m_u(z) = K_u(z, \theta). \tag{6.71}$$

By (6.54) and (6.71), we have ($z \in \mathbb{C}^+$, $|z|$ large)

$$\begin{aligned}
B_u(1_{\mathcal{A}_{\mathrm{odd}}}, z) &= E_\theta(\cos(u\theta)1_{\mathcal{A}_{\mathrm{odd}}}(1 + y\cos(u\theta)m_u(z))^{-1}) \\
&= E_\theta(\cos(u\theta)(1 + y\cos(u\theta)m_u(z))^{-1})1_{\mathcal{A}_{\mathrm{odd}}} \\
&= \frac{1}{2\pi} \int_0^{2\pi} \frac{\cos(u\theta)}{1 + y\cos(u\theta)m_u(z)} 1_{\mathcal{A}_{\mathrm{odd}}}. \tag{6.72}
\end{aligned}$$

Hence, by (6.70) and (6.72) and, for $z \in \mathbb{C}^+$ and $|z|$ large, the Stieltjes transform of the LSD of $\frac{1}{2n}Z(P_u + P_u^*)Z^*$ satisfies,

$$\begin{aligned}
m_u(z) &= \varphi_{\mathrm{odd}}(B(1_{\mathcal{A}_{\mathrm{odd}}}, z) - z)^{-1} \\
&= -\varphi_{\mathrm{odd}}\Big(z - \frac{1}{2\pi}\int_0^{2\pi} \frac{\cos(u\theta)\, d\theta}{1 + ym_u(z)\cos(u\theta)} 1_{\mathcal{A}_{\mathrm{odd}}}\Big)^{-1} \\
&= -\Big(z - \frac{1}{2\pi}\int_0^{2\pi} \frac{\cos(u\theta)\, d\theta}{1 + ym_u(z)\cos(u\theta)}\Big)^{-1}. \tag{6.73}
\end{aligned}$$

Now by contour integration, it can be shown that

$$\frac{1}{2\pi}\int_0^{2\pi} \frac{\cos(u\theta)\, d\theta}{1 + ym_u(z)\cos(u\theta)} = \frac{1}{ym_u(z)} - \frac{2}{y^2m_u^2(z)}\frac{1}{\omega_1 - \omega_2}$$

where ω_1 and ω_2 are two roots of $\omega^2 + 2(ym_u(z))^{-1}\omega + 1 = 0$ with $|\omega_1| > 1$, $|\omega_2| < 1$ and $(\omega_1 - \omega_2)^{-2} = \frac{y^2m_u^2(z)}{4(1 - y^2m_u^2(z))}$. Therefore, by (6.73), for $z \in \mathbb{C}^+$ and $|z|$ large, we have

$$\begin{aligned}
\frac{-1}{m_u(z)} &= z - \frac{1}{ym_u(z)} + \frac{2(\omega_1 - \omega_2)^{-1}}{y^2m^2(z)} \quad \text{or} \\
1 &= ((1 - y) - zym_u(z))^2(1 - y^2m_u^2(z)).
\end{aligned}$$

Hence, (6.66) is established for $z \in \mathbb{C}^+$ and $|z|$ large. Using analyticity of $m_u(z)$, (6.66) continues to hold for all $z \in \mathbb{C}^+$. $\qquad\square$

Recall that in Theorems 4.3.1 and 4.3.2, we stated the LSD of $n^{-1}ZZ^*$ and $n^{-1}A^{1/2}ZZ^*A^{1/2}$ where A is a $p \times p$ symmetric, non-negative definite matrix. Recall the class $U(\delta)$ in (4.26). Among other things, there we assumed that $\{\varepsilon_{i,j}\} \in U(0)$.

Now if we are willing to work under the stronger assumption (A1) and norm bounded A, then those conclusions follow from Theorems 6.3.1 and 6.4.1. If one carefully follows the proofs of Theorems 4.3.1 and 4.3.2 given in Bai and Silverstein [2009], he/she can see that these are first proved under (A1) and when A is norm bounded. Then to relax these assumptions, appropriate truncations on the entries of Z and on the ESD of A are used. The same truncation arguments can also be used to justify the following two corollaries. Recall the class of matrices $\mathcal{NND}$ in (4.27).

Corollary 6.5.4. *Suppose (A1) holds and $p, n = n(p) \to \infty$, $p/n \to y > 0$. Suppose $\{A_p\} \in \mathcal{NND}$, norm bounded and has LSD F^A. Then the almost sure LSD of $n^{-1}A_p^{1/2}ZZ^*A_p^{1/2}$ is given by (4.37). The same LSD result continues to hold if we relax the norm bounded assumption on A and, instead of (A1), assume $\{\varepsilon_{i,j} : i, j \geq 1\} \in U(0)$.*

Proof. We shall prove only the first part. Since the proof of the second part involves standard truncation as discussed, we shall omit it.

To prove the first part, we again use Corollary 6.5.1. So assume $z \in \mathbb{C}^+$ and $|z|$ large. Note that $n^{-1}A_p^{1/2}ZZ^*A_p^{1/2} = \Delta_0$ with

$$\psi_0 = A_p^{1/2}, \quad \psi_j = 0, \ \forall j \geq 1. \tag{6.74}$$

Suppose a has the distribution F^A. As A_p is symmetric and non-negative definite, $a^{1/2}$ is meaningful. By (6.74) and (6.48), $\eta_0 = a^{1/2}$ and $\eta_j = 0$, $\forall j \geq 1$. Therefore, (6.53) reduces to

$$\lambda = a^{1/2}, \quad h(\lambda, \theta) = a^{1/2}a^{1/2} = a. \tag{6.75}$$

By (6.51) and (6.52), we have for $z \in \mathbb{C}^+$, $|z|$ large (since $u = 0$),

$$
\begin{aligned}
K_0(z, \theta) &= \varphi_{\mathrm{odd}}(a(B_0(a^{1/2}, z) - z)^{-1}) \\
&= \int \frac{a dF^A(a)}{B_0(a^{1/2}, z) - z} = K(z), \text{ say,} \tag{6.76} \\
m_U(z) &\quad \varphi_{\mathrm{odd}}((B_0(a^{1/2}, z) - z)^{-1}) \\
&= \int \frac{dF^A}{B_0(a^{1/2}, z) - z}. \tag{6.77}
\end{aligned}
$$

Now by (6.54) and (6.76), we have for $z \in \mathbb{C}^+$, $|z|$ large (since $u = 0$),

$$
\begin{aligned}
B_0(a^{1/2}, z) &= E_\theta(\cos(0\theta)a(1 + y\cos(0\theta)K(z)^{-1}) \\
&= E_\theta(a(1 + yK(z))^{-1}) = \frac{a}{1 + yK(z)}. \tag{6.78}
\end{aligned}
$$

Hence, by (6.76), (6.77) and (6.78), we have for $z \in \mathbb{C}^+$, $|z|$ large,

$$
\begin{aligned}
zm_0(z) &= \int \frac{z dF^A}{B_0(a^{1/2}, z) - z} = \int \frac{B_0(a^{1/2}, z) dF^A}{B_0(a^{1/2}, z) - z} - 1 \\
&= \frac{1}{1 + yK(z)} \int \frac{a dF^A}{B_0(a^{1/2}, z) - z} - 1 = \frac{K(z)}{1 + yK(z)} - 1. \\
&= \frac{1}{y} \frac{yK(z)}{1 + yK(z)} - 1 = -\frac{1}{y} \frac{1}{1 + yK(z)} + \frac{1}{y} - 1. \quad (6.79)
\end{aligned}
$$

Therefore, by (6.79) and (6.78), we have for $z \in \mathbb{C}^+$, $|z|$ large,

$$
a(zym_0(z) + y - 1) = -\frac{a}{1 + yK(z)} = -B_0(a^{1/2}, z). \quad (6.80)
$$

Now substituting the value of $B_0(a^{1/2}, z)$ obtained in (6.80) into (6.77), we have

$$
m_0(z) = -\int \frac{dF^A}{z - a(zym_0(z) + y - 1)} \quad z \in \mathbb{C}^+, \ |z| \text{ large.}
$$

Therefore, for $z \in \mathbb{C}^+$ and $|z|$ large, (4.37) is proved. Using analyticity, (4.37) continues to hold for all $z \in \mathbb{C}^+$. $\qquad \square$

We now give an alternative free probability proof of Theorem 4.3.1. Recall the Marčenko-Pastur law MP_y with parameter y from Section 4.3.1.

Corollary 6.5.5. *Suppose (A1) holds and $p, n = n(p) \to \infty$, $p/n \to y > 0$. Then the almost sure LSD of $n^{-1}ZZ^*$ exists and it is the MP_y law whose moment sequence and Stieltjes transform respectively satisfy (4.34) and (4.35). The result continues to hold if instead of (A1), we assume $\{\varepsilon_{i,j} : i, j \geq 1\} \in U(0)$.*

Proof. Again we shall prove only the first part. We have already established (4.37) for general A_p in Corollary 6.5.4. Put $A_p^{1/2} = I_p$, where I_p is as in (1.9). Then (4.35) follows immediately.

Next we show (4.34) using Theorem 6.3.1. Let $B_1 = I_p$ and $B_2 = I_n$. Then note that $n^{-1}ZZ^* = n^{-1}B_1 Z B_2 Z^* B_1$. Moreover, $\bar{B}_1 \to a_0, \underline{B}_2 \to c_0$, where a_0 and c_0 are both Bernoulli random variables with success probabilities $y(1 + y)^{-1}$ and $(1 + y)^{-1}$ respectively. Let s be a semi-circle variable and suppose s, a_0 and c_0 are free. Observe that, by (6.22),

$$
\lim \frac{1}{p} E\mathrm{Tr}(n^{-1}ZZ^*)^h = y^{-1}(1 + y)\bar{\varphi}[((1 + y)a_0 s c_0 s a_0)^h], \ \forall h \geq 1.
$$

Hence, by (5.36), the h-th moment of the LSD of $n^{-1}ZZ^*$ is given by

$$
\frac{(1 + y)^{h+1}}{y} \sum_{\pi \in NC(h)} \bar{\varphi}_\pi[a_0^2, \dots, a_0^2] \ \bar{\varphi}_{K(\pi)}[c_0, \dots, c_0]. \quad (6.81)
$$

Note that if $\pi \in NC(h)$ has k blocks, then

$$\begin{aligned}
\bar{\varphi}_\pi[a_0^2, a_0^2, \ldots, a_0^2] &= \bar{\varphi}_\pi[a_0, a_0, \ldots, a_0] = y^k(1+y)^{-k}, \\
\bar{\varphi}_\pi[c_0, c_0, \ldots, c_0] &= (1+y)^{-k}.
\end{aligned}$$

By Property 4 of Kreweras complement in Section 5.2, if $\pi \in NC(h)$ has k blocks then $K(\pi)$ has $(h - k + 1)$ many blocks. Therefore, (6.81) equals

$$\sum_{k=1}^h \#\{\pi \in NC(h) : \pi \text{ has } k \text{ blocks}\}\, y^{k-1} = \sum_{k=1}^h \frac{1}{k}\binom{h-1}{k-1}\binom{h}{k-1}y^{k-1},$$

which is indeed the h-th moment of the Marčenko-Pastur law (see (4.34)). For the last equality see page 144 of Nica and Speicher [2006]. This completes the proof of Corollary 6.5.5. □

For the next corollary, we need the following definition.

Definition 6.5.1. *A random variable X is said to follow a free Bessel(s, t) if its h-th order moment is given by*

$$EX^h = \sum_{k=1}^h \frac{1}{k}\binom{h-1}{k-1}\binom{sh}{k-1}t^k \quad \forall h \geq 1. \tag{6.82}$$

For more details on the importance of this distribution in free probability, see Banica et al. [2011].

Corollary 6.5.6. *Suppose (A1) and $p, n = n(p) \to \infty$, $p/n \to y > 0$ hold. Then for each $u \geq 1$, LSD of $p^{-2}ZP_uZ^*ZP_{-u}Z^*$ exists almost surely and is the free Bessel$(2, y^{-1})$ law whose h-th moment satisfies*

$$\beta_h = \sum_{k=1}^h \frac{1}{k}\binom{h-1}{k-1}\binom{2h}{k-1}y^{-k}, \ h \geq 1. \tag{6.83}$$

Proof. To establish (6.83), we again use Theorem 6.3.1. Let $B_1 = I_p$ and $B_2 = P_u$. Then

$$p^{-2}ZP_uZ^*ZP_{-u}Z^* = (n/p)^2 n^{-2} B_1 Z B_2 Z^* B_1 Z B_2^* Z^* B_1.$$

Note that $\bar{B}_1 \to a_0$ where a_0 is a Bernoulli random variable with success probability $y(1+y)^{-1}$. Also $(\bar{B}_2, \bar{B}_2^*) \to (c, c^*)$, where c and c^* commute and

$$\bar{\varphi}(c^k c^{*l}) = \frac{1}{1+y} I(k = l). \tag{6.84}$$

Let s be the semi-circle variable and suppose s, a_0 and $\{c, c^*\}$ are free. Observe that, by (6.22), for all $h \geq 1$,

$$\lim \frac{1}{p} E \text{Tr}\left(p^{-2}ZP_uZ^*ZP_{-u}Z^*\right)^h = \frac{1+y}{y^{2h+1}}\bar{\varphi}((1+y)^2 a_0 s c s a_0^2 s c s a_0).$$

Recall $NCE(2n)$ from (5.15). By (5.36), the h-th moment of the LSD of $p^{-2}ZP_uZ^*ZP_{-u}Z^*$ is given by

$$\frac{(1+y)^{2h+1}}{y^{2h+1}} \sum_{\pi \in NC(2h)} \bar{\varphi}_{K(\pi)}[a_0, \dots, a_0] \; \bar{\varphi}_\pi[c, c^*, \dots, c, c^*]. \tag{6.85}$$

Note that

$$\bar{\varphi}_\pi[c, c^*, c, c^* \dots, c, c^*] = \begin{cases} 0 & \text{if } \pi \in NC(2h) - NCE(2h) \\ (1+y)^k, & \text{if } \pi \in NCE(2h) \text{ has } k \text{ many blocks.} \end{cases} \tag{6.86}$$

Also note that by Property 4 of Kreweras complement in Section 5.2, $K(\pi)$ has $2h+1-k$ blocks and hence $\bar{\varphi}_{K(\pi)}[a_0, a_0, \dots, a_0] = y^{2h+1-k}(1+y)^{2h+1-k}$. Therefore (6.85) equals

$$y^{-2h} \sum_{k=1}^{h} \#\{\pi \in NCE(2h) : \pi \text{ has } k \text{ blocks}\} \, y^{2h+1-k-1},$$

$$= \sum_{k=1}^{h} \frac{1}{k}\binom{h-1}{k-1}\binom{2h}{k-1} y^{-k},$$

where the last equality follows from Lemma 4.1 of Edelman [1980]. The final expression is indeed the h-th moment of the free Bessel$(2, y^{-1})$ law. This proves Corollary 6.5.6. $\qquad\qquad\qquad\qquad\qquad\qquad\qquad\qquad\qquad\square$

Exercises

1. Establish the bounds in (6.41).
2. Show that $K_u(z, \theta)$ and $B_u(\lambda, z)$ exist for sufficiently large $|z|$.
3. Establish Theorem 6.4.1(b) by observing (6.36).

Chapter 7

GENERALIZED COVARIANCE MATRIX II

In Chapter 6, we discussed the NCP convergence and LSD for the class of matrices $\{\mathbb{P}_{l,(u_{l,1},u_{l,2},\ldots,u_{l,k_l})}\}$ defined in (6.1) when $p, n = n(p) \to \infty$, $p/n \to y > 0$. There we used asymptotic freeness of Wigner and deterministic matrices after embedding matrices of different orders into larger square matrices of the same order.

In this chapter, we are interested in the case where $p, n = n(p) \to \infty$ but $p/n \to 0$. In this case, the embedding technique that we used in Chapter 6, does not work since the growth of p and n are not comparable. At the same time, if we recall the statements of Theorems 4.3.3–4.3.5, we conclude that very different scaling as well as some centering would be needed to get non-degenerate limits.

Taking a cue from these results, define the centered and scaled matrices

$$\mathcal{R}_{l,(u_{l,1},\ldots,u_{l,k_l})} = (n/p)^{1/2}(\mathbb{P}_{l,(u_{l,1},\ldots,u_{l,k_l})} - \mathbb{G}_{l,k_l}), \text{ where} \qquad (7.1)$$

$$\mathbb{G}_{l,k_l} = \Big(\prod_{i=1}^{k_l} n^{-1}\mathrm{Tr}\,(A_{l,2i})\Big)\prod_{i=0}^{k_l} A_{l,2i+1} \qquad (7.2)$$

are the *centering matrices*.

Let

$$\mathcal{V}_p = \mathrm{Span}\{\mathcal{R}_{l,(u_{l,1},\ldots,u_{l,k_l})} : l, k_l \geq 1\}. \qquad (7.3)$$

Note that $\mathcal{V}_p$ forms a $*$-algebra. We shall see that the sequence of NCP $(\mathcal{V}_p, p^{-1}E\mathrm{Tr})$ converges and the limit NCP can be expressed in terms of some free variables. In addition, the LSD of any symmetric polynomial in $\{\mathcal{R}_{l,(u_{l,1},\ldots,u_{l,k_l})}\}$ exists and can be expressed in terms of these free variables. We also derive the Stieltjes transform of these LSD. Finally we then present several applications of these results to specific models. In Chapters 8, 9, and 10, these results will be used for statistical inference in high-dimensional time series.

7.1   Preliminaries

7.1.1   Assumptions

We first list all the assumptions that are required for the convergence of $(\mathcal{V}_p, p^{-1}E\mathrm{Tr})$ as $p, n = n(p) \to \infty$, $p/n \to 0$. Some of these have already appeared in Chapter 6. For convenience of the reader, we state them again.

Let $Z_u = ((\varepsilon_{u,i,j}))_{p \times n}$, $1 \leq u \leq U$ be $p \times n$ independent matrices (see Definition 4.3.1). Therefore, $\{\varepsilon_{u,i,j}\}$ are independently distributed with $E(\varepsilon_{u,i,j}) = 0$, $E(\varepsilon_{u,i,j})^2 = 1$. Recall the classes $\mathcal{L}$ and $C(\delta, p)$ respectively from (4.23) and (4.25). We assume that

(A1) $((\varepsilon_{u,i,j})) \in \mathcal{L} \cup C(\delta, p)$ for some $\delta > 0$ and for all $1 \leq u \leq U$.

Recall that (A1) was also assumed in Chapter 6. It will be weakened later for some corollaries and applications. If there is only one u i.e., if $U = 1$, we will write $\varepsilon_{i,j}$ and Z respectively for $\varepsilon_{1,i,j}$ and Z_1.

Now we move to the assumptions on the deterministic matrices $\{B_i\}$. The following assumption on $\{B_{2i-1}\}$ are borrowed from Chapter 6.

(A2) $\{B_{2i-1} : 1 \leq i \leq K\}$ are norm bounded $p \times p$ matrices and $(\mathrm{Span}(B_{2i-1}, B^*_{2i-1} : 1 \leq i \leq K), p^{-1}\mathrm{Tr})$ converges.

Recall the following notation from (6.13):

$$(\mathrm{Span}\{B_{2i-1}, B^*_{2i-1} : i \leq K\}, p^{-1}\mathrm{Tr}) \to$$
$$(\mathcal{A}_{\mathrm{odd}} = \mathrm{Span}\{b_{2i-1}, b^*_{2i-1} : i \leq K\}, \varphi_{\mathrm{odd}}). \qquad (7.4)$$

For the even indexed matrices $\{B_{2i}\}$, in Chapter 6, we had assumed their joint convergence (see Assumption (A3) there). In the present scenario it suffices to stipulate a more relaxed assumption:

(A3a) $\{B_{2i} : 1 \leq i \leq L\}$ are $n \times n$ matrices with bounded spectral norms. For all $1 \leq i, i' \leq L$, $\epsilon_1, \epsilon_2 = 1$ or $*$, we assume

$$(a) \ -\infty < \lim_{n \to \infty} n^{-1}\mathrm{Tr}(B^{\epsilon_1}_{2i}) < \infty, \quad (b) \ -\infty < \lim_{n \to \infty} n^{-1}\mathrm{Tr}(B^{\epsilon_1}_{2i} B^{\epsilon_2}_{2i'}) < \infty.$$
$$(7.5)$$

Thus, $\{B_{2i}\}$ may not converge jointly. Only moments of polynomials of degree 1 and 2 are assumed to converge.

7.1.2   Centering and Scaling

To see the necessity of the appropriate centering and scaling on matrix polynomials, let us consider the following example.

Example 1. Let $H = n^{-1}A_1 Z_1 A_2 Z^*_1 A^*_1$, where $A_1, A^*_1 \in \{B_{2i-1}, B^*_{2i-1}\}$, $A_2 \in \{B_{2i}, B^*_{2i}\}$ and $A_2 = A^*_2$. Recall the convergence in (7.4). Let $\{a_1, a^*_1\} \in \{b_{2i-1}, b^*_{2i-1}\}$ denote the limits of $\{A_1, A^*_1\}$. Let

$$d_0 = \lim n^{-1}\mathrm{Tr}(A_2). \qquad (7.6)$$

By (A3a), the right side of (7.6) exists and is finite. Using simple algebra, under (A1), it is easy to see that

$$\lim \frac{1}{p} E\mathrm{Tr}(H) \;=\; \lim \frac{1}{p}\mathrm{Tr}(A_1 A_1^*)\,\lim \frac{1}{n}\mathrm{Tr}(A_2)$$

$$=\; \varphi_{\mathrm{odd}}(d_0 a_1 a_1^*),\ \text{by (A2), (A3a) and (7.6)},\qquad (7.7)$$

and

$$\lim \frac{1}{p} E\mathrm{Tr}(H^2) \;=\; \lim \frac{1}{p}\mathrm{Tr}(A_1 A_1^*)^2 \left(\lim \frac{1}{n}\mathrm{Tr}(A_2)\right)^2$$

$$+\lim \frac{p}{n}\left(\lim \frac{1}{p}\mathrm{Tr}(A_1 A_1^*)\right)^2 \lim \frac{1}{n}\mathrm{Tr}(A_2^2)$$

$$=\; \varphi_{\mathrm{odd}}[(d_0 a_1 a_1^*)^2],$$
$$\text{by (A2), (A3a) and (7.6) and, as } p/n \to 0.\quad (7.8)$$

Similarly, under (A1), (A2), (A3a) and if $p/n \to 0$, we have

$$\lim \frac{1}{p} E\mathrm{Tr}(H^h) = \varphi_{\mathrm{odd}}[(d_0 a_1 a_1^*)^h],\ \forall h > 2. \qquad (7.9)$$

Therefore, H converges to $d_0 a_1 a_1^*$. Consider the matrix $G = n^{-1}\mathrm{Tr}(A_2)A_1 A_1^*$. Note that by (A2) and (A3a), G also converges to $d_0 a_1 a_1^*$. Therefore, there is no contribution of the random matrix Z_1 in the limit of H. This is not desirable because such results cannot be used in any statistical application.

To get a non-trivial limit of H, we need appropriate centering and scaling. Since $G \to d_0 a_1 a_1^*$, by (7.7), the appropriate centering for H is G. To find the suitable scaling, consider the following computation.

$$\lim p^{-1} E\mathrm{Tr}((H - G)^2) \;=\; \lim p^{-1} E\mathrm{Tr}(H^2)$$
$$+ \lim p^{-1}\mathrm{Tr}(G^2) - 2\lim p^{-1} E\mathrm{Tr}(HG). \quad (7.10)$$

Now, under (A1), it is easy to see that

$$\frac{1}{p} E\mathrm{Tr}(H^2) \;=\; \left(\frac{1}{p}\mathrm{Tr}(A_1 A_1^*)^2\right)\left(\frac{1}{n}\mathrm{Tr}(A_2)\right)^2$$

$$+ \frac{p}{n}\left(\frac{1}{p}\mathrm{Tr}(A_1 A_1^*)\right)^2 \frac{1}{n}\mathrm{Tr}(A_2^2) + O(1/n),$$

$$p^{-1}\mathrm{Tr}(G^2) \;=\; \left(\frac{1}{n}\mathrm{Tr}(A_2)\right)^2 \frac{1}{p}\mathrm{Tr}(A_1 A_1^*)^2,$$

$$p^{-1}\mathrm{Tr}(HG) \;=\; \left(\frac{1}{n}\mathrm{Tr}(A_2)\right)^2 \frac{1}{p}\mathrm{Tr}(A_1 A_1^*)^2.$$

$$(7.11)$$

Therefore, by (7.10), we have

$$p^{-1} E\mathrm{Tr}((H - G)^2) = \frac{p}{n}\left(\frac{1}{p}\mathrm{Tr}(A_1 A_1^*)\right)^2 \frac{1}{n}\mathrm{Tr}(A_2^2) + O(n^{-1}). \qquad (7.12)$$

Hence, an appropriate scaling for $(H - G)$ is $\sqrt{np^{-1}}$ and under (A1), (A2), (A3a) and $p/n \to 0$, we have

$$\lim p^{-1} E\mathrm{Tr}(\sqrt{np^{-1}}(H - G))^2 = d_1 \varphi_{\mathrm{odd}}[(a_1 a_1^*)^2], \quad \text{where} \quad d_1 = \lim \frac{1}{n}\mathrm{Tr}(A_2^2).$$

Moreover, one can easily see that

$$\lim \frac{1}{p} E\mathrm{Tr}(\sqrt{np^{-1}}(H - G))^4 = 2d_1^2 \varphi_{\mathrm{odd}}[(a_1 a_1^*)^2](\varphi_{\mathrm{odd}}(a_1 a_1^*))^2. \quad (7.13)$$

Therefore, the limit of $\sqrt{np^{-1}}(H - G)$ is not trivial. A precise description of the limit will emerge from the next example where we shall identify the contribution of Z_1 in the limit via freeness.

7.1.3 Main idea

To see how freeness comes into the picture and hence how it motivates the limiting NCP of $(\mathcal{V}_p, p^{-1}E\mathrm{Tr})$, let us consider the following example.

Example 2. Consider the following four polynomials

$$\pi_1 = \sqrt{np^{-1}}\,(S_1 - G_1)\,, \quad \pi_2 = \sqrt{np^{-1}}\,(S_2 - G_2)\,, \quad (7.14)$$

$$\pi_3 = \sqrt{np^{-1}}\,(S_3 - G_3)\,, \quad \pi_4 = \sqrt{np^{-1}}(S_1 S_2 - G_1 G_2) \quad (7.15)$$

where

$$\begin{aligned} S_1 &= n^{-1} A_1 Z_1 A_2 Z_1^* A_3, \ S_2 = n^{-1} A_1 Z_2 A_2 Z_2^* A_3, \\ S_3 &= n^{-1} A_5 Z_1 A_6 Z_1^* A_7, \quad (7.16) \\ G_1 &= n^{-1}\mathrm{Tr}(A_2) A_1 A_3, \ G_2 = n^{-1}\mathrm{Tr}(A_2) A_1 A_3, \\ G_3 &= n^{-1}\mathrm{Tr}(A_6) A_5 A_7, \quad (7.17) \end{aligned}$$

and $A_1, A_3, A_5, A_7 \in \{B_{2i-1}, B_{2i-1}^* : i \geq 1\}$, $A_2, A_6 \in \{B_{2i}, B_{2i}^* : i \geq 1\}$. Suppose $\{A_i\}$ are norm bounded matrices. Now we investigate the following.

1. Convergence of $(\mathrm{Span}(\pi_1, \pi_1^*), p^{-1}E\mathrm{Tr})$: to see how the limit of a single polynomial π_1 can be expressed in terms of free variables.

2. Joint convergence of $(\mathrm{Span}\{\pi_1, \pi_2, \pi_3, \pi_1^*, \pi_2^*, \pi_3^*\}, p^{-1}E\mathrm{Tr})$: to see how several independent Z_u matrices interact in the limit.

3. Convergence of $(\mathrm{Span}(\pi_4, \pi_4^*), p^{-1}E\mathrm{Tr})$: to see how the limit of a polynomial involving more that one (Z, Z^*) pair can be described.

1. Convergence of $(\mathrm{Span}(\pi_1, \pi_1^*), p^{-1}E\mathrm{Tr})$. As discussed in Definition 5.1.4, convergence of π_1 is equivalent to the convergence of $p^{-1}E\mathrm{Tr}(\Pi(\pi_1, \pi_1^*))$ for all polynomials Π. Using simple matrix algebra, under (A1), one can easily see that

$$\lim \frac{1}{p} E\mathrm{Tr}(\pi_1) = 0,$$

$$\lim \frac{1}{p} E\mathrm{Tr}(\pi_1^2) = \lim \frac{1}{n} \mathrm{Tr}(A_2^2) \left(\lim \frac{1}{p}\mathrm{Tr}(A_1 A_3)\right)^2, \qquad (7.18)$$

$$\lim \frac{1}{p} E\mathrm{Tr}(\pi_1 \pi_1^*) = \lim \frac{1}{n} \mathrm{Tr}(A_2 A_2^*) \lim \frac{1}{p}\mathrm{Tr}(A_1 A_1^*)$$

$$\times \lim \frac{1}{p}\mathrm{Tr}(A_3 A_3^*). \qquad (7.19)$$

All the above limits are finite once we use Assumptions (A2) and (A3a).

Recall (7.4). Let $\{a_1, a_3, a_1^*, a_3^*\} \in \mathrm{Span}\ \{b_{2i-1}, b_{2i-1}^* : i \geq 1\}$ denote the limit of $\{A_1, A_3, A_1^*, A_3^*\}$. Also by (A3a),

$$\lim n^{-1}\mathrm{Tr}(A_2^{\epsilon_1} A_2^{\epsilon_2}) < \infty, \ \forall \epsilon_1, \epsilon_2 = 1, *. \qquad (7.20)$$

Recall the free cumulant κ_l of order l in (5.22). By enlarging the NCP of $\{b_{2i-1}, b_{2i-1}^*\}$ if necessary, let w_1 be a variable which is free of $\{a_1, a_3, a_1^*, a_3^*\}$ and whose all marginal free cumulants of order greater than two are 0 and the first two free cumulants satisfy

$$\kappa_1(w_1^{\epsilon_1}) = 0, \ k_2(w_1^{\epsilon_1}, w_1^{\epsilon_2}) = \lim n^{-1}\mathrm{Tr}(A_2^{\epsilon_1} A_2^{\epsilon_2}), \ \forall \epsilon_1, \epsilon_2 = 1, *. \qquad (7.21)$$

Denote the state of the above enlarged NCP by φ_0. Therefore, the restriction of φ_0 on $\{b_{2i-1}, b_{2i-1}^*\}$ is φ_{odd}. Using the algorithm for computing moments of free variables given in Section 5.4, one can easily see that

$$\varphi_0(a_1 w_1 a_3) = 0, \ \text{by (5.34), and} \qquad (7.22)$$

$$\varphi_0[(a_1 w_1 a_3)^2] = (\varphi_{\mathrm{odd}}(a_1 a_3))^2 k_2(w_1, w_1) \ \text{by (5.34)}$$

$$= (\varphi_{\mathrm{odd}}(a_1 a_3))^2 \lim \frac{1}{n}\mathrm{Tr}(A_2^2), \ \text{by (7.21)}$$

$$= \lim \frac{1}{n}\mathrm{Tr}(A_2^2) \left(\lim \frac{1}{p}\mathrm{Tr}(A_1 A_3)\right)^2,$$

$$\varphi(a_1 w_1 a_3 a_3^* w_1^* a_1^*) = \varphi_{\mathrm{odd}}(a_1 a_1^*)\varphi_{\mathrm{odd}}(a_3 a_3^*)k_2(w_1, w_1^*), \ \text{by (5.34)}$$

$$= \varphi_{\mathrm{odd}}(a_1 a_1^*) \ \varphi_{\mathrm{odd}}(a_3 a_3^*) \lim \frac{1}{n}\mathrm{Tr}(A_2 A_2^*)$$

$$= \lim \frac{1}{n}\mathrm{Tr}(A_2 A_2^*) \lim \frac{1}{p}\mathrm{Tr}(A_1 A_1^*)$$

$$\times \lim \frac{1}{p}\mathrm{Tr}(A_3 A_3^*). \qquad (7.23)$$

Therefore by (7.18), (7.19), (7.22), and (7.23),

$$\lim \frac{1}{p} E\mathrm{Tr}(\pi_1) = \varphi_0(a_1 w_1 a_3), \qquad (7.24)$$

$$\lim p^{-1} E\mathrm{Tr}(\pi_1^2) = \varphi_0[(a_1 w_1 a_3)^2] \ \text{and} \qquad (7.25)$$

$$\lim p^{-1} E\mathrm{Tr}(\pi_1 \pi_1^*) = \varphi_0((a_1 w_1 a_3)(a_1 w_1 a_3)^*). \qquad (7.26)$$

Similarly, one can show that, for $T \geq 1$ and $\epsilon_1, \ldots, \epsilon_T = 1, *$, we have

$$\lim p^{-1} E\text{Tr}(\pi_1^{\epsilon_1} \cdots \pi_T^{\epsilon_T}) = \varphi_0((a_1 w_1 a_3)^{\epsilon_1} \cdots (a_1 w_1 a_3)^{\epsilon_T}). \qquad (7.27)$$

Therefore,

$$(\text{Span}\{\pi_1, \pi_1^*\}, \frac{1}{p} E\text{Tr}) \to (\text{Span}\{\alpha_1, \alpha_1^*\}, \varphi_0), \text{ where } \alpha_1 = a_1 w_1 a_3. \qquad (7.28)$$

Note that the right side of the above equations (7.24), (7.25), and (7.27) can be in principle computed by using the distribution of $\{a_1, a_3, a_1^*, a_3^*\}$, the distribution of w_1 and, the freeness of these two collections.

Then under (A1), (A2), and (A3a), one can similarly see that,

$$(\text{Span}\{\pi_2, \pi_2^*\}, p^{-1} E\text{Tr}) \quad \to \quad (\text{Span}\{\alpha_2, \alpha_2^*\}, \varphi_0), \quad \alpha_2 := a_1 w_2 a_3, \qquad (7.29)$$

$$(\text{Span}\{\pi_2, \pi_2^*\}, p^{-1} E\text{Tr}) \quad \to \quad (\text{Span}\{\alpha_3, \alpha_3^*\}, \varphi_0), \quad \alpha_3 := a_5 w_3 a_7 \qquad (7.30)$$

where $\{a_5, a_7, a_5^*, a_7^*\}$ is the limit of $\{A_5, A_7, A_5^*, A_7^*\}$ and w_2, w_3 have exactly the same free cumulants given in (7.21) as w_1 except A_2 is replaced by A_6 for w_3. We are yet to unearth the relation between w_1, w_2 and w_3. This is done next.

2. Joint convergence of $(\text{Span}\{\pi_1, \pi_2, \pi_3, \pi_1^*, \pi_2^*, \pi_3^*\}, p^{-1} E\text{Tr})$. Suppose the marginal cumulants of $\{w_1, w_2, w_3\}$ are as before and the joint cumulants are as follows.

$$\begin{aligned} \kappa_r(w_{i_1}^{\epsilon_1}, \ldots, w_{i_r}^{\epsilon_r}) &= 0, \ \forall r > 2, \ i_1, \ldots i_r = 1, 2, 3, \ \epsilon_1, \ldots, \epsilon_r = 1, *. \\ \kappa_2(w_1^{\epsilon_1}, w_2^{\epsilon_2}) &= \kappa_2(w_2^{\epsilon_1}, w_3^{\epsilon_2}) = 0 \ \text{ and} \\ \kappa_2(w_1^{\epsilon_1}, w_3^{\epsilon_2}) &= \lim n^{-1} \text{Tr}(A_2^{\epsilon_1} A_6^{\epsilon_2}). \end{aligned} \qquad (7.31)$$

Using arguments similar to those in the marginal cases, under (A1), (A2), and (A3a), one can show that

$$\begin{aligned} \lim \frac{1}{p} E\text{Tr}(\pi_1 \pi_2) &= \lim \frac{1}{p} E\text{Tr}(\pi_2 \pi_3) = 0 = \varphi_0(\alpha_1 \alpha_2) = \varphi_0(\alpha_2 \alpha_3), \\ \lim \frac{1}{p} E\text{Tr}(\pi_1 \pi_3) &= \lim \frac{1}{n} \text{Tr}(A_2 A_6) \lim \frac{1}{p} \text{Tr}(A_1 A_7) \lim \frac{1}{p} \text{Tr}(A_3 A_5) \\ &= \varphi_0(\alpha_1 \alpha_3). \end{aligned}$$

Moreover, one can indeed show that for $T \geq 1$, $\epsilon_1, \epsilon_2, \ldots, \epsilon_T = 1, *$,

$$\frac{1}{p} E\text{Tr}(\pi_{i_1}^{\epsilon_1} \pi_{i_2}^{\epsilon_2} \cdots \pi_{i_T}^{\epsilon_t}) = \varphi_0(\alpha_{i_1}^{\epsilon_1} \alpha_{i_2}^{\epsilon_2} \cdots \alpha_{i_T}^{\epsilon_T}), \ i_1, i_2, \ldots \in \{1, 2, 3\}. \qquad (7.32)$$

Hence,

$$(\text{Span}\{\pi_1, \pi_2, \pi_3, \pi_1^*, \pi_2^*, \pi_3^*\}, \frac{1}{p} E\text{Tr}) \to (\text{Span}\{\alpha_1, \alpha_2, \alpha_3, \alpha_1^*, \alpha_2^*, \alpha_3^*\}, \varphi_0)$$

$$(7.33)$$

where $\{a_1, a_3, a_5\}$, $\{w_1, w_2, w_3\}$ are free and the joint free cumulants of the latter are as in (7.31).

3. Convergence of $(\mathrm{Span}(\pi_4), p^{-1}E\mathrm{Tr})$. Let

$$g_1 = \lim n^{-1}\mathrm{Tr}(A_2)a_1a_3, \quad g_2 = \lim n^{-1}\mathrm{Tr}(A_6)a_5a_7.$$

Note that

$$(\mathrm{Span}\{G_1, G_2, G_1^*, G_2^*\}, \frac{1}{p}E\mathrm{Tr}) \to (\mathrm{Span}\{g_1, g_2, g_1^*, g_2^*\}, \varphi_{\mathrm{odd}}). \tag{7.34}$$

Recall π_1 and π_3 respectively from (7.14) and (7.15). To understand the convergence of π_4, observe that

$$\pi_4 = \pi_1 G_2 + G_1\pi_2 + \sqrt{pn^{-1}}\pi_1\pi_2. \tag{7.35}$$

Hence, by the previous example,

$$
\begin{aligned}
\pi_4 \quad &\to \quad \alpha_1 g_2 + g_1\alpha_2 + 0.\alpha_1\alpha_2 \text{ (since } p/n \to 0) \\
&= \quad a_1 w_1 a_3 g_2 + g_1 a_5 w_3 a_7 \\
&= \quad a_1 w_1 a_3 \lim \frac{1}{n}\mathrm{Tr}(A_6)\, a_5 a_7 + \lim \frac{1}{n}\mathrm{Tr}(A_2)\, a_1 a_3 a_5 w_3 a_7 \quad (7.36)
\end{aligned}
$$

where (w_1, w_3) are as in the previous example.

It is to be noted that even though we started with one random matrix Z, in (7.36) we ended up with two w_i variables. The expression in (7.36) can be visualized as follows. Ignoring centering and scaling, consider the approximate relation

$$\pi_4 \approx S_1 S_2 \approx A_1(Z_1 A_2 Z_1^*)A_3 A_5 (Z_1 A_6 Z_1^*)A_7. \tag{7.37}$$

Each pair (Z_i, Z_i^*) gives rise to a w variable. For example, the first pair gives a w_1 and it contributes $a_1 w_1 a_3 \lim n^{-1}\mathrm{Tr}(A_6)a_5a_7$. Similarly, the second pair gives a w_3 and it contributes $\lim n^{-1}\mathrm{Tr}(A_2)a_1a_3a_5w_3a_7$. Then the limit is the *sum* of these two (see (7.36)). Later we shall refer to variables on the left and right of any w as c and $\tilde{c}$ variables respectively.

From the above example it is intuitively apparent why the centered polynomials $\{\sqrt{np^{-1}}(\mathbb{P} - \mathbb{G}) = \mathcal{R}\}$ should converge jointly and what their limits would be. We make these ideas precise in the next section.

7.2   NCP convergence

To describe the limit, define a family of variables $\mathcal{T} = \{w_{u,l,i} : u, l, i \geq 1\}$ (note that $w_{u,l,i}$ is attached to the matrix $A_{l,2i}$ and Z_u-index u), where for all $l_j, u_j, i_j \geq 1$, $\epsilon_j = 1, *, \forall j \geq 1$,

$$\kappa_r(w_{u_j, l_j, i_j}^{\epsilon_j} : j \leq r) = \begin{cases} \lim \frac{1}{n}\mathrm{Tr}(A_{l_1, 2i_1}^{\epsilon_1}, A_{l_2, 2i_2}^{\epsilon_2}), & \text{if } r = 2 \text{ and } u_1 = u_2 \\ 0, & \text{if } r \neq 2 \text{ or } u_1 \neq u_2, \end{cases}$$

and they are free of $\{b_{2i-1}, b_{2i-1}^* : 1 \leq i \leq K\}$. That is, $\mathcal{A}_u = \{w_{u,l,i} : l, i \geq 1\}$, $1 \leq u \leq U$ are free. The above sequence of free cumulants naturally defines a state on $\text{Span}\{w_{u,l,i} : u, l, i \geq 1\}$, say φ_w.

Two special cases are worth mentioning. Recall Definitions 5.3.1 and 5.3.2 of the semi-circle family and the circular family of non-commutative variables. If B_{2i}, $i \geq 1$ are self-adjoint, then each $w_{u,l,j}$ can be taken to be self-adjoint and $\mathcal{T}$ would be a semi-circle family. On the other hand, if $\lim n^{-1}\text{Tr}(B_{2i}^2) = \lim n^{-1}\text{Tr}(B_{2i}^{*2}) = 0$, $\forall i \geq 1$, then $\mathcal{T}$ would be a circular family.

Now we formalize the definition of the left and the right variables c and $\tilde{c}$. Motivated by the ideas given towards the end of Section 7.1.3, in general, for $\mathcal{V}_p$, let for all $l \geq 1$ and $1 \leq j \leq k_l$,

$$c_{l,j} = \prod_{i=1}^{j-1} \lim \frac{1}{n}\text{Tr}(A_{l,2i}) \prod_{i=0}^{j-1} a_{l,2i+1}, \tag{7.38}$$

$$\tilde{c}_{l,j} = \prod_{i=j+1}^{k_l} \lim \frac{1}{n}\text{Tr}(A_{l,2i}) \prod_{i=j}^{k_l} a_{l,2i+1}, \tag{7.39}$$

$$\alpha_{l,(u_{l,1}, u_{l,2}, \ldots, u_{l,k_l})} = \sum_{j=1}^{k_l} c_{l,j} w_{u_{l,j}, l, j} \tilde{c}_{l,j}. \tag{7.40}$$

Recall the NCP $(\mathcal{A}_{\text{odd}}, \varphi_{\text{odd}})$ from (6.13). Let,

$$(\mathcal{B}, \varphi_0) = \text{free product of } (\mathcal{A}_{\text{odd}}, \varphi_{\text{odd}}) \text{ and } (\text{Span}\{w_{u,l,i} : u, l, i \geq 1\}, \varphi_w). \tag{7.41}$$

Consider the following $*$-sub-algebra of $\mathcal{B}$,

$$\mathcal{V} = \text{Span}\big(\alpha_{l,(u_{l,1}, u_{l,2}, \ldots, u_{l,k_l})} : u_{l,j} \geq 1, l \geq 1\big).$$

Now we have all the ingredients to state the joint convergence theorem. Proof of this theorem is very technical. We provide the proof later in Section A.3 of the Appendix. This result is the cornerstone to obtain the LSD of symmetric polynomials in $\{\mathcal{R}_{l,(u_{l,1}, \ldots, u_{l,k_l})}\}$ in the next section.

Theorem 7.2.1. *(Bhattacharjee and Bose [2016c])) Suppose Assumptions (A1), (A2), and (A3a) hold and $p, n = n(p) \to \infty$, $p/n \to 0$. Then*

(a) $(\mathcal{V}_p, Ep^{-1}\text{Tr}) \to (\mathcal{V}, \varphi_0)$.

(b) $(\text{Span}\{\sqrt{np^{-1}}(n^{-1}Z_{u_j}B_{2j}Z_{u_j}^* - n^{-1}\text{Tr}(B_{2j})) : u_j \geq 1, j \geq 1\}, p^{-1}E\text{Tr})$ and $(\text{Span}(B_{2i-1} : 1 \leq i \leq K), p^{-1}\text{Tr})$ are asymptotically free.

7.3　LSD of symmetric polynomials

The following theorem guarantees the existence of the LSD of any symmetric polynomial in $\{\mathcal{R}_{l,(u_{l,1}, u_{l,2}, \ldots, u_{l,k_l})}\}$.

Theorem 7.3.1. *(Bhattacharjee and Bose [2016c]) Suppose Assumptions (A1), (A2), and (A3a) hold and $p, n = n(p) \to \infty$, $p/n \to 0$. Then the LSD of any self-adjoint polynomial $\mathbb{P}(\mathcal{R}_{l,(u_{l,1}, u_{l,2}, \ldots, u_{l,k_l})} : 1 \le l \le r)$ in $\mathcal{V}_p$ exists with probability 1 and it is given by $\mathbb{P}(\alpha_{l,(u_{l,1}, u_{l,2}, \ldots, u_{l,k_l})} : 1 \le l \le r)$.*

Proof. To prove the theorem, by Lemma 4.1.2, we need to establish the conditions (M1), (M4), and (C) as described in the moment method. The (M1) condition is immediate from Theorem 7.2.1. Proof of (M4) and (C) go along the same lines as the proof of (M4) and (C) in the proof Theorem 6.3.1. We omit the similar and tedious technical details. Hence, the proof of Theorem 7.3.1 is complete. □

7.4 Stieltjes transform

By utilizing Assumptions (A2) and (A3a), we can verify that the self-adjoint elements in $\mathcal{V}$ have moments with nice bounds. Hence, they uniquely define proper probability distributions of usual bounded random variables. In principle we know how to calculate the moments of these variables. On the other hand, as we have seen, the LSD results in the literature are mostly in terms of Stieltjes transform. To show how these are linked, we now establish a general Stieltjes transform result. Let

$$\gamma = \sum_{j=1}^{r}(a_j w_j c_j + c_j^* w_j^* a_j^*). \tag{7.42}$$

Here $\{w_j, w_j^*\}$ is a family of non-commutative variables which satisfy

$$\kappa_r(w_{j_i}^{\epsilon_i} : 1 \le i \le r) = \begin{cases} b_{j_1, j_2, \epsilon_1, \epsilon_2}, & \text{if } r = 2 \\ 0, & \text{if } r \ne 2, \end{cases} \tag{7.43}$$

for all $j_i \ge 1$, $\epsilon_i = 1, *$, $i \ge 1$. The variables $\{a_j, c_j\} \subset \{b_{2i-1}\}$ and $\{a_j\}$ is some permutation of $\{c_j\}$. Further $\{w_j, w_j^*, j \le r\}$ and $\{a_j, a_j^*, j \le r\}$ are free. Recall the state φ_0 in (7.41), whose restriction on $\mathcal{A}_{odd}$ is φ_{odd}.

The general Stieltjes transform formula given below is quite messy. However, the formulae will be significantly simplified in special cases of interest to us. We shall deal with some special cases in the next section.

Theorem 7.4.1. *For $z \in \mathbb{C}^+$, $|z|$ large, the Stieltjes transform of γ is given by*

$$\begin{aligned} m_\gamma(z) &= -\varphi_0((z + \beta(z, a))^{-1}), & (7.44) \\ &= -\varphi_{odd}((z + \beta(z, a))^{-1}) & (7.45) \end{aligned}$$

where,

$$(z + \beta(z, a))^{-1} = z^{-1} \sum_{i=0}^{\infty} z^{-i}(-\beta(z, a))^i$$

and $\beta(z, a)$ is given by

$$\beta(z, a) = -\sum_{j_1, j_2 = 1}^{r} \left[b_{j_1, j_2, 1, 1} c_{j_1} a_{j_2} \varphi_0 \left(\frac{a_{j_1} c_{j_2}}{z + \beta(z, a)} \right) \right.$$

$$+ b_{j_1, j_2, 1, *} c_{j_1} c_{j_2}^* \varphi_0 \left(\frac{a_{j_1} a_{j_2}^*}{z + \beta(z, a)} \right)$$

$$- b_{j_1, j_2, *, 1} a_{j_1}^* a_{j_2} \varphi_0 \left(\frac{c_{j_1}^* c_{j_2}}{z + \beta(z, a)} \right)$$

$$\left. + b_{j_1, j_2, *, *} a_{j_1}^* c_{j_2}^* \varphi_0 \left(\frac{c_{j_1}^* a_{j_2}^*}{z + \beta(z, a)} \right) \right]$$

$$= \quad same\ expression\ with\ \varphi_0\ replaced\ by\ \varphi_{\text{odd}}.$$

Using the same arguments as in Lemma 6.4.1, it is easy to see that the power series above are all meaningful for large $|z|$.

Proof of Theorem 7.4.1. Throughout, $|z|$ is assumed to be sufficiently large for any relevant expression to be meaningful.

For all $i \geq 1$, define

$$R_i = \sum_{j_1, j_2 = 1}^{r} \left[b_{j_1, j_2, 1, 1} c_{j_1} a_{j_2} \varphi_0 (a_{j_1} c_{j_2} \gamma^{i-1}) + b_{j_1, j_2, 1, *} c_{j_1} c_{j_2}^* \varphi_0 (a_{j_1} a_{j_2}^* \gamma^{i-1}) \right.$$

$$\left. + b_{j_1, j_2, *, 1} a_{j_1}^* a_{j_2} \varphi_0 (c_{j_1}^* c_{j_2} \gamma^{i-1}) + b_{j_1, j_2, *, *} a_{j_1}^* c_{j_2}^* \varphi_0 (c_{j_1}^* a_{j_2}^* \gamma^{i-1}) \right], \quad (7.46)$$

$$\beta(z, a) = -\sum_{i=1}^{\infty} z^{-i} R_i. \quad (7.47)$$

Note that $\varphi_0(\gamma^{2h-1}) = 0$ and $\varphi_0(R_{2h}) = 0$, $\forall h \geq 1$. By (5.43), we have

$$\varphi_0(\gamma^{2h}) = \sum_{\epsilon_1, \ldots, \epsilon_{2h}} \sum_{j_1, \ldots, j_{2h}} \varphi_0 \left(\prod_{k=1}^{2h} a_{j_k}^{\epsilon_k} w_{j_k}^{\epsilon_k} c_{j_k}^{\epsilon_k} \right)$$

$$= \sum_{\substack{\epsilon_1, \ldots, \epsilon_{2h} \\ j_1, \ldots, j_{2h}}} \sum_{\pi \in NC_2(2h)} \varphi_{0K(\pi)} (c_{j_1} a_{j_2}, c_{j_2} a_{j_3}, \ldots, c_{j_{2h}} a_{j_1})$$

$$\times k_\pi (w_{j_1}, \ldots, w_{j_{2h}}). \quad (7.48)$$

For any subset A of $NC(n)$, by contribution of A in $\varphi_0(\gamma^{2h})$, we mean

$$\sum_{\substack{\epsilon_1, \ldots, \epsilon_{2h} \\ j_1, \ldots, j_{2h}}} \sum_{\pi \in A} \varphi_{0K(\pi)} (c_{j_1} a_{j_2}, c_{j_2} a_{j_3}, \ldots, c_{j_{2h}} a_{j_1}) \, k_\pi (w_{j_1}, \ldots, w_{j_{2h}}).$$

To simplify (7.48), consider the following decomposition of $NC_2(2h)$.

$$NC_2(2h) = \cup_{i=1}^{h} C_{i,h}, \text{ where}$$

$$C_{i,h} = \text{set of all } \sigma \in NC_2(2h) \text{ such that } \{1, 2i\} \in \sigma.$$

Note that the contribution of $\{\{1, 2h\}, \{2, 3\}, \{4, 5\}, \ldots, \{2h - 2, 2h - 1\}\} \in \mathcal{C}_{h,h}$ to right side of (7.48), is $\varphi_0(R_1^h)$. Now,

$$\varphi_0(\gamma^2) \quad = \quad \text{contribution of } \mathcal{C}_{1,1} \text{ in } \varphi_0(\gamma^2) = \varphi_0(R_1).$$

Again,

$$\begin{aligned}
\varphi_0(\gamma^4) \quad &= \quad \text{contribution of } \mathcal{C}_{1,2} \text{ in } \varphi_0(\gamma^4) + \text{contribution of } \mathcal{C}_{2,2} \text{ in } \varphi_0(\gamma^4) \\
&= \quad \varphi_0(R_3 + R_1^2).
\end{aligned}$$

Next,

$$\begin{aligned}
\varphi_0(\gamma^6) \quad &= \quad \text{contribution of } \mathcal{C}_{1,3} \text{ in } \varphi_0(\gamma^6) + \text{contribution of } \mathcal{C}_{2,3} \text{ in } \varphi_0(\gamma^6) \\
&\qquad + \text{contribution of } \mathcal{C}_{3,3} \text{ in } \varphi_0(\gamma^6) \\
&= \quad \varphi_0(R_5 + R_1 R_3 + (R_3 R_1 + R_1^3)).
\end{aligned}$$

Now, let us define the set of all ordered partitions of the integer K into t blocks as follows.

$$S_{K,t} = \{(i_1, \ldots, i_t) : i_1, \ldots, i_t \in \mathbb{N}, \sum_{j=1}^{t} i_j = K\}, \ 1 \leq t \leq K.$$

Then, one can show easily by induction on h that

$$\varphi_0(\gamma^{2h}) = \varphi_0\Big(\sum_{t=1}^{h} \sum_{i_1, \ldots, i_t \in S_{2h-t,t}} \prod_{j=1}^{t} R_{i_j} \Big), \forall h \geq 1.$$

We omit the tedious details. Hence, using the power series expansion (4.16) for Stieltjes transform, we have

$$\begin{aligned}
m_\gamma(z) \quad &= \quad \varphi_0((\gamma - z)^{-1}) = -z^{-1} \sum_{h=0}^{\infty} z^{-2h} \varphi_0(\gamma^{2h}) \qquad (7.49) \\
&= \quad -\varphi_0\Big(z^{-1} \sum_{h=0}^{\infty} z^{-2h} \sum_{t=1}^{h} \sum_{i_1, i_2, \ldots, i_t \in S_{2h-t,t}} \prod_{j=1}^{t} R_{i_j} \Big) \\
&= \quad -\varphi_0\Big(z^{-1} \sum_{t=0}^{\infty} z^{-t} \sum_{h=t}^{\infty} \sum_{i_1, i_2, \ldots, i_t \in S_{2h-t,t}} \prod_{j=1}^{t} z^{-i_j} R_{i_j} \Big) \\
&= \quad -\varphi_0\Big(z^{-1} \sum_{t=0}^{\infty} z^{-t} \Big(\sum_{i=1}^{\infty} z^{-i} R_i \Big)^t \Big) \\
&= \quad -\varphi_0\Big(z^{-1} \sum_{t=0}^{\infty} z^{-t} (-\beta(z, a))^t \Big) \\
&= \quad -\varphi_0((z + \beta(z, a))^{-1}). \qquad (7.50)
\end{aligned}$$

Similarly, one can easily show by induction on h and the assumption $\{a_j : j \leq r\} = \{c_j : j \leq r\}$, that

$$
\begin{aligned}
R_{2h+1} = \sum_{j_1,j_2=1}^{r} \Bigg[& b_{j_1,j_2,1,1} a_{j_1} c_{j_2} \varphi_0\big(c_{j_1} a_{j_2} \sum_{t=1}^{h} \sum_{i_1,\ldots,i_t \in S_{2h-t,t}} \prod_{j=1}^{t} R_{i_j}\big) \\
& + b_{j_1,j_2,1,*} a_{j_1} a_{j_2}^* \varphi_0\big(c_{j_1} c_{j_2}^* \sum_{t=1}^{h} \sum_{i_1,\ldots,i_t \in S_{2h-t,t}} \prod_{j=1}^{t} R_{i_j}\big) \\
& + b_{j_1,j_2,*,1} c_{j_1}^* c_{j_2} \varphi_0\big(a_{j_1}^* a_{j_2} \sum_{t=1}^{h} \sum_{i_1,\ldots,i_t \in S_{2h-t,t}} \prod_{j=1}^{t} R_{i_j}\big) \\
& + b_{j_1,j_2,*,*} c_{j_1}^* a_{j_2}^* \varphi_0\big(a_{j_1}^* c_{j_2}^* \sum_{t=1}^{h} \sum_{i_1,\ldots,i_t \in S_{2h-t,t}} \prod_{j=1}^{t} R_{i_j}\big) \Bigg].
\end{aligned}
$$

Now (7.46) is immediate from the power series expansion of $\beta(z,a)$ in (7.47) and using calculations similar to (7.49)–(7.50). Hence, the proof of Theorem 7.4.1 is complete. $\qquad\square$

7.5   Corollaries

The following corollaries and remarks discuss some special cases. They follow from Theorems 7.2.1, 7.3.1, and 7.4.1. Recall the classes $\mathcal{NND}$ and $\mathcal{N}$ respectively from (4.27) and (4.46).

Corollary 7.5.1. *Let $Z_{p \times n}$ be an independent matrix whose entries satisfy (A1). Let A_p and B_n be norm bounded deterministic matrices. Suppose $\{A_p\} \in \mathcal{NND}$ with LSD F^A. Suppose $\{B_n\} \in \mathcal{N}$ and $\lim n^{-1} \operatorname{Tr}(B^2) = d_2$. Let (a,s) be free in some NCP $(\mathcal{B}, \varphi_0)$ where a is distributed as F^A and s is a standard semi-circle variable. Suppose $p, n = n(p) \to \infty$, $p/n \to 0$. Then*

(a) $\left(Span\{ \sqrt{\tfrac{n}{p}}(\tfrac{1}{n} A^{1/2} Z B Z^* A^{1/2} - \tfrac{1}{n}\operatorname{Tr}(B)A)\}, \tfrac{1}{p} E\operatorname{Tr} \right)$
converges to $\left(Span\{a^{1/2}\sqrt{d_2}sa^{1/2}\}, \varphi_0 \right)$

(b) The LSD of $\sqrt{np^{-1}}(n^{-1}A^{1/2} ZBZ^* A^{1/2} - n^{-1}\operatorname{Tr}(B)A)$ exists almost surely and it is distributed as $a^{1/2}\sqrt{d_2}sa^{1/2}$ whose Stieltjes transform satisfies the pair of equations (4.47) and (4.48). Here (a,s) is as in (a) above.

Proof. As the assumptions on A_p and B_n respectively imply that Assumptions (A2) and (A3a) hold, (a) follows immediately from Theorem 7.2.1 (a). Also the first part of (b) follows from Theorem 7.3.1.

To verify the claim of Stieltjes transform, note that by (7.42), we have $a^{1/2}\sqrt{d_2}sa^{1/2} = \gamma$, where for all $j > 1$,

$$
a_1 = a_1^* = c_1 = c_1^* = \frac{1}{\sqrt{2}} a^{1/2},
$$

$$
w_1 = w_1^* = s, \quad a_j = c_j = w_j = 0. \tag{7.51}
$$

Also by (7.43)

$$
\begin{aligned}
b_{j_1,j_2,\epsilon_1,\epsilon_2} &= \kappa_2(w_{j_1}^{\epsilon_1}, w_{j_2}^{\epsilon_2}) \\
&= \begin{cases} \kappa_2(\sqrt{d_2}s, \sqrt{d_2}s) = d_2, & \forall \epsilon_1, \epsilon_2 = 1 \text{ or } *, \ j_1 = j_2 = 1 \\ 0, & \text{otherwise.} \end{cases}
\end{aligned}
$$

Therefore, for $z \in \mathbb{C}^+$, $|z|$ large , (7.46) reduces to

$$
\begin{aligned}
\beta(z,a) &= -4\big[d_2 \frac{a}{2} \varphi_0\big(\frac{a}{2}(z + \beta(z,a))^{-1}\big)\big] \\
&= -d_2 a \varphi_0(a(z + \beta(z,a))^{-1}) = d_2 a g(z), \text{ say,} \qquad (7.52)
\end{aligned}
$$

where

$$
\begin{aligned}
g(z) &= -\varphi_0(a(z + \beta(z,a))^{-1}) \\
&= -\varphi_0(a(z + d_2 a g(z))^{-1}) = -\int_{\mathbb{R}} \frac{t \, dF^A(t)}{z + d_2 t g(z)}.
\end{aligned}
$$

Hence, (4.48) is established. Now, for $z \in \mathbb{C}^+$ and $|z|$ large, (7.44) reduces to

$$
\begin{aligned}
m_\gamma(z) &= -\varphi_0((z + \beta(z,a))^{-1}) \\
&= -\varphi_0((z + d_2 a g(z))^{-1}) = -\int_{\mathbb{R}} \frac{dF^A(t)}{z + d_2 t g(z)}.
\end{aligned}
$$

This established (4.47) for large $|z|$. Since both sides of (4.47) are analytic, it continues to hold for all $z \in \mathbb{C}^+$. □

Recall the classes $\mathcal{L}_4$ and $U(\delta)$ defined in (4.23) and (4.26). Consider the following weak assumptions on the entries of Z.

(A4) $\{\varepsilon_{i,j}\} \in \mathcal{L}_4 \cap U(\delta)$ for some $\delta > 0$.

(A5) $\{\varepsilon_{i,j} : i, j \geq 1\}$ are *i.i.d.* with mean 0, variance 1 and $E|\varepsilon_{i,j}|^4 < \infty$.

Remark 7.5.1. *The LSD result in Corollary 7.5.1(b) continues to hold if we replace (A1) by (A4) or (A5). This relaxation is possible by first observing that Theorem 7.3.1 is applicable to appropriately truncated variables and then using a suitable metric to estimate the distance between the ESD of the original and the truncated version. We omit the tedious details of this argument, specially because the proof of Corollary 8.3.1 (c) given later is also along the same lines. Incidentally, if both A and B are taken to be the identity matrices, we recover Theorem 4.3.3 of Bai and Yin [1988]. In addition, we can drop the norm boundedness assumption on A_p by truncating the ESD of the matrix A. For details of this argument see Section 3.1 of Wang and Paul [2014]. Consequences of that are outlined in the exercises.*

Recall the compound free Poisson distribution from Definition 5.5.1.

Corollary 7.5.2. *Suppose all the assumptions in Corollary 7.5.1 hold. Then the LSD of*

$$\sqrt{np^{-1}}(n^{-1}ZBZ^* - \frac{1}{n}\mathrm{Tr}(B))\sqrt{np^{-1}}(n^{-1}ZBZ^* - \frac{1}{n}\mathrm{Tr}(B))$$

is the compound free Poisson distribution with rate 1, and jump distribution same as the distribution of d_2a.

Proof. From the discussions around Definition 5.5.1, it is clear that if a semi-circle variable s and another variable a are freely independent, then for any constant $c > 0$, $\sqrt{cs}a\sqrt{cs}$ has the compound free Poisson distribution with rate 1 and jump distribution ca. Therefore, Corollary 7.5.2 is immediate since by Theorem 7.2.1(a),

$$\sqrt{np^{-1}}(n^{-1}ZBZ^* - \frac{1}{n}\mathrm{Tr}(B)) \to \sqrt{d_2}s,$$

where s is the standard semi-circle variable and by Theorem 7.2.1(b), s and a are freely independent. $\qquad\square$

The following corollary will be used later in Chapter 8, when we deal with the LSD of $\hat{\Gamma}_u + \hat{\Gamma}_u^*$. Recall $\{\Delta_u\}$ from (6.3). We use the same assumptions on $\{\psi_j\}$ as in Corollary 6.5.1. For convenience of the reader, we state it again.

Suppose $\{\psi_j\} \subset \{B_{2i-1}, B_{2i-1}^*\}$ i.e., we assume:

(B) $\{\psi_j\}$ are norm bounded and

$$(\mathrm{Span}\{\psi_j, \psi_j^* : j \geq 0\}, p^{-1}\mathrm{Tr}) \to (\mathrm{Span}\{\eta_j, \eta_j^* : j \geq 0\}, \varphi_{\mathrm{odd}}) \text{ (say).} \quad (7.53)$$

Recall the NCP $(\mathcal{A}_{\mathrm{odd}}, \varphi_{\mathrm{odd}})$ from (7.4). Clearly the NCP in the right side of (7.53) is a $*$-sub-algebra of $(\mathcal{A}_{\mathrm{odd}}, \varphi_{\mathrm{odd}})$. Recall that φ_0 is the state corresponding to the free product given in (7.41). Therefore, by Definition 5.3.4, the restriction of φ_0 on $\mathcal{A}_{\mathrm{odd}}$ is φ_{odd}.

To describe the Stieltjes transform below, for $x = (x_1, \ldots, x_q)$, $x_l \in \mathcal{A}_{\mathrm{odd}}$ $\forall l$, we define

$$\Psi(x, \theta) = (\sum_{l=0}^{q} x_l e^{il\theta})(\sum_{l=0}^{q} x_l^* e^{-il\theta}) = \sum_{l_1, k_1=0}^{q} x_{l_1} x_{k_1}^* e^{i(l_1-k_1)\theta}. \quad (7.54)$$

For $x = (x_1, \ldots, x_q), y = (y_1, y_2, \ldots, y_q)$, $x_l, y_l \in \mathcal{A}_{\mathrm{odd}}$ $\forall l$, we define

$$R_u(x, y) = \frac{1}{2\pi} \int_0^{2\pi} \cos^2(u\theta)\Psi(x, \theta)\Psi(y, \theta)d\theta \quad (7.55)$$

$$= \sum_{l_1, l_2, k_1, k_2=0}^{q} x_{l_1} x_{k_1}^* y_{l_2} y_{k_2}^* \frac{1}{2\pi} \int_0^{2\pi} \cos^2(u\theta) e^{i(l_1-k_1+l_2-k_2)\theta}d\theta.$$

This simplfies to

$$R_u(x,y) = \frac{1}{2} \sum_{l_1,l_2,k_1,k_2=0}^{q} x_{l_1} x_{k_1}^* y_{l_2} y_{k_2}^* I(l_1 - k_1 + l_2 - k_2 = 0)$$

$$+\frac{1}{4} \sum_{l_1,l_2,k_1,k_2=0}^{q} x_{l_1} x_{k_1}^* y_{l_2} y_{k_2}^* I(l_1 - k_1 + l_2 - k_2 = 2u)$$

$$+\frac{1}{4} \sum_{l_1,l_2,k_1,k_2=0}^{q} x_{l_1} x_{k_1}^* y_{l_2} y_{k_2}^* I(l_1 - k_1 + l_2 - k_2 = -2u).$$

For $x = (x_1, \ldots, x_q)$, $x_l \in \mathcal{A}_{\text{odd}}$ for all l and $\eta = (\eta_1, \eta_2, \ldots, \eta_q)$ ($\{\eta_j\}$ are as in (7.53)), we define

$$\beta_u(z,x) = -\varphi_0(R_u(x,\eta)(z + \beta_u(z,\eta))^{-1} | x) \qquad (7.56)$$

where the right side above is defined as

$$-\frac{1}{2} \sum_{l_1,l_2,k_1,k_2=0}^{q} x_{l_1} x_{k_1}^* \varphi_0(\eta_{l_2} \eta_{k_2}^* (z + \beta_u(z,\eta))^{-1}) I(l_1 - k_1 + l_2 - k_2 = 0)$$

$$-\frac{1}{4} \sum_{l_1,l_2,k_1,k_2=0}^{q} x_{l_1} x_{k_1}^* \varphi_0(\eta_{l_2} \eta_{k_2}^* (z + \beta_u(z,\eta))^{-1}) I(l_1 - k_1 + l_2 - k_2 = 2u)$$

$$-\frac{1}{4} \sum_{l_1,l_2,k_1,k_2=0}^{q} x_{l_1} x_{k_1}^* \varphi_0(\eta_{l_2} \eta_{k_2}^* (z + \beta_u(z,\eta))^{-1}) I(l_1 - k_1 + l_2 - k_2 = -2u).$$

Now we are ready to state the following corollary. Recall $\{\Gamma_u\}$ from (3.4).

Corollary 7.5.3. *Suppose Assumptions (A1), (B) and (7.53) hold and $p, n = n(p) \to \infty$, $p/n \to 0$. Then the almost sure LSD of $\frac{1}{2}\sqrt{np^{-1}}(\Delta_u + \Delta_u^* - \Gamma_u - \Gamma_u^*)$ exists and its Stieltjes transform is given by*

$$m_u(z) = -\varphi_0((z + \beta_u(z,\eta))^{-1}), \quad z \in \mathbb{C}^+, |z| \text{ large}, \qquad (7.57)$$

where $\beta_u(z,\eta)$ satisfies (7.56).

Proof. First note that $\{\frac{1}{2}\sqrt{np^{-1}}(\Delta_u + \Delta_u^* - \Gamma_u - \Gamma_u^*)\}$ satisfy the form (7.1) with $\{B_{2i}\} = \{P_i\}$ where $P_i = ((I(j - k = i)))_{1 \le j,k \le n}$. Moreover, under (B) and (7.53), $\{\psi_j\}$ satisfy Assumption (A2) and (7.4). Also note that the matrices $\{P_u : u = 0, \pm 1, \pm 2, \ldots\}$ satisfy Assumption (A3a) and (6.56). Hence, by Theorem 7.3.1, the LSD of $\{\frac{1}{2}\sqrt{np^{-1}}(\Delta_u + \Delta_u^* - \Gamma_u - \Gamma_u^*)\}$ is given by

$$\gamma = \sum_{j,k=0}^{q} \eta_j w_{u,j,k} \eta_k^*, \text{ where } w_{u,j,k}^* = w_{u,k,j}, \text{ and}$$

$$\kappa_r(w_{u,j_l,k_l} : 1 \leq l \leq r)$$

$$= \begin{cases} \lim \frac{1}{n} \mathrm{Tr}\left(\frac{(P_{j_1-k_1+u}+P_{j_1-k_1-u})}{2} \frac{(P_{j_2-k_2+u}+P_{j_2-k_2-u})}{2} \right), & \text{if } r = 2 \\ 0, & r \neq 2 \end{cases}$$

by (6.56)
$$= \begin{cases} \frac{1}{8\pi} \int_0^{2\pi} e^{i(j_1-k_1+j_2-k_2+2u)\theta} + \frac{1}{8\pi} \int_0^{2\pi} e^{i(j_1-k_1+j_2-k_2-2u)\theta} \\ + \frac{1}{4\pi} \int_0^{2\pi} e^{i(j_1-k_1+j_2-k_2)\theta}, & \text{if } r = 2 \\ 0, & r \neq 2, \end{cases}$$

and, $\{\eta_j\}$ which are as in (7.53)) and $\{w_{u,j,k}\}$ are free.

Now Theorem 7.4.1 can be applied to get the Stieltjes transform of γ. For brevity, we use the notation $\overset{*}{\sum} = \overset{q}{\underset{j_1,j_2,k_1,k_2=1}{\sum}}$.

By (7.46), $-\beta_u(z,x)$ equals

$$\frac{1}{8\pi} \int_0^{2\pi} \overset{*}{\sum} e^{2iu\theta} e^{i(j_1-k_1+j_2-k_2)\theta} x_{j_1}^* x_{k_1} \varphi_0 \left(\eta_{j_2} \eta_{k_2}^* (z + \beta(z,\eta))^{-1} \right) d\theta$$

$$+ \frac{1}{8\pi} \int_0^{2\pi} e^{-2iu\theta} \overset{*}{\sum} e^{i(j_1-k_1+j_2-k_2)\theta} x_{j_1}^* x_{k_1} \varphi_0 \left(\eta_{j_2} \eta_{k_2}^* (z + \beta(z,\eta))^{-1} \right) d\theta$$

$$+ \frac{1}{4\pi} \int_0^{2\pi} \overset{*}{\sum} e^{i(j_1-k_1+j_2-k_2)\theta} x_{j_1}^* x_{k_1} \varphi_0 \left(\eta_{j_2} \eta_{k_2}^* (z + \beta(z,\eta))^{-1} \right) d\theta$$

$$= \frac{1}{8\pi} \int_0^{2\pi} (e^{2iu\theta} + e^{-2iu\theta} + 2) \Psi(x,\theta) \varphi_0 \left(\Psi(\eta,\theta)(z + \beta(z,\eta))^{-1} \right) d\theta$$

$$= \varphi_0 \left(\frac{1}{2\pi} \int_0^{2\pi} \cos^2(u\theta) \Psi(x,\theta) \Psi(\eta,\theta)(z + \beta(z,\eta))^{-1} d\theta \Big| x \right)$$

$$= \varphi_0 \left(R_u(x,\eta)(z + \beta(z,\eta))^{-1} \Big| x \right).$$

Hence, (7.56) is proved. Now by (7.44), (7.57) holds for large $|z|$. $\qquad\square$

Exercises

1. Establish the relations in (7.9), (7.12) and (7.13).

2. Show that (7.27), (7.29), (7.30) and (7.32) hold.

3. Proof of Theorem 7.2.1 was given only for the case $U = 1$. Verify that the proof goes through when $U > 1$.

4. Verify conditions (M4) and Carleman's condition (C) in Theorem 7.3.1.

5. Show that the power series that appeared in Theorem 7.4.1 are meaningful.

6. Show that the LSD result in Corollary 7.5.1(b) continues to hold if we replace (A1) by (A4) or (A5) and drop the norm boundedness assumption on A_p. As a consequence, check that Theorem 4.3.5 of Wang et al. [2017] follows. Show further that if B is taken to be the identity matrix, we recover Theorem 4.3.4 of Bao [2012].

Part III

Part III

Chapter 8

SPECTRA OF AUTOCOVARIANCE
MATRIX I

Sample autocovariance matrices are important in high-dimensional linear time series. In addition to just $\hat{\Gamma}_u$ and $\hat{\Gamma}_u^*$, we may also be interested in functions of these. For example, if we wish to study the *singular values* of $\hat{\Gamma}_u$, we need to consider $\hat{\Gamma}_u \hat{\Gamma}_u^*$. Likewise, as we may recall, in the one-dimensional case, all tests for white noise are based on quadratic functions of autocovariances. The analogous objects in our model are quadratic polynomials in autocovariances. Thus, we are naturally led to the consideration of matrix polynomials of autocovariances.

This chapter focuses on the LSD of polynomials in sample autocovariance matrices $\{\hat{\Gamma}_u\}$ for the infinite-dimensional moving average processes. However, due to the reasons discussed in the previous chapters, we restrict ourselves to only symmetric polynomials. We make use of the general results developed in Chapters 4–7 to deal with all symmetric polynomials in $\{\hat{\Gamma}_u\}$. We show that under the most reasonable conditions on the parameter matrices $\{\psi_j\}$, the LSD of any symmetric polynomial in $\{\hat{\Gamma}_u\}$ exists for both the cases $p/n \to y \in (0, \infty)$ and $p/n \to 0$. In the latter case, we will need appropriate scaling constants and centering matrices. We describe the limits in terms of some free variables and derive formulae for their Stieltjes transform in some specific cases. In the next chapter, we will extend these results when we have more than one independent time series and also consider joint polynomials. In Chapters 10 and 11, we shall use these results in statistical applications.

There does not seem to be any general LSD results known for non-symmetric polynomials. We provide a few simulations to convince the reader that LSD results in these cases should also hold.

8.1   Assumptions

We first list the assumptions. Recall the following classes of independent variables from (4.23)–(4.25):

$$\mathcal{L}_r \;=\; \text{set of all collections of independent random variables} \quad (8.1)$$
$$\{\epsilon_{i,j} : i,j \geq 1\} \text{ such that } \sup_{i,j} E|\varepsilon_{i,j}|^r < \infty,$$

$$\mathcal{L} \;=\; \bigcap_{r=1}^{\infty} \mathcal{L}_r, \qquad\qquad\qquad\qquad\qquad\qquad (8.2)$$

$$C(\delta,p) \;=\; \text{set of all collections } \{\varepsilon_{i,j}\} \text{ such that } \forall i,j$$
$$P(|\varepsilon_{i,j}| \leq \eta_p p^{\frac{1}{2+\delta}}) = 1, \text{ for some } \eta_p \downarrow 0 \text{ as } p \to \infty. \qquad (8.3)$$

Consider the following assumption on $\{\varepsilon_{i,j}\}$.

(B1) $\{\varepsilon_{i,j}\}$ are independently distributed with mean 0 and variance 1.

(B2) $\{\varepsilon_{i,j}\}$ are i.i.d. random variables with mean 0, variance 1 and $E(\varepsilon_{i,j})^4 < \infty$.

(B3) $\{\varepsilon_{i,j}\} \in \mathcal{L} \cup C(\delta,p)$ for all $p \geq 1$ and for some $\delta > 0$.

Later we will relax $(B3)$ for specific polynomials.

We consider the same assumption on $\{\psi_j\}$ as in Corollaries 6.5.1 and 7.5.3. For convenience, we state it again. Recall the collection of some $p \times p$ matrices $\{B_{2i-1}\}$ which satisfy Assumption (A2) in Chapters 6 and 7.

Suppose $\{\psi_j\} \subset \{B_{2i-1}, B_{2i-1}^*\}$ i.e., we assume:

(B) $\{\psi_j\}$ are norm bounded and converge jointly.

Suppose

$$(\text{Span}\{\psi_j, \psi_j^* : j \geq 0\}, p^{-1}\text{Tr}) \to (\text{Span}\{\eta_j, \eta_j^* : j \geq 0\}, \varphi_{\text{odd}}), \quad (8.4)$$
$$(\text{Span}\{\bar{\psi}_j, \bar{\psi}_j^* : j \geq 0\}, (n+p)^{-1}\text{Tr}) \to (\text{Span}\{\bar{\eta}_j, \bar{\eta}_j^* : j \geq 0\}, \bar{\varphi}_{\text{odd}}). \quad (8.5)$$

Recall the NCP $(\mathcal{A}_{\text{odd}}, \varphi_{\text{odd}})$ and $(\bar{\mathcal{A}}_{\text{odd}}, \bar{\varphi}_{\text{odd}})$ respectively from (6.13) and (6.14). Clearly the NCP at right side of (8.4) and (8.5) are sub-algebras of $(\mathcal{A}_{\text{odd}}, \varphi_{\text{odd}})$ and $(\bar{\mathcal{A}}_{\text{odd}}, \bar{\varphi}_{\text{odd}})$. Moreover, for any polynomial Π,

$$\varphi_{\text{odd}}(\Pi(\eta_j, \eta_j^* : j \geq 0)) \;=\; \frac{1+y}{y}\bar{\varphi}_{\text{odd}}(\Pi(\bar{\eta}_j, \bar{\eta}_j^* : j \geq 0))$$
$$=\; \frac{1+y}{y}\bar{\varphi}(\Pi(\bar{\eta}_j, \bar{\eta}_j^* : j \geq 0)).$$

8.2   LSD when $p/n \to y \in (0,\infty)$

We shall first deal with the MA(q) model when $q < \infty$. The MA(∞) model will be analysed using the same ideas but there will be some added technical difficulties. This will done later.

8.2.1 $MA(q)$, $q < \infty$

Consider the infinite-dimensional $MA(q)$ process

$$X_{t,p} = \sum_{j=0}^{q} \psi_j \varepsilon_{t-j}, \ t \geq 1, \ q < \infty. \tag{8.6}$$

To describe the LSD of any symmetric polynomial in $\{\hat{\Gamma}_u, \hat{\Gamma}_u^*\}$ for (8.6), we need the matrices $\{\Delta_u\}$ and $\{P_i\}$ defined respectively in (6.3) and just before (6.3). The matrices $\{\hat{\Gamma}_u\}$ can be approximated by $\{\Delta_u\}$ for LSD purposes (see (6.3)). Since $\{\Delta_u\}$ is of the form (6.1) with $\{B_{2i}\} = \{P_i : i = 0, \pm1, \pm2, \ldots\}$, by (6.16), (6.55), and (6.56), we now have

$$\bar{\mathcal{A}}_{\text{even}} = \text{Span}\{\underline{c}_u, \underline{c}_u^* = \underline{c}_{-u} : u = 0, \pm1, \ldots\}, \text{ and}$$

$$(\text{Span}\{\underline{P}_u, \underline{P}_u^* : u = 0, \pm1, \ldots\}, (n+p)^{-1}\text{Tr}) \to (\bar{\mathcal{A}}_{\text{even}}, \bar{\varphi}_{\text{even}}),$$

where for all $T \geq 1$ and $i_1, i_2, \ldots, i_T = 0, \pm1, \pm2, \ldots$, we have

$$\bar{\varphi}_{\text{even}}(\prod_{j=1}^{T} \underline{c}_{i_j}) = \lim \frac{1}{n+p}\text{Tr}(\prod_{j=1}^{T} \underline{P}_{i_j}) = \frac{1}{1+y}I(\sum_{j=1}^{T} i_j = 0). \tag{8.7}$$

Recall the NCP $(\text{Span}\{s_u\}, \varphi_s)$ of free semi-circle variables, defined at the beginning of Section 6.2. Let $s \in \{s_u\}$ be any typical standard semi-circle variable and $(\text{Span}\{s\}, \varphi_s)$ be the NCP generated by s with moment sequence $\{\varphi_s(s^k) = \beta_k\}$ where $\{\beta_k\}$ is given in (4.19).

Recall the NCP $(\mathcal{A}, \bar{\varphi})$ defined in (6.17), where $\{\bar{\eta}_j, \bar{\eta}_j^*\}$, $\{\underline{c}_j, \underline{c}_j^*\}$ and s are free.

Consider the following polynomials in $\mathcal{A}$

$$\bar{\gamma}_{uq} = (1+y) \sum_{j,j'=0}^{q} \bar{\eta}_j s \underline{c}_{j-j'+u} s \bar{\eta}_{j'}^*, \ \forall u, q \geq 0. \tag{8.8}$$

Then we have the following Theorem.

Theorem 8.2.1. *(Bhattacharjee and Bose [2016a]) Consider the model (8.6). Suppose (B1), (B3), and (B) hold and $p, n = n(p) \to \infty$, $p/n \to y \in (0, \infty)$. Then the LSD of any symmetric polynomial $\Pi(\hat{\Gamma}_u, \hat{\Gamma}_u^* : u \geq 0)$ exists almost surely and it is uniquely determined by the moment sequence*

$$\lim p^{-1} E\text{Tr}(\Pi(\hat{\Gamma}_u, \hat{\Gamma}_u^* : u \geq 0))^k = \frac{1+y}{y}\bar{\varphi}[(\Pi(\bar{\gamma}_{uq}, \bar{\gamma}_{uq}^* : u \geq 0))^k]. \tag{8.9}$$

Proof. To prove the above theorem, we use the moment method and Lemma 4.1.2. Recall $\{\Delta_u\}$ from (6.3):

$$\Delta_u = \frac{1}{n} \sum_{j,j'} \psi_j Z P_{j-j'+u} Z^* \psi_{j'}^*. \tag{8.10}$$

The following lemma describes the approximation of $\{\hat{\Gamma}_u\}$ by $\{\Delta_u\}$. Its proof is given in Section A.4 of the Appendix.

Lemma 8.2.1. *Consider the model (8.6). Suppose (B1), (B3), and (B) hold and $p, n = n(p) \to \infty$, $p/n \to y > 0$. Then the following statements are true.*

(a) For any polynomial Π,

$$\lim p^{-1} E \mathrm{Tr}(\Pi(\hat{\Gamma}_u, \hat{\Gamma}_u^* : u \ge 0)) = \lim p^{-1} E \mathrm{Tr}(\Pi(\Delta_u, \Delta_u^* : u \ge 0)). \quad (8.11)$$

(b) Let, for $1 \le i \le T$, m_i be polynomials. Let for all $1 \le i \le T$,

$$
\begin{aligned}
\mathcal{P}_i &= \mathrm{Tr}(m_i(\Delta_u, \Delta_u^* : u \ge 0)), & \mathcal{P}_i^0 &= E\mathcal{P}_i, \\
\tilde{\mathcal{P}}_i &= \mathrm{Tr}(m_i(\hat{\Gamma}_u, \hat{\Gamma}_u^* : u \ge 0)), & \tilde{\mathcal{P}}_i^0 &= E\tilde{\mathcal{P}}_i.
\end{aligned}
$$

Then we have

$$\lim E(\prod_{i=1}^{T}(\tilde{\mathcal{P}}_i - \tilde{\mathcal{P}}_i^0)) = \lim E(\prod_{i=1}^{T}(\mathcal{P}_i - \mathcal{P}_i^0)). \quad (8.12)$$

Now we continue the proof of Theorem 8.2.1. By Theorem 6.2.1, for any polynomial Π, and for all $k \ge 1$,

$$\lim \frac{1}{p} E \mathrm{Tr}(\Pi(\Delta_u, \Delta_u^* : u \ge 0))^k = \frac{1+y}{y} \bar{\varphi}(\Pi(\bar{\gamma}_{uq}, \bar{\gamma}_{uq}^* : u \ge 0))^k. \quad (8.13)$$

Hence, by Lemma 8.2.1(a), for all $k \ge 1$,

$$\lim \frac{1}{p} E \mathrm{Tr}(\Pi(\hat{\Gamma}_u, \hat{\Gamma}_u^* : u \ge 0))^k = \frac{1+y}{y} \bar{\varphi}(\Pi(\bar{\gamma}_{uq}, \bar{\gamma}_{uq}^* : u \ge 0))^k. \quad (8.14)$$

This establishes (M1).

By Lemmas 6.3.1 and 8.2.1(b), we have for any polynomial Π

$$E\left[\frac{1}{p}\mathrm{Tr}\left(\Pi(\hat{\Gamma}_u, \hat{\Gamma}_u^* : u \ge 0)\right)^h - E\left(\frac{1}{p}\mathrm{Tr}\left(\Pi(\hat{\Gamma}_u, \hat{\Gamma}_u^* : u \ge 0)\right)^h\right)\right]^4 = O(p^{-4}). \quad (8.15)$$

and hence (M4) is established.

Proof of Carleman's condition (C) is essentially same as its counterpart in Theorem 6.3.1. This completes the proof of Theorem 8.2.1. □

8.2.2 $MA(\infty)$

Now consider the $MA(\infty)$ process

$$X_t = \sum_{j=0}^{\infty} \psi_j \varepsilon_{t-j}, \ \forall t. \quad (8.16)$$

Recall $||\cdot||_2$ defined in (1.8). We make the following additional assumption on $\{\psi_j\}$.

(B4) $\sum_{j=0}^{\infty} \sup_p ||\psi_j||_2 < \infty$.

The approach is now essentially to truncate the series and consider the corresponding MA(q) model and then let $q \to \infty$. To describe the limit, which will be in terms of infinite sums of non-commuting variables, we need the following lemma. Its proof is given in Section A.8 of the Appendix.

Lemma 8.2.2. *Suppose (B4) holds. Then for any polynomial $\Pi(\bar\gamma_{uq}, \bar\gamma_{uq}^* : u \geq 0)$, $\lim_{q\to\infty} \bar\varphi(\Pi(\bar\gamma_{uq}, \bar\gamma_{uq}^* : u \geq 0))$ exists and is finite.*

Now consider the NCP $(\mathcal{A}_\infty, \bar\varphi_\infty)$ where

$$\mathcal{A}_\infty = \mathrm{Span}\{\bar\gamma_{u\infty}, \bar\gamma_{u\infty}^* : u \geq 0\} \tag{8.17}$$

and for any polynomial $\Pi(\bar\gamma_{u\infty}, \bar\gamma_{u\infty}^* : u \geq 0)$,

$$\bar\varphi_\infty(\Pi(\bar\gamma_{u\infty}, \bar\gamma_{u\infty}^* : u \geq 0)) = \lim_{q\to\infty} \bar\varphi(\Pi(\bar\gamma_{uq}, \bar\gamma_{uq}^* : u \geq 0)). \tag{8.18}$$

The existence of the limit at right side of (8.18) is guaranteed by Lemma 8.2.2. Now we have the following theorem.

Theorem 8.2.2. *(Bhattacharjee and Bose [2016a]) Consider the model (8.16) with $q = \infty$. Suppose Assumptions (B1), (B3), (B), and (B4) hold and $p, n = n(p) \to \infty$, $p/n \to y \in (0, \infty)$. Then the LSD of any symmetric polynomial $\Pi(\hat\Gamma_u, \hat\Gamma_u^* : u \geq 0)$ in $\{\hat\Gamma_u, \hat\Gamma_u^*\}$ exists almost surely and it is uniquely determined by the moment sequence*

$$\lim p^{-1} E\mathrm{Tr}(\Pi(\hat\Gamma_u, \hat\Gamma_u^* : u \geq 0))^k = \frac{1+y}{y} \bar\varphi_\infty[\Pi(\bar\gamma_{u\infty}, \bar\gamma_{u\infty}^* : u \geq 0))^k]. \tag{8.19}$$

To prove the theorem, we need the following Lemma, whose proof is given in Section A.9 of the Appendix.

Lemma 8.2.3. *Suppose (B4) holds. Then we have the following results.*

(a) For any symmetric polynomial $\Pi(\bar\gamma_{uq}, \bar\gamma_{uq}^* : u \geq 0)$, there exists a unique probability measure F_q on $\mathbb{R}$ such that

$$\int x^K dF_q = \frac{1+y}{y} \bar\varphi[(\Pi(\bar\gamma_{uq}, \bar\gamma_{uq}^* : u \geq 0))^K], \ \forall K \geq 1. \tag{8.20}$$

(b) For any symmetric polynomial $\Pi(\bar\gamma_{u\infty}, \bar\gamma_{u\infty}^* : u \geq 0)$, there exists a unique probability measure F on $\mathbb{R}$ such that

$$\int x^K dF = \frac{1+y}{y} \bar\varphi_\infty(\Pi[(\bar\gamma_{u\infty}, \bar\gamma_{u\infty}^* : u \geq 0))^K], \ \forall K \geq 1. \tag{8.21}$$

(c) F_q converges weakly to F as $q \to \infty$.

Proof of Theorem 8.2.2. Let $F, G : \mathbb{R} \to [0, 1]$ be two distribution functions. Define the *Lévy distance* between F and G as

$$L(F, G) = \inf\{\epsilon > 0|\ F(x - \epsilon) - \epsilon \leq G(x) \leq F(x + \epsilon) + \epsilon \ \ \forall x \in \mathbb{R}\}. \quad (8.22)$$

Let,

$$
\begin{aligned}
F_{p,q} &= \text{ESD of } \Pi(\hat{\Gamma}_u, \hat{\Gamma}_u^* : u \geq 0) \text{ for the MA}(q) \text{ process}, \quad (8.23)\\
F_{p,\infty} &= \text{ESD of } \Pi(\hat{\Gamma}_u, \hat{\Gamma}_u^* : u \geq 0) \text{ for the MA}(\infty) \text{ process}. \quad (8.24)
\end{aligned}
$$

It is enough to show that $\lim_{p \to \infty} L(F_{p,\infty}, F_\infty) = 0$, almost surely.

All the inequalities, equalities and limits below are in the almost sure sense. Note that

$$L(F_{p,\infty}, F_\infty) \leq L(F_{p,\infty}, F_{p,q}) + L(F_{p,q}, F_q) + L(F_q, F_\infty). \quad (8.25)$$

Now, taking limit as $p \to \infty$ on both sides of (8.25), by Theorem 8.2.1,

$$\lim_{p \to \infty} L(F_{p,\infty}, F_\infty) \leq \lim_{p \to \infty} L(F_{p,\infty}, F_{p,q}) + L(F_q, F_\infty), \ \forall q \geq 0. \quad (8.26)$$

Taking limit as $q \to \infty$ on both sides of (8.26), by Lemma 8.2.3

$$
\begin{aligned}
\lim_{p \to \infty} L(F_{p,\infty}, F_\infty) &\leq \lim_{q \to \infty} \lim_{p \to \infty} L(F_{p,\infty}, F_{p,q}) + \lim_{q \to \infty} L(F_q, F_\infty)\\
&\leq \lim_{q \to \infty} \lim_{p \to \infty} L(F_{p,\infty}, F_{p,q}). \quad (8.27)
\end{aligned}
$$

Hence, the proof of Theorem 8.2.2 will be complete once we prove that right side of (8.27) is 0. This proof is very technical and hence we defer it to Lemma A.10.2 in the Appendix. This completes the proof of Theorem 8.2.2. □

8.2.3 *Application to specific cases*

This section consists of applications of Theorems 8.2.1 and 8.2.2 to some specific cases when $p/n \to y \in (0, \infty)$.

Recall the class of independent random variables defined in (4.26):

$$
\begin{aligned}
U(\delta) = {} & \text{all collections of independent random variables } \{\varepsilon_{i,j}\}\\
& \text{such that for all } \eta > 0\\
& \lim \frac{\eta^{-(2+\delta)}}{np} \sum_{i=1}^{p} \sum_{j=1}^{n} E(|\varepsilon_{i,j}|^{2+\delta} I(|\varepsilon_{i,j}| > \eta p^{\frac{1}{2+\delta}})) = 0. \quad (8.28)
\end{aligned}
$$

Also recall the class (8.1).

Corollary 8.2.1. (a) *Consider the model (8.6). Suppose Assumptions (B1), (B3), and (B) hold and $p, n = n(p) \to \infty$, $p/n \to y \in (0, \infty)$. Then the LSD of $\frac{1}{2}(\hat{\Gamma}_u + \hat{\Gamma}_u^*)$ exists almost surely and its Stieltjes transform is given by (6.51)–(6.54).*

(b) *Under the additional assumption (B4), the above result in (a) holds for the model (8.16) once we replace q by ∞.*

(c) *The above results in (a) and (b) hold if instead of (B3) we assume (B2) or $\{\varepsilon_{i,j} : i, j \geq 1\} \in \mathcal{L}_{2+\delta} \cap U(\delta)$ for some $\delta > 0$.*

(d) *Consider the model (8.6). Then the almost sure LSD of $\frac{1}{2}(\hat{\Gamma}_u + \hat{\Gamma}_u^*)$ are identical whenever $u > q$ and are different for $u \leq q$.*

Proof. (a) Existence of the LSD of $\frac{1}{2}(\hat{\Gamma}_u + \hat{\Gamma}_u^*)$ is immediate from Theorem 8.2.1. Recall the matrices $\{\Delta_u\}$ in (8.10). By Theorems 6.3.1 and 8.2.1, LSD of $\frac{1}{2}(\hat{\Gamma}_u + \hat{\Gamma}_u^*)$ and $\frac{1}{2}(\Delta_u + \Delta_u^*)$ are identical and therefore their limiting Stieltjes transforms are same. By Corollary 6.5.1, the latter is given by (6.51)–(6.54). Hence, the same is true for $\frac{1}{2}(\hat{\Gamma}_u + \hat{\Gamma}_u^*)$ and the proof of (a) is complete.

(b) This is immediate from Theorem 8.2.2.

(c) This proof needs appropriate truncation on the support of $\{\varepsilon_{i,j}\}$ and is very technical. We defer the proof to Section A.5 in the Appendix.

(d) By Theorem 8.2.1, the LSD of $\frac{1}{2}(\hat{\Gamma}_u + \hat{\Gamma}_u^*)$ depends on u only through the distribution of $\{c_{j_1 - j_2 + u}, c_{j_2 - j_1 - u} : 0 \leq j_1, j_2 \leq q\}$. But by (8.7), these classes have identical distribution for $u > q$. Therefore, the LSD of $\frac{1}{2}(\hat{\Gamma}_u + \hat{\Gamma}_u^*)$ are identical for $u > q$.

At the same time, by Theorem 8.2.1 and Lemma 5.4.2(a), for $u \leq q$ we have

$$
\lim \frac{1}{p} E\mathrm{Tr}(\hat{\Gamma}_u + \hat{\Gamma}_u^*) = \frac{1+y}{y} \bar{\varphi}(\bar{\gamma}_{uq} + \bar{\gamma}_{uq}^*)
$$

$$
= \frac{(1+y)^2}{y} \bar{\varphi}\Big(\sum_{j=0}^{q-u} \bar{\eta}_j \bar{\eta}_{j+u}^* + \sum_{j=0}^{q-u} \bar{\eta}_j^* \bar{\eta}_{j+u} \Big). \quad (8.29)
$$

These vary as u varies. Thus, LSD of $\frac{1}{2}(\hat{\Gamma}_u + \hat{\Gamma}_u^*)$ are different for $u \leq q$. □

The Stieltjes transform expression in (6.51)–(6.54) can be simplified further under the following restrictive assumption on $\{\psi_j\}$. This assumption is taken from Wang et al. [2017].

(WAP) $\{\psi_j\}$ are Hermitian, simultaneously diagonalizable and norm bounded. There are continuous functions $f_j : \mathbb{R}^m \to \mathbb{R}$ and a $p \times p$ unitary matrix U such that $U \psi_j U^* = \mathrm{diag}(f_j(\alpha_1), f_j(\alpha_2), \dots, f_j(\alpha_p))$, $\alpha_j \in \mathbb{R}^m$ for all j and some positive integer m. The measure $p^{-1} \sum_{i=1}^{p} \delta_{\alpha_i}$ converges weakly to a compactly supported probability distribution F on $\mathbb{R}^m$.

The following corollary is immediate once we observe that under (WAP),

$\varphi_{\text{odd}}(\cdot)$ in (6.51)–(6.54) reduces to $\int \cdot \, dF$. For a direct proof using the method of Stieltjes transform, see Liu et al. [2015].

Corollary 8.2.2. *Consider the model (8.16). Suppose Assumptions (B2) and (WAP) hold, $\sum_{j=0}^{\infty} |f_j(\alpha)| < \infty$, $\forall \alpha \in \mathbb{R}^m$ and $p, n = n(p) \to \infty$, $p/n \to y \in (0, \infty)$. Then for each $u \geq 1$, the LSD of $\frac{1}{2}(\hat{\Gamma}_u + \hat{\Gamma}_u^*)$ exists almost surely. The limiting Stieltjes transform $m_u(z)$, $z \in \mathbb{C}^+$ satisfies*

$$m_u(z) = \int_{\mathbb{R}^m} \left(\frac{1}{2\pi} \int_0^{2\pi} \frac{\cos(u\theta')h_1(\alpha, \theta')d\theta'}{1 + y\cos(u\theta')K_u(z, \theta')} - z \right)^{-1} dF(\alpha), \text{ where}$$

(8.30)

$$K_u(z, \theta) = \int_{\mathbb{R}^m} \left(\frac{1}{2\pi} \int_0^{2\pi} \frac{\cos(u\theta')h_1(\alpha, \theta')d\theta'}{1 + y\cos(u\theta')K_u(z, \theta')} - z \right)^{-1} h_1(\alpha, \theta) dF(\alpha),$$

(8.31)

$$h_1(\alpha, \theta) = |\sum_{j=0}^{\infty} e^{ij\theta} f_j(\alpha)|^2, \quad \alpha \in \mathbb{R}^m.$$

(8.32)

The following corollary describes the LSD of $\frac{1}{2}(\hat{\Gamma}_u + \hat{\Gamma}_u^*)$ for the particular case when $\psi_j = \lambda_j I_p$, $\forall j$. Using the method of Stieltjes transform, part (d) has been proved directly by Pfaffel and Schlemm [2011].

Corollary 8.2.3. *(a) Consider the model (8.6). Suppose Assumptions (B1) and (B3) hold and $p, n = n(p) \to \infty$, $p/n \to y \in (0, \infty)$. Let $\psi_j = \lambda_j I_p$, $\forall j$. Then for each $u \geq 1$, the almost sure LSD of $\frac{1}{2}(\hat{\Gamma}_u + \hat{\Gamma}_u^*)$ exists and the limiting Stieltjes transform $m_u(z)$ satisfies (only one solution yields a valid Stieltjes transform)*

$$z = -\frac{1}{m_u(z)} + \frac{1}{2\pi} \int_0^{2\pi} \frac{d\theta}{ym_u(z) + f^{-1}(\theta)}, \quad z \in \mathbb{C}^+ \text{ where}$$

(8.33)

$$f(\theta) = \cos(u\theta)|\sum_{k=0}^{q} \lambda_k e^{ik\theta}|^2.$$

(8.34)

(b) Under the additional assumption $\sum_{j=0}^{\infty} |\lambda_j| < \infty$, the result in (a) holds for the model (8.16) once we replace q by ∞.

(c) The results in (a) and (b) continue to hold if instead of (B3) we assume $\{\varepsilon_{i,j}\} \in \mathcal{L}_{2+\delta} \cap U(\delta)$ for some $\delta > 0$.

(d) The results in (a) and (b) continue to hold if instead of (B1) and (B3) we assume (B2).

Proof. (a) Recall the definition of free cumulant κ in (5.22). Note that by Theorems 6.3.1 and 8.2.1, the LSD of $\frac{1}{2}(\hat{\Gamma}_u + \hat{\Gamma}_u^*)$ and $\frac{1}{2}(\Delta_u + \Delta_u^*)$ are identical and therefore, their free cumulants are also same. By Corollary 6.5.2, free cumulants of the LSD of $\frac{1}{2}(\hat{\Gamma}_u + \hat{\Gamma}_u^*)$ are given by

$$\kappa_{ur} = \frac{1}{2y\pi} \int_0^{2\pi} (yf(\theta))^r d\theta, \quad \forall r \geq 1.$$

(8.35)

Recall the free cumulant generating function $C(\cdot)$ in (5.30). Note that $|\kappa_{ur}| \le C^r$, $\forall r \ge 1$ and some $C > 0$. Therefore, for $z \in \mathbb{C}^+$, $|z|$ small,

$$C(z) = 1 + \sum_{r=1}^{\infty} \kappa_{ur} z^r = 1 + \frac{1}{2y\pi} \int_0^{2\pi} \frac{yz f(\theta) d\theta}{1 - yz f(\theta)}. \tag{8.36}$$

Note that $m_u(z) \to 0$ as $|z| \to \infty$. Hence, using the relation (5.31) between the generating functions for moments and free cumulants, for some $K > 0$ and all $|z| > K$, $z \in \mathbb{C}^+$

$$-z m_u(z) = C(-m_u(z)) = 1 - \frac{1}{2y\pi} \int_0^{2\pi} \frac{y m_u(z) f(\theta) d\theta}{1 + y m_u(z) f(\theta)}. \tag{8.37}$$

Upon simplifying, (8.37) reduces to (8.33) for large $|z|$. Using analyticity, (8.33) holds for all $z \in \mathbb{C}^+$. We omit the details. Hence, (a) is proved.

(b) follows from Corollary 8.2.1(b) as under $\psi_j = \lambda_j I_p$, $\forall j$, (B4) reduces to $\sum_{j=0}^{\infty} |\lambda_j| < \infty$.

(c) is immediate from Corollary 8.2.1(c).

(d) is immediate as (B2) imples $\{\varepsilon_{i,j}\} \in \mathcal{L}_{2+\delta} \cap U(\delta)$ for some $\delta > 0$. $\qquad \square$

The following corollary describes the LSD of $\hat{\Gamma}_u \hat{\Gamma}_u^*$.

Corollary 8.2.4. (a) *Consider the model (8.6). Suppose Assumptions (B1), (B3), and (B) hold and $p, n = n(p) \to \infty$, $p/n \to y \in (0, \infty)$. Then the LSD of $\hat{\Gamma}_u \hat{\Gamma}_u^*$ exists almost surely.*

(b) *Under the additional assumption (B4), the result in (a) holds for the model (8.16) once we replace q by ∞.*

(c) *The results in (a) and (b) hold if instead of (B3) we assume (B2) or $\{\varepsilon_{i,j} : i, j \ge 1\} \in \mathcal{L}_4 \cap U(\delta)$ for some $\delta > 0$.*

(d) *The almost sure LSD of $\hat{\Gamma}_u \hat{\Gamma}_u^*$ in (a) are identical whenever $u > q$ and are different for $u \le q$.*

Proof. (a) and (b) follow immediately from Theorems 8.2.1 and 8.2.2. For (c) we need truncation on $\{\varepsilon_{i,j}\}$ and the arguments are very technical. We provide the details in Section A.6 in the Appendix. The first part of (d) is true for the same reason as given in the proof of Corollary 8.2.1(d). For the second part, by Theorem 8.2.1 and Lemma 5.4.2(a), for all $u \le q$ note that

$$\begin{aligned}
\lim \frac{1}{p} E\mathrm{Tr}(\hat{\Gamma}_u \hat{\Gamma}_u^*) &= \frac{1+y}{y} \bar{\varphi}(\bar{\gamma}_{uq} \bar{\gamma}_{uq}^*) \\
&= \frac{(1+y)^3}{y} \sum_{i,j=0}^{q-u} \bar{\varphi}(\bar{\eta}_i \bar{\eta}_{i+u}^* \bar{\eta}_j \bar{\eta}_{j+u}^*) \\
&\quad + \frac{(1+y)^3}{y} \sum_{\substack{0 \le i, j, j' \le u \\ 0 \le i+j'-j \le q}} \bar{\varphi}(\bar{\eta}_j \bar{\eta}_{i+j'-j}^*) \bar{\varphi}(\bar{\eta}_{j'} \bar{\eta}_i^*). \tag{8.38}
\end{aligned}$$

These expressions are different for different values of u. Therefore, the LSD of $\hat{\Gamma}_u\hat{\Gamma}_u^*$ are different for $0 \leq u \leq q$. Hence, (d) is proved. This completes the proof of Corollary 8.2.4.   □

Now consider the simplest model, the MA(0) process, defined in (3.6):

$$X_{t,p} = \varepsilon_t, \quad \text{for all } t. \tag{8.39}$$

For convenience, let us write X_t for $X_{t,p}$. Let $\varepsilon_{t,i}$ be the i-th element of ε_t. Also recall the Marčenko–Pastur law MP_y with parameter $y > 0$ with the moment sequence (4.34). The following corollary describes the LSD of $\hat{\Gamma}_0$, $\{\hat{\Gamma}_u+\hat{\Gamma}_u^*\}_{u\geq 1}$ and $\{\hat{\Gamma}_u\hat{\Gamma}_u^*\}_{u\geq 1}$ for the MA(0) process. Direct proofs of Parts (a) and (b) can also be found in Bai and Silverstein [2009] and Jin et al. [2014], respectively.

Corollary 8.2.5. *Consider the model (8.39). Suppose Assumption (B1) holds and $p, n = n(p) \to \infty$, $p/n \to y \in (0, \infty)$. Then the following hold.*

(a) Suppose Assumption (B3) is satisfied or $\{\varepsilon_{i,j} : i, j \geq 1\} \in U(0)$. Then the almost sure LSD of $\hat{\Gamma}_0$ is the MP_y law.

(b) Suppose Assumption (B3) is satisfied or $\{\varepsilon_{i,j} : i, j \geq 1\} \in U(\delta)$ for some $\delta > 0$. Then for each $u \geq 1$, the almost sure LSD of $\frac{1}{2}(\hat{\Gamma}_u + \hat{\Gamma}_u^*)$ exists and its limiting Stieltjes transform is given by (6.66).

(c) Suppose Assumption (B3) is satisfied or $\{\varepsilon_{i,j} : i, j \geq 1\} \in \mathcal{L}_4 \cap U(\delta)$ for some $\delta > 0$. Then for each $u \geq 1$, the almost sure LSD of $\hat{\Gamma}_u\hat{\Gamma}_u^*$ exists and its moment sequence is given by (6.83).

Proof. We prove all the above results under (B3). To prove the results under the corresponding alternative assumptions, we need appropriate truncation on $\{\varepsilon_{i,j}\}$ and follow the same arguments as in the proof of Corollary 8.2.1 (c) given later in Section A.5 in the Appendix. We shall not provide the details of the truncation arguments.

Recall the matrix Z in (4.3.1). Now note that, under (B1) and (B3), by Theorems 6.3.1 and 8.2.1, we have the following:

(a) LSDs of $n^{-1}ZZ^*$ and $\hat{\Gamma}_0$ are identical;

(b) LSDs of $\frac{1}{2n}Z(P_u + P_{-u})Z^*$ and $\frac{1}{2}(\hat{\Gamma}_u + \hat{\Gamma}_u^*)$ are identical;

(c) LSDs of $p^{-2}ZP_uZ^*ZP_{-u}Z^*$ and $\hat{\Gamma}_u\hat{\Gamma}_u^*$ are identical.

Therefore, by Corollaries 6.5.3 , 6.5.5 and 6.5.6, the result follows.   □

8.3   LSD when $p/n \to 0$

Now we shall discuss the LSD of symmetric polynomials in $\{\hat{\Gamma}_u\}$ for the case $p, n(p) \to \infty$ such that $p/n \to 0$. Consider the collection of non-commutative variables $\{w_{u,j_1,j_2}\}$, whose free cumulants are as follows:

For all $j_1, j_2, \ldots \geq 1$ and $u_1, u_2, \ldots = 0, \pm 1, \pm 2, \ldots,$

$$w^*_{u,j_1,j_2} = w_{-u,j_2,j_1}, \tag{8.40}$$

$$\kappa_2(w_{u_1,j_1,j_2}, w_{u_2,j_3,j_4}) = \frac{1}{2\pi} \int_0^{2\pi} e^{i(j_1 - j_2 + u_1)\theta} e^{i(j_3 - j_4 + u_2)\theta} \, d\theta$$

$$= \begin{cases} 1, & \text{if } j_1 - j_2 + j_3 - j_4 = -(u_1 + u_2) \\ 0, & \text{otherwise,} \end{cases} \tag{8.41}$$

$$\kappa_r(w_{u_i,j_{2i-1},j_{2i}} : 1 \leq i \leq r) = 0, \; \forall r \neq 2. \tag{8.42}$$

As mentioned in Section 7.2, the above sequence of free cumulants naturally define a state, say φ_w, on $\mathrm{Span}\{w_{u,l,i} : u,l,i \geq 1\}$. This is because, the moments and free cumulants are in one-to-one correspondence (see (5.22) and (5.24) in Chapter 4).

Recall $(\mathcal{A}_{\mathrm{odd}}, \varphi_{\mathrm{odd}})$ and $\{\eta_j, \eta_j^*\}$ respectively in (6.13) and (8.4). Also recall the free product $(\mathcal{B}, \varphi_0)$ of $(\mathcal{A}_{\mathrm{odd}}, \varphi_{\mathrm{odd}})$ and $(\mathrm{Span}\{w_{u,l,i} : u,l,i \geq 1\}, \varphi_w)$ in (7.41). Therefore, $\{w_{u,j_1,j_2}\}$ and $\{\eta_j, \eta_j^*\}$ are free in $(\mathcal{B}, \varphi_0)$.

Consider the following polynomial from $\mathcal{B}$:

$$S_{uq} = \sum_{j_1,j_2=0}^{q} \eta_{j_1} w_{u,j_1,j_2} \eta_{j_2}^*, \quad \forall u, q \geq 0. \tag{8.43}$$

Recall the population autocovariance matrices $\{\Gamma_u\}$ defined in (3.1):

$$\Gamma_u = \sum_{j=0}^{q-u} \psi_j \psi_{j+u}^*, \quad \forall u \geq 0.$$

Note that, by (8.4),

$$(\mathrm{Span}\{\Gamma_u, \Gamma_u^* : u \geq 0\}, p^{-1}\mathrm{Tr}) \to (\mathrm{Span}\{G_{uq}, G_{uq}^* : u \geq 0\}, \varphi_{\mathrm{odd}}), \tag{8.44}$$

where

$$G_{uq} = \sum_{j=0}^{q-u} \eta_j \eta_{j+u}^*, \quad \forall u, q \geq 0. \tag{8.45}$$

Now we will state the LSD for general self-adjoint polynomials $\Pi(\hat{\Gamma}_u, \hat{\Gamma}_u^* : u \geq 0)$. To describe the limit write it in the form

$$\Pi(\hat{\Gamma}_u, \hat{\Gamma}_u^* : u \geq 0) = \sum_{l=1}^{T} \beta_l \left(\prod_{i=1}^{k_l} \hat{\Gamma}_{u_{l,i}}^{\epsilon_{l,i}} \right), \tag{8.46}$$

where $\epsilon_{l,i} \in \{1, *\}$ and $u_{l,i} \in \{0, 1, 2, \ldots\}$.

Then we have the following theorem.

Theorem 8.3.1. *(Bhattacharjee and Bose [2016c]) Consider the model (8.6). Suppose (B1), (B3), and (B) hold and $p, n(p) \to \infty$, $p/n \to 0$. Then the LSD of the self-adjoint polynomial*

$$\sqrt{np^{-1}}\big(\Pi(\hat{\Gamma}_u, \hat{\Gamma}_u^* : u \geq 0) - \Pi(\Gamma_u, \Gamma_u^* : u \geq 0)\big) \tag{8.47}$$

exists almost surely and is distributed as

$$\sum_{l=1}^{T} \beta_l \Big(\sum_{\substack{\nu_{l,i} \in \{0,1\} \\ \sum_i \nu_{l,i} = 1}} \prod_{i=1}^{k_l} G_{u_{l,i}q}^{\epsilon_{l,i}(1-\nu_{l,i})} S_{u_{l,i}q}^{\epsilon_{l,i}\nu_{l,i}} \Big) \tag{8.48}$$

where $\{S_{uq}\}$ and $\{G_{uq}\}$ are as in (8.43) and (8.45) with $q = \infty$.

Proof. As in the proof of Theorem 8.2.1, we use the moment method and Lemma 4.1.2. Recall $\{\Delta_u\}$ given in (6.3). The following lemma describes the approximation of $\{\hat{\Gamma}_u\}$ by $\{\Delta_u\}$ and is an analogue of Lemma 8.2.1 given earlier for the $p/n \to y > 0$ case. Proof of this lemma follows the same arguments as the proof of Lemma 8.2.1 and therefore we omit it.

Lemma 8.3.1. *Consider the model (8.6). Suppose (B1), (B3), and (B) hold and $p, n(p) \to \infty$, $p/n \to 0$. Then the following statements are true.*

(a) For any polynomial Π,

$$\lim p^{-1} E\mathrm{Tr}\big(\sqrt{np^{-1}}(\Pi(\hat{\Gamma}_u, \hat{\Gamma}_u^* : u \geq 0) - \Pi(\Gamma_u, \Gamma_u^* : u \geq 0))\big)$$
$$= \lim \frac{1}{p} E\mathrm{Tr}\big(\sqrt{np^{-1}}(\Pi(\Delta_u, \Delta_u^* : u \geq 0) - \Pi(\Gamma_u, \Gamma_u^* : u \geq 0))\big). \tag{8.49}$$

(b) Let, for $1 \leq i \leq T$, m_i be polynomials. Let for all $1 \leq i \leq T$,

$$\mathcal{P}_i = \mathrm{Tr}\big(\sqrt{np^{-1}}(m_i(\Delta_u, \Delta_u^* : u \geq 0) - m_i(\Gamma_u, \Gamma_u^* : u \geq 0))\big), \quad \mathcal{P}_i^0 = E\mathcal{P}_i,$$
$$\tilde{\mathcal{P}}_i = \mathrm{Tr}\big(\sqrt{np^{-1}}(m_i(\hat{\Gamma}_u, \hat{\Gamma}_u^* : u \geq 0) - m_i(\Gamma_u, \Gamma_u^* : u \geq 0))\big), \quad \tilde{\mathcal{P}}_i^0 = E\tilde{\mathcal{P}}_i.$$

Then we have

$$\lim E\big(\prod_{i=1}^{T}(\tilde{\mathcal{P}}_i - \tilde{\mathcal{P}}_i^0)\big) = \lim E\big(\prod_{i=1}^{T}(\mathcal{P}_i - \mathcal{P}_i^0)\big). \tag{8.50}$$

Now it is easy to see that, exactly like the proof of Theorem 8.2.1, here (M1), (M4) and (C) hold by application of Theorems 7.2.1, 7.3.1 and Lemma 8.3.1. This completes the proof of Theorem 8.3.1. □

We now consider the case $q = \infty$. Proof of the next Theorem uses the same arguments as in the proof of Theorem 8.2.2 and hence we omit it.

Theorem 8.3.2. *(Bhattacharjee and Bose [2016c]) Consider the model (8.16). Suppose (B1), (B3), (B), and (B4) hold and $p, n(p) \to \infty$, $p/n \to 0$. Let $\Pi(\hat{\Gamma}_u, \hat{\Gamma}_u^* : u \geq 0)$ be decomposed as in (8.46). Then LSD of the self-adjoint polynomial*

$$\sqrt{np^{-1}}\big(\Pi(\hat{\Gamma}_u, \hat{\Gamma}_u^* : u \geq 0) - \Pi(\Gamma_u, \Gamma_u^* : u \geq 0)\big) \tag{8.51}$$

exists almost surely and is distributed as

$$\sum_{l=1}^{T} \beta_l \Big(\sum_{\substack{\nu_{l,i} \in \{0,1\} \\ \sum_i \nu_{l,i}=1}} \prod_{i=1}^{k_l} G_{u_{l,i}\infty}^{\epsilon_{l,i}(1-\nu_{l,i})} S_{u_{l,i}\infty}^{\epsilon_{l,i}\nu_{l,i}} \Big). \tag{8.52}$$

where $\{S_{u\infty}\}$ and $\{G_{u\infty}\}$ are as in (8.43) and (8.45).

8.3.1 Application to specific cases

We now list some consequences of Theorems 8.3.1 and 8.3.2.

The following corollary describes the LSD of $\hat{\Gamma}_u + \hat{\Gamma}_u^*$. Recall the classes $\mathcal{L}_r$ and $U(\delta)$ defined respectively in (8.1) and (8.28).

Corollary 8.3.1. *(a) Consider the model (8.6). Suppose (B1), (B3), and (B) hold and $p, n(p) \to \infty$, $p/n \to 0$. Then the LSD g_{uq} of $\sqrt{np^{-1}}\big(2^{-1}(\hat{\Gamma}_u + \hat{\Gamma}_u^*) - 2^{-1}(\Gamma_u + \Gamma_u^*)\big)$ exists almost surely and*

$$g_{uq} \overset{\mathcal{D}}{=} \frac{1}{2}\Big(\sum_{j_1, j_2=0}^{q} \eta_{j_1} w_{u,j_1,j_2} \eta_{j_2}^* + \sum_{j_1, j_2=0}^{q} \eta_{j_2} w_{-u,j_2,j_1} \eta_{j_1}^* \Big). \tag{8.53}$$

The Stieltjes transform of g_{uq} is given by (7.54)–(7.57).

(b) Under the additional assumption (B4), the result in (a) holds for the model (8.16) once we replace q by ∞.

(c) The results in (a) and (b) hold if instead of (B3) we assume (B2) or if $\{\varepsilon_{i,j} : i,j \geq 1\} \in \mathcal{L}_{2+\delta} \cap U(\delta)$ for some $\delta > 0$.

(d) Consider the model (8.6). Then the almost sure LSD of $\sqrt{np^{-1}}\big(2^{-1}(\hat{\Gamma}_u + \hat{\Gamma}_u^*) - 2^{-1}(\Gamma_u + \Gamma_u^*)\big)$ are identical whenever $u > q$ and are different for $u \leq q$.

Proof. (a) Existence of the LSD of $\sqrt{np^{-1}}\big(2^{-1}(\hat{\Gamma}_u + \hat{\Gamma}_u^*) - 2^{-1}(\Gamma_u + \Gamma_u^*)\big)$ and relation (8.53) are immediate from Theorem 8.3.1 once we put $T = 2, k_1 = k_2 = \beta_1 = \beta_2 = 1$, $u_{1,1} = u_{2,1} = u$, $\epsilon_{1,1} = 1, \epsilon_{2,1} = *$. By Theorems 7.3.1 and 8.3.1, this LSD and LSD of $\sqrt{np^{-1}}\big(2^{-1}(\Delta_u + \Delta_u^*) - 2^{-1}(\Gamma_u + \Gamma_u^*)\big)$ are identical and therefore their limiting Stieltjes transforms are same. But by Corollary 7.5.3, the Stieltjes transform of this LSD is given by (7.54)–(7.57). Hence, the proof of (a) is complete.

(b) This is immediate from Theorem 8.3.2.

(c) This proof needs appropriate truncation on the support of $\{\varepsilon_{i,j}\}$ and is very technical. We will provide the truncation arguments in Section A.7 of the Appendix.

(d) For each u, the distribution of g_{uq} depends only on the distribution of $\{w_{u,j_1,j_2}, w_{-u,j_2,j_1} : 1 \le j_1, j_2 \le q\}$. By (8.41), the distribution of $\{w_{u,j_1,j_2}, w_{-u,j_2,j_1} : 1 \le j_1, j_2 \le q\}$ is characterized by the second order cumulants,

$$\kappa_2(w_{v,j_1,j_2}, w_{w,k_1,k_2}) = \begin{cases} 1, & \text{if } j_1 - j_2 + k_1 - k_2 = \pm 2u \text{ or } 0; \, v, w = \pm u, \\ 0, & \text{otherwise.} \end{cases}$$

For an MA(q) process, $-2q \le j_1 - j_2 + k_1 - k_2 \le 2q$. Hence, for $u > q$, $j_1 - j_2 + k_1 - k_2 = \pm 2u$ can never happen. Therefore, for $u > q$, the distribution of $\{w_{u,j_1,j_2}, w_{-u,j_2,j_1} : 1 \le j_1, j_2 \le q\}$ does not depend on u and hence g_{uq}, $u \ge q$, are identically distributed. Now, note that for $u \le q$

$$\varphi_0(g_{uq}^2) = \sum_{\substack{0 \le j_1, j_2, k_1 \le q \\ 0 \le j_1 + k_1 - j_2 - 2u \le q}} \varphi_0(\eta_{j_2}^* \eta_{k_1}) \varphi_0(\eta_{j_1} \eta_{j_1 + k_1 - j_2 - 2u}^*)$$

$$+ \sum_{\substack{0 \le j_1, j_2, k_1 \le q \\ 0 \le j_1 + k_1 - j_2 + 2u \le q}} \varphi_0(\eta_{j_2}^* \eta_{k_1}) \varphi_0(\eta_{j_1} \eta_{j_1 + k_1 - j_2 + 2u}^*)$$

$$+ \sum_{\substack{0 \le j_1, j_2, k_1 \le q \\ 0 \le j_1 + k_1 - j_2 \le q}} \varphi_0(\eta_{j_2}^* \eta_{k_1}) \varphi_0(\eta_{j_1} \eta_{j_1 + k_1 - j_2}^*).$$

These are different as u varies. Therefore, the distribution of g_{uq} are different for $0 \le u \le q$. Hence, (d) is proved. □

The Stieltjes transform expression in (7.54)–(7.57) can be simplified further under suitable assumptions on $\{\psi_j\}$. The following corollary is immediate once we observe that under (WAP), $\varphi_0(\cdot)$ in (7.54)–(7.57) reduces to $\int \cdot \, dF$. For a direct Stieltjes transform proof of Part (b), see Wang et al. [2017].

Corollary 8.3.2. (a) Consider the model (8.16)). Suppose (B1), (B3), and (WAP) hold, $\sum_{j=0}^{\infty} |f_j(\alpha)| < \infty \; \forall \alpha \in \mathbb{R}^m$ and $p, n(p) \to \infty$, $p/n \to 0$. Then the LSD of $\frac{1}{2}(\hat{\Gamma}_u + \hat{\Gamma}_u^*)$ exists almost surely and the limiting Stieltjes transform $m_u(z)$ satisfies

$$m_u(z) = -\int_{\mathbb{R}^m} \frac{dF(\alpha)}{z + \beta_u(z, \alpha)}, \quad z \in \mathbb{C}^+, \; where \tag{8.54}$$

$$\beta_u(z, x) = -\int_{\mathbb{R}^m} \frac{R_u(x, \alpha) dF(\alpha)}{z + \beta_u(z, \alpha)}, \quad z \in \mathbb{C}^+, x \in \mathbb{R}^m, \tag{8.55}$$

$$R_u(x, y) = \frac{1}{2\pi} \int_0^{2\pi} \cos^2(u\theta) h_1(x, \theta) h_1(y, \theta) d\theta, \quad x, y \in \mathbb{R}^m, \tag{8.56}$$

and $h_1(\cdot, \cdot)$ is as in (8.32).

(b) The result in (a) continues to hold if we assume (B2) instead of (B1) and (B3).

The following corollary describes the LSD of $\hat{\Gamma}_u \hat{\Gamma}_u^*$. Its proof is similar to the proof of Corollary 8.2.4 and is left as exercise.

Corollary 8.3.3. (a) Consider the model (8.6). Suppose (B1), (B3) and (B) hold and $p, n = n(p) \to \infty$, $p/n \to 0$. Then the LSD of $\sqrt{np^{-1}}(\hat{\Gamma}_u \hat{\Gamma}_u^* - \Gamma_u \Gamma_u^*)$ exists almost surely.

(b) Under the additional assumption (B4), the result in (a) holds for the model (8.16) once we replace q by ∞.

(c) The results in (a) and (b) hold if instead of (B3) we assume (B2) or if $\{\varepsilon_{i,j} : i, j \geq 1\} \in \mathcal{L}_4 \cap U(\delta)$ for some $\delta > 0$.

(d) The almost sure LSD of $\sqrt{np^{-1}}(\hat{\Gamma}_u \hat{\Gamma}_u^* - \Gamma_u \Gamma_u^*)$ in (a) are identical whenever $u > q$ and are different for $u \leq q$.

Example 8.3.1. Suppose $X_t = \varepsilon_t + \lambda \varepsilon_{t-1}$, $\forall t$ where $\lambda \in \mathbb{R}$ is fixed. Suppose (B1) and (B3) hold. Then the almost sure LSD of $\sqrt{np^{-1}}(\hat{\Gamma}_1 \hat{\Gamma}_1^* - \lambda^2 I_p)$ is distributed as $\lambda\sqrt{2((1+\lambda^2)^2 + \lambda^2 + 2)}s$ where s is a standard semi-circle variable.

Proof. Recall $\Pi(\hat{\Gamma}_u, \hat{\Gamma}_u^* : u \geq 0)$ in (8.46). Note that $\hat{\Gamma}_u \hat{\Gamma}_u^* = \Pi(\hat{\Gamma}_u, \hat{\Gamma}_u^* : u \geq 0)$ with $T = 1, \beta_1 = 1, k_1 = 2, \epsilon_{1,1} = 1, \epsilon_{1,2} = *, u_{1,1} = 1, u_{1,2} = 1$. By (8.4), $\eta_0 = 1_{\mathcal{A}_{\mathrm{odd}}}, \eta_1 = \lambda 1_{\mathcal{A}_{\mathrm{odd}}}, \eta_j = 0 \ \forall j > 1$. Therefore, by (8.43),

$$S_{11} = w_{1,0,0} + \lambda^2 w_{1,1,1} + \lambda w_{1,1,0} + \lambda w_{1,0,1}, \tag{8.57}$$

where, by (8.40)–(8.42),

$$
\begin{aligned}
\kappa_2(w_{1,0,0}, w_{1,0,0}^*) &= \kappa_2(w_{1,1,1}, w_{1,1,1}^*) \\
\kappa_2(w_{1,1,0}, w_{1,1,0}^*) &= \kappa_2(w_{1,0,1}, w_{1,0,1}^*) = 1, \tag{8.58} \\
\kappa_2(w_{1,0,0}^*, w_{1,1,1}) &= \kappa_2(w_{1,0,0}, w_{1,1,1}^*) = 1 \\
\kappa_2(w_{1,1,0}, w_{1,1,0}) &= \kappa_2(w_{1,0,1}, w_{1,0,1}) = 1, \tag{8.59}
\end{aligned}
$$

and all other joint free cumulants of $(w_{1,0,0}, w_{1,1,1}, w_{1,1,0}, w_{1,0,1})$ are zero. By (8.45)

$$G_{11} = G_{11}^* = \lambda 1_{\mathcal{A}_{\mathrm{odd}}}. \tag{8.60}$$

Hence, by Theorem 8.3.1, the almost sure LSD of $\sqrt{np^{-1}}(\hat{\Gamma}_1 \hat{\Gamma}_1^* - \lambda^2 I_p)$ is distributed as (in view of (8.48))

$$S_{11}G_{11}^* + G_{11}S_{11}^* = g_1 + g_1^*$$

where

$$g_1 = \lambda[w_{1,0,0} + \lambda^2 w_{1,1,1} + \lambda w_{1,1,0} + \lambda w_{1,0,1}] \tag{8.61}$$

Let w_1 and w_2 be circular elements with $\kappa_2(w_1, w_1^*) = \kappa_2(w_2, w_2^*) = 1$ and s_3 be a standard semi-circle variable. Moreover, suppose w_1, w_2 and w_3 are free. Then by (8.58) and (8.59), g_1 has same distribution as $\lambda(1 + \lambda^2)w_1 + \lambda^2 w_2 + \lambda^2 s_3$. Therefore, $g_1 + g_1^*$ has same distribution as $\lambda(1+\lambda^2)(w_1 + w_1^*) + \lambda^2(w_2 + w_2^*) + 2\lambda^2 s_3$. Now, note that $(w_1 + w_1^*)$ is a self-adjoint element with all marginal cumulants zero except $\kappa_2(w_1 + w_1^*, w_1 + w_1^*) = 2$. Therefore, by Definition 5.3.1, $(w_1 + w_1^*)$ is distributed as $\sqrt{2}s_1$, where s_1 is a standard semi-circle variable. Similarly, $(w_2 + w_2^*)$ is distributed as $\sqrt{2}s_2$, where s_2 is a standard semi-circle variable. Therefore, $g_1 + g_1^*$ has same distribution as $\sqrt{2}\lambda((1 + \lambda^2)s_1 + \lambda s_2 + \sqrt{2}\lambda s_3)$ where s_1, s_2 and s_3 are free standard semi-circle variables. However this has identical distribution with $\lambda\sqrt{2((1 + \lambda^2)^2 + \lambda^2 + 2)}s$ where s is a standard semi-circle variable. This completes the proof of Example 8.3.1.   □

8.4   Non-symmetric polynomials

The autocovariance matrices $\{\hat{\Gamma}_u\}$ themselves are not symmetric for $u > 0$ and hence the theorems in Sections 8.2 and 8.3 do not apply. The study of the limit spectrum of non-hermitian matrices is extremely difficult and very few results are known for general non-hermitian sequences of matrices. As far as we know, there is no general LSD results known for non-symmetric polynomials in $\{\hat{\Gamma}_u : u > 0\}$. Nevertheless, as the results of our simulations given in Figures 8.1–8.3 suggest, there are LSD results in these cases waiting to be discovered.

Recall I_p and J_p respectively from (1.9) and (1.10). Let $A_p = 0.5I_p + 0.5J_p$ and $B_p = 0.7I_p + 0.3J_p$. Suppose $\varepsilon_t \sim \mathcal{N}(0, I_p)$. Consider the following models:

Model A: $X_t = \varepsilon_t$,

Model B: $X_t = \varepsilon_t + A_p\varepsilon_{t-1}$,

Model C: $X_t = \varepsilon_t + A_p\varepsilon_{t-1} + B_p\varepsilon_{t-1}$.

We simulated from these models and drew the scatter plots of eigenvalues of $\hat{\Gamma}_1$ (appropriately centered and scaled) for different choices of (p, n) which correspond to the cases $y = 0, 0.5, 1$ and 2 in Figures 8.1–8.3.

Exercises

1. Suppose $X_t = \varepsilon_t + \lambda_1\varepsilon_{t-1} + \lambda_2\varepsilon_{t-2}$ for all t, $\lambda_1, \lambda_2 \in \mathbb{R}$ and $p/n \to y > 0$. Under appropriate assumptions, establish existence of LSD of the following matrices:
   (a) $\hat{\Gamma}_1 + \hat{\Gamma}_2 + \hat{\Gamma}_1^* + \hat{\Gamma}_2^*$,
   (b) $\hat{\Gamma}_1\hat{\Gamma}_2 + \hat{\Gamma}_2^*\hat{\Gamma}_1^*$,
   (c) $\hat{\Gamma}_1\hat{\Gamma}_2^* + \hat{\Gamma}_2\hat{\Gamma}_1^*$,
   (d) $\hat{\Gamma}_1 + \hat{\Gamma}_2 + \hat{\Gamma}_1\hat{\Gamma}_2 + \hat{\Gamma}_1^* + \hat{\Gamma}_2^* + \hat{\Gamma}_2^*\hat{\Gamma}_1^*$.

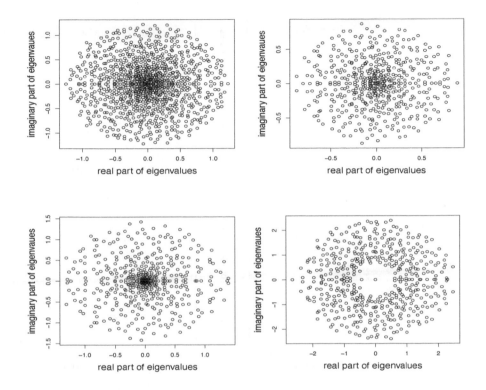

Figure 8.1   *Model A. ESD of (i)* $\sqrt{np^{-1}}\hat{\Gamma}_1$, $n = p^2$, *Row 1, Column 1 (ii)* $\hat{\Gamma}_1$, $n = 2p$, *Row 1, Column 2 (iii)* $\hat{\Gamma}_1$, $n = p$, *Row 2, Column 1 (iv)* $\hat{\Gamma}_1$, $n = p/2$, *Row 2, Column 2. In all cases* $p = 1000$ *(multiple eigenvalues are plotted only once).*

Also describe these LSD in terms of polynomials in free independent variables. Moreover, find LSD of the above matrices when λ_1 is a $p \times p$ matrix with (i, j)-th entry $\theta^{|i-j|}$ and $\theta \in (0, 1)$. Do the appropriate modification when $p/n \to 0$.

2. Find the LSD of $\hat{\Gamma}_4\hat{\Gamma}_4^* + \hat{\Gamma}_5\hat{\Gamma}_5^*$ and $\hat{\Gamma}_5\hat{\Gamma}_5^* + \hat{\Gamma}_6\hat{\Gamma}_6^*$ for Model C in Section 8.4 when $p/n \to y > 0$. Show that these two LSD are identical. What happens when $p/n \to 0$?

3. Give a complete proof of Lemma 8.2.1(b).

4. Prove the approximation of $\hat{\Gamma}_u$ by Δ_u in Lemma 8.2.1(b) when $p/n \to y \in (0, \infty)$.

5. Verify Carleman's condition (C) in the proof of Theorem 8.2.1.

6. Show that the expression in (8.38) are indeed different for $0 \le u \le q$.

7. Prove the approximation of $\hat{\Gamma}_u$ by Δ_u in Lemma 8.3.1 when $p/n \to 0$.

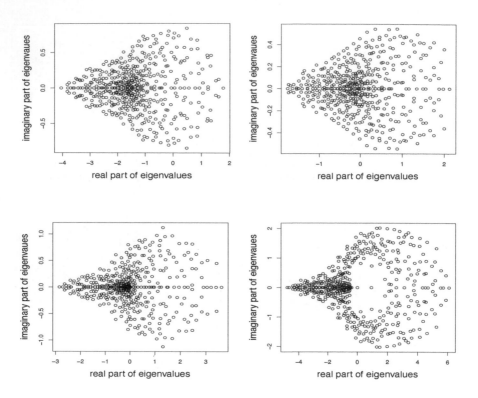

Figure 8.2   *Model B. ESD of (i) $\sqrt{np^{-1}}(\hat{\Gamma}_1 - \Gamma_1)$, $n = p^2$, Row 1, Column 1, (ii) $\hat{\Gamma}_1$, $n = 2p$, Row 1, Column 2 (iii) $\hat{\Gamma}_1$, $n = p$, Row 2, Column 1 (iv) $\hat{\Gamma}_1$, $n = p/2$, Row 2, Column 2. In all cases, $p = 1000$ (multiple eigenvalues are plotted only once).*

8. Prove Theorem 8.3.2 on the LSD of symmetric polynomials in sample autocovariance matrices under $MA(\infty)$ process when $p/n \to 0$.

9. Give a proof of Corollary 8.3.2(b).

10. Establish Corollary 8.3.3 on the LSD of $\sqrt{np^{-1}}(\hat{\Gamma}_u\hat{\Gamma}_u^* - \Gamma_u\Gamma_u^*)$ when $p/n \to 0$.

11. Show that (B2) implies $\{\varepsilon_{i,j} : i, j \geq 1\} \in \mathcal{L}_{2+\delta} \cap U(\delta)$ for some $\delta > 0$.

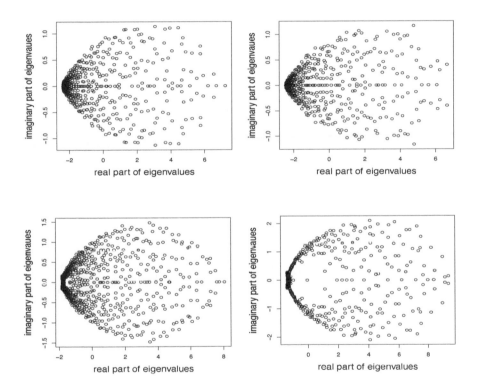

Figure 8.3 *Model C. ESD of (i) $\sqrt{np^{-1}}(\hat{\Gamma}_1 - \Gamma_1)$, $n = p^2$, Row 1, Column 1, (ii) $\hat{\Gamma}_1$, $n = 2p$, Row 1, Column 2 (iii) $\hat{\Gamma}_1$, $n = p$, Row 2, Column 1 (iv) $\hat{\Gamma}_1$, $n = p/2$, Row 2, Column 2. In all cases, $p = 1000$ (multiple eigenvalues are plotted only once)*

Chapter 9

SPECTRA OF AUTOCOVARIANCE MATRIX II

For statistical inference when more than one time series are involved, we need to extend the results of the previous chapter. It is the goal of this chapter to state and prove LSD results for symmetric polynomials in sample autocovariance matrices of two independent infinite-dimensional MA processes. These results will be used in Chapter 10 to construct graphical tests for comparing two independent MA processes.

So suppose that we have observations on the two independent MA(q) processes

$$X_t = \sum_{j=0}^{q} \psi_j \varepsilon_{t-j} \text{ and } Y_t = \sum_{j=0}^{q} \phi_j \xi_{t-j} \ \forall t \geq 1 \qquad (9.1)$$

where $\varepsilon_t = (\varepsilon_{t,1}, \varepsilon_{t,2}, \ldots, \varepsilon_{t,p})'$ and $\xi_t = (\xi_{t,1}, \xi_{t,2}, \ldots, \xi_{t,p})'$ are $p \times 1$ independent vectors and $p, n \to \infty$. Let $\hat{\Gamma}_{uX}$ and $\hat{\Gamma}_{uY}$ be the sample autocovariance matrices of order u respectively for the processes $\{X_t\}$ and $\{Y_t\}$.

Now we can construct symmetric polynomials which involve both these sequences. For example, one can consider $(\hat{\Gamma}_{1X} + \hat{\Gamma}_{1X}^* + \hat{\Gamma}_{2Y} + \hat{\Gamma}_{2Y}^*)$ or $(\hat{\Gamma}_{1X}\hat{\Gamma}_{2Y} + \hat{\Gamma}_{2Y}^*\hat{\Gamma}_{1X}^*)$. From the experience of Chapter 8, it is easy to guess that the LSD of such symmetric polynomials would exist in the two regimes (1) $p, n = n(p) \to \infty$, $p/n \to y \in (0, \infty)$ and (2) $p, n = n(p) \to \infty$, $p/n \to 0$. Here also we need additional centering and scaling for the latter case. It is to be noted that the results for two time series are easily extendable to the case where we have more than two processes of the form (9.1).

9.1 Assumptions

Let $Z_X = ((\varepsilon_{t,j}))_{p \times n}$ and $Z_Y = ((\xi_{t,j}))_{p \times n}$ be the two independent matrices which are also mutually independent. Recall the collection of matrices $\{P_i\}$ defined just before (6.3). Define the matrices

$$\Delta_{uX} = \frac{1}{n} \sum_{j,j'=0}^{q} \psi_j Z_X P_{j-j'+u} Z_X^* \psi_{j'}^* \text{ and}$$

$$\Delta_{uY} \;=\; \frac{1}{n}\sum_{j,j'=0}^{q}\phi_j Z_Y P_{j-j'+u} Z_Y^* \phi_{j'}^*. \qquad (9.2)$$

The joint convergence of $\{\Delta_{uX}, \Delta_{uX}^*, \Delta_{uY}, \Delta_{uY}^* : u \geq 0\}$ follows easily from the results of Chapters 6 and 7, respectively, for the regimes (1) and (2). Under suitable assumptions, the LSD of any symmetric polynomial in $\{\hat{\Gamma}_{uX}, \hat{\Gamma}_{uX}^*, \hat{\Gamma}_{uY}, \hat{\Gamma}_{uY}^* : u \geq 0\}$ is identical with the LSD of the same symmetric polynomial in $\{\Delta_{uX}, \Delta_{uX}^*, \Delta_{uY}, \Delta_{uY}^* : u \geq 0\}$. The details will be provided in Lemma 9.2.1 and the proof of Theorem 9.2.1.

We first list the assumptions that are needed. Recall the classes of independent variables $\mathcal{L}_r$, $\mathcal{L}$ and $C(\delta, p)$ from (8.1), (8.2), and (8.3). Suppose $\{\varepsilon_{i,j}\}$ and $\{\xi_{i,j}\}$ both satisfy (B1)-(B3) from Section 8.1 i.e.,

(B5) $\{\varepsilon_{i,j}, \xi_{i,j}\}$ are independently distributed with mean 0 and variance 1.

(B6) $\{\varepsilon_{i,j}, \xi_{i,j}\}$ are i.i.d. with mean 0, variance 1 and $E\big[(\varepsilon_{i,j})^4 + (\xi_{i,j})^4\big] < \infty$.

(B7) $\{\varepsilon_{i,j}, i \leq p, \; j \leq n\}$, $\{\xi_{i,j}, i \leq p, \; j \leq n\} \in \mathcal{L} \cup C(\delta, p)$ for all $p \geq 1$ and for some $\delta > 0$.

Later we will relax $(B7)$ for specific polynomials.

Recall the collection of some $p \times p$ matrices $\{B_{2i-1}\}$ which satisfy Assumption (A2) of Chapters 6 and 7.

Suppose $\{\psi_j, \phi_j : j \geq 0\} \subset \{B_{2i-1}, B_{2i-1}^*\}$ i.e., we assume:

(D) $\{\psi_j, \phi_j : j \geq 0\}$ are norm bounded and converge jointly.

Suppose

$$(\mathrm{Span}\{\psi_j, \psi_j^*, \phi_j, \phi_j^* : j \geq 0\}, p^{-1}\mathrm{Tr})$$
$$\to (\mathrm{Span}\{\eta_j, \eta_j^*, \theta_j, \theta_j^* : j \geq 0\}, \varphi_{\mathrm{odd}}), \qquad (9.3)$$
$$(\mathrm{Span}\{\bar{\psi}_j, \bar{\psi}_j^*, \bar{\phi}_j, \bar{\phi}_j^* : j \geq 0\}, (n+p)^{-1}\mathrm{Tr})$$
$$\to (\mathrm{Span}\{\bar{\eta}_j, \bar{\eta}_j^*, \bar{\theta}_j, \bar{\theta}_j^* : j \geq 0\}, \bar{\varphi}_{\mathrm{odd}}). \qquad (9.4)$$

Recall the NCP $(\mathcal{A}_{\mathrm{odd}}, \varphi_{\mathrm{odd}})$ and $(\bar{\mathcal{A}}_{\mathrm{odd}}, \bar{\varphi}_{\mathrm{odd}})$ respectively from (6.13) and (6.14). Clearly the NCP at right side of (9.3) and (9.4) are sub-algebras of $(\mathcal{A}_{\mathrm{odd}}, \varphi_{\mathrm{odd}})$ and $(\bar{\mathcal{A}}_{\mathrm{odd}}, \bar{\varphi}_{\mathrm{odd}})$. Moreover, for any polynomial Π,

$$\varphi_{\mathrm{odd}}(\Pi(\eta_j, \eta_j^*, \theta_j, \theta_j^* : j \geq 0)) \;=\; \frac{1+y}{y}\bar{\varphi}_{\mathrm{odd}}(\Pi(\bar{\eta}_j, \bar{\eta}_j^*, \bar{\theta}_j, \bar{\theta}_j^* : j \geq 0))$$
$$=\; \frac{1+y}{y}\bar{\varphi}(\Pi(\bar{\eta}_j, \bar{\eta}_j^*, \bar{\theta}_j, \bar{\theta}_j^* : j \geq 0)).$$

9.2   LSD when $p/n \to y \in (0, \infty)$

9.2.1 $MA(q)$, $q < \infty$

First assume that we have $MA(q)$ processes, $q < \infty$. By (6.16), (6.55) and (6.56), we now have

$$\bar{\mathcal{A}}_{\text{even}} = \text{Span}\{\underline{c}_u, \underline{c}_u^* - \underline{c}_{-u} : u = 0, \pm 1, \ldots\}, \text{ and}$$
$$(\text{Span}\{\underline{P}_u, \underline{P}_u^* : u = 0, \pm 1, \ldots\}, (n+p)^{-1}\text{Tr}) \to (\bar{\mathcal{A}}_{\text{even}}, \bar{\varphi}_{\text{even}}),$$

where for all $T \geq 1$ and $i_1, i_2, \ldots, i_T = 0, \pm 1, \pm 2, \ldots$, we have

$$\bar{\varphi}_{\text{even}}\left(\prod_{j=1}^{T} \underline{c}_{i_j}\right) = \lim \frac{1}{n+p}\text{Tr}\left(\prod_{j=1}^{T} \underline{P}_{i_j}\right) = \frac{1}{1+y}I\left(\sum_{j=1}^{T} i_j = 0\right). \tag{9.5}$$

Recall the NCP (Span$\{s_u\}, \varphi_s$) of free semi-circle variables, defined at the beginning of Section 6.2. Let $s_1, s_2 \in \{s_u\}$ be two typical standard semi-circle variables and for $i = 1, 2$, (Span$\{s_i\}, \varphi_s$) be the NCP generated by s_i with moment sequence $\{\varphi_s(s_i^k) = \beta_k\}$ where $\{\beta_k\}$ are given in (4.19).

Recall the NCP $(\mathcal{A}, \bar{\varphi})$ defined in (6.17), where $\{\bar{\eta}_j, \bar{\eta}_j^*, \bar{\theta}_j, \bar{\theta}_j^*\}, \{\underline{c}_j, \underline{c}_j^*\}, s_1$ and s_2 are free.

For all $u, q \geq 0$, define the polynomials in $\mathcal{A}$

$$\bar{\gamma}_{X,uq} = (1+y)\sum_{j,j'=0}^{q} \bar{\eta}_j s_1 \underline{c}_{j-j'+u} s_1 \bar{\eta}_{j'}^*, \tag{9.6}$$

$$\bar{\gamma}_{Y,uq} = (1+y)\sum_{j,j'=0}^{q} \bar{\theta}_j s_2 \underline{c}_{j-j'+u} s_2 \bar{\theta}_{j'}^*. \tag{9.7}$$

Then we have the following Theorem.

Theorem 9.2.1. *Consider the model (9.1) with $q < \infty$. Suppose (B5), (B7), and (D) hold and $p, n = n(p) \to \infty$, $p/n \to y \in (0, \infty)$. Then the LSD of any symmetric polynomial $\Pi(\hat{\Gamma}_{uX}, \hat{\Gamma}_{uX}^*, \hat{\Gamma}_{uY}, \hat{\Gamma}_{uY}^* : u \geq 0)$ exists almost surely and it is uniquely determined by the moment sequence*

$$\lim p^{-1}E\text{Tr}(\Pi(\hat{\Gamma}_{uX}, \hat{\Gamma}_{uX}^*, \hat{\Gamma}_{uY}, \hat{\Gamma}_{uY}^* : u \geq 0))^k \tag{9.8}$$
$$= \frac{1+y}{y}\bar{\varphi}[(\Pi(\bar{\gamma}_{X,uq}, \bar{\gamma}_{X,uq}^*, \bar{\gamma}_{Y,uq}, \bar{\gamma}_{Y,uq}^* : u \geq 0))^k].$$

Proof. As in the proof of Theorem 8.2.1, we use the moment method and Lemma 4.1.2. The following lemma describes the approximation of $\{\hat{\Gamma}_{uX}, \hat{\Gamma}_{uY}\}$ by $\{\Delta_{uX}, \Delta_{uY}\}$. Its proof is similar to the proof of Lemma 8.2.1 and is left as an exercise.

Lemma 9.2.1. *Consider the model (9.1) with $q < \infty$. Suppose (B5), (B7), and (D) hold and $p, n = n(p) \to \infty$, $p/n \to y > 0$. Then the following statements are true.*

(a) For any polynomial Π,

$$\lim p^{-1}E\text{Tr}(\Pi(\hat{\Gamma}_{uX}, \hat{\Gamma}_{uX}^*, \hat{\Gamma}_{uY}, \hat{\Gamma}_{uY}^* : u \geq 0)) \tag{9.9}$$
$$= \lim p^{-1}E\text{Tr}(\Pi(\Delta_{uX}, \Delta_{uX}^*, \Pi(\Delta_{uY}, \Delta_{uY}^* : u \geq 0)).$$

(b) For $1 \leq i \leq T$, let m_i be polynomials and let

$$
\begin{aligned}
\mathcal{P}_i &= \mathrm{Tr}(m_i(\Delta_{uX}, \Delta_{uX}^*, \Delta_{uY}, \Delta_{uY}^* : u \geq 0)), \quad \mathcal{P}_i^0 = E\mathcal{P}_i, \\
\tilde{\mathcal{P}}_i &= \mathrm{Tr}(m_i(\hat{\Gamma}_{uX}, \hat{\Gamma}_{uX}^*, \hat{\Gamma}_{uY}, \hat{\Gamma}_{uY}^* : u \geq 0)), \quad \tilde{\mathcal{P}}_i^0 = E\tilde{\mathcal{P}}_i.
\end{aligned}
$$

Then we have

$$
\lim E(\prod_{i=1}^{T}(\tilde{\mathcal{P}}_i - \tilde{\mathcal{P}}_i^0)) = \lim E(\prod_{i=1}^{T}(\mathcal{P}_i - \mathcal{P}_i^0)). \tag{9.10}
$$

Now we continue with the proof of Theorem 9.2.1. By Theorem 6.2.1, for any polynomial Π, and for all $k \geq 1$,

$$
\lim \frac{1}{p} E\mathrm{Tr}(\Pi(\Delta_{uX}, \Delta_{uX}^*, \Delta_{uY}, \Delta_{uY}^* : u \geq 0))^k \tag{9.11}
$$

$$
= \frac{1+y}{y} \bar{\varphi}(\Pi(\bar{\gamma}_{X,uq}, \bar{\gamma}_{X,uq}^*, \bar{\gamma}_{Y,uq}, \bar{\gamma}_{Y,uq}^* : u \geq 0))^k.
$$

Hence, by Lemma 9.2.1(a), for all $k \geq 1$,

$$
\lim \frac{1}{p} E\mathrm{Tr}[((\Pi(\hat{\Gamma}_{uX}, \hat{\Gamma}_{uX}^*, \hat{\Gamma}_{uY}, \hat{\Gamma}_{uY}^* : u \geq 0))^k] \tag{9.12}
$$

$$
= \frac{1+y}{y} \bar{\varphi}[(\Pi(\bar{\gamma}_{X,uq}, \bar{\gamma}_{X,uq}^*, \bar{\gamma}_{Y,uq}, \bar{\gamma}_{Y,uq}^* : u \geq 0))^k].
$$

This establishes (M1).

By Lemmas 6.3.1 and 9.2.1(b), we have for any polynomial Π

$$
E\left[\frac{1}{p}\mathrm{Tr}\big(\Pi(\hat{\Gamma}_{uX}, \hat{\Gamma}_{uX}^*, \hat{\Gamma}_{uY}, \hat{\Gamma}_{uY}^* : u \geq 0)\big)^h \right. \tag{9.13}
$$

$$
\left. -E(\frac{1}{p}\mathrm{Tr}\big(\Pi(\hat{\Gamma}_{uX}, \hat{\Gamma}_{uX}^*, \hat{\Gamma}_{uY}, \hat{\Gamma}_{uY}^* : u \geq 0))^h\big)\right]^4 = O(p^{-4})
$$

and hence (M4) is established.

Proof of Carleman's condition (C) is essentially same as its counter part in Theorem 6.3.1. This completes the proof of Theorem 9.2.1. $\square$

9.2.2   MA(∞)

Now consider the processes in (9.1) with $q = \infty$. Recall $||\cdot||_2$ defined in (1.8). We now introduce the assumption

(B8) $\sum_{j=0}^{\infty}\left[\sup_p ||\psi_j||_2 + \sup_p ||\phi_j||_2\right] < \infty$.

Suppose (B8) holds. Then for any polynomial $\Pi(\bar{\gamma}_{X,uq}, \bar{\gamma}_{X,uq}^*, \bar{\gamma}_{Y,uq}, \bar{\gamma}_{Y,uq}^* : u \geq 0)$, $\lim_{q\to\infty} \bar{\varphi}(\Pi(\bar{\gamma}_{X,uq}, \bar{\gamma}_{X,uq}^*, \bar{\gamma}_{Y,uq}, \bar{\gamma}_{Y,uq}^* : u \geq 0))$ exists and is finite. Consider the NCP $(\mathcal{A}_\infty, \bar{\varphi}_\infty)$ where

$$
\mathcal{A}_\infty = \mathrm{Span}\{\bar{\gamma}_{X,u\infty}, \bar{\gamma}_{X,u\infty}^*, \bar{\gamma}_{Y,u\infty}, \bar{\gamma}_{Y,u\infty}^* : u \geq 0\} \tag{9.14}
$$

and for any polynomial $\Pi(\bar{\gamma}_{X,u\infty}, \bar{\gamma}^*_{X,u\infty}, \bar{\gamma}_{Y,u\infty}, \bar{\gamma}^*_{Y,u\infty} : u \geq 0)$,

$$\bar{\varphi}_\infty(\Pi(\bar{\gamma}_{X,u\infty}, \bar{\gamma}^*_{X,u\infty}, \bar{\gamma}_{Y,u\infty}, \bar{\gamma}^*_{Y,u\infty} : u \geq 0)) \tag{9.15}$$
$$= \lim_{q \to \infty} \bar{\varphi}(\Pi(\bar{\gamma}_{X,uq}, \bar{\gamma}^*_{X,uq}, \bar{\gamma}_{Y,uq}, \bar{\gamma}^*_{Y,uq} : u \geq 0)).$$

Now we have the following theorem. Its proof is similar to the proof of Theorem 8.2.2 and is left as an exercise.

Theorem 9.2.2. *Consider the model (9.1) with* $q = \infty$. *Suppose (B5), (B7), (D) and (B8) hold and* $p, n = n(p) \to \infty$, $p/n \to y \in (0, \infty)$. *Then the LSD of any symmetric polynomial* $\Pi(\hat{\Gamma}_{uX}, \hat{\Gamma}^*_{uX}, \hat{\Gamma}_{uY}, \hat{\Gamma}^*_{uY} : u \geq 0)$ *in* $\{\hat{\Gamma}_{uX}, \hat{\Gamma}^*_{uX}, \hat{\Gamma}_{uY}, \hat{\Gamma}^*_{uY}\}$ *exists almost surely and it is uniquely determined by the moment sequence*

$$\lim p^{-1} E \mathrm{Tr}[(\Pi(\hat{\Gamma}_{uX}, \hat{\Gamma}^*_{uX}, \hat{\Gamma}_{uY}, \hat{\Gamma}^*_{uY} : u \geq 0))^k] \tag{9.16}$$
$$= \frac{1+y}{y} \bar{\varphi}_\infty[(\Pi(\bar{\gamma}_{X,u\infty}, \bar{\gamma}^*_{X,u\infty}, \bar{\gamma}_{Y,u\infty}, \bar{\gamma}^*_{Y,u\infty} : u \geq 0))^k].$$

Now we shall discuss application of Theorems 9.2.1 and 9.2.2 to some specific cases. Recall the classes of independent random variables defined in (8.28):

$$U(\delta) = \text{all collections of independent random variables } \{\varepsilon_{i,j}\}$$
$$\text{such that for all } \eta > 0$$
$$\lim \frac{\eta^{-(2+\delta)}}{np} \sum_{i=1}^p \sum_{j=1}^n E(|\varepsilon_{i,j}|^{2+\delta} I(|\varepsilon_{i,j}| > \eta p^{\frac{1}{2+\delta}})) - 0. \tag{9.17}$$

Justification of the following examples is similar to the proofs of Corollaries 8.2.1 and 8.2.4. We leave them as exercises.

Example 9.2.1. *(a)* Consider the model (9.1) with $q < \infty$. Suppose (B5), (B7), and (D) hold and $p, n = n(p) \to \infty$, $p/n \to y \in (0, \infty)$. Then the LSD of $(\hat{\Gamma}_{uX} + \hat{\Gamma}^*_{uX}) \pm (\hat{\Gamma}_{uY} + \hat{\Gamma}^*_{uY})$ exist almost surely.

(b) Under the additional assumption (B8), the above result in (a) holds for the model (9.1) with $q = \infty$.

(c) The above results in (a) and (b) hold if, instead of (B7), we assume (B6) or, $\{\varepsilon_{i,j} : i, j \geq 1\}, \{\xi_{i,j} : i, j \geq 1\} \in \mathcal{L}_{2+\delta} \cap U(\delta)$ for some $\delta > 0$.

Example 9.2.2. *(a)* Consider the model (9.1) with $q < \infty$. Suppose (B5), (B7), and (D) hold and $p, n = n(p) \to \infty$, $p/n \to y \in (0, \infty)$. Then the LSD of $\hat{\Gamma}_{uX}\hat{\Gamma}^*_{uX} \pm \hat{\Gamma}_{uY}\hat{\Gamma}^*_{uY}$ exists almost surely.

(b) Under the additional assumption (B8), the above result in (a) holds for the model (9.1) with $q = \infty$.

(c) The above results in (a) and (b) hold if, instead of (B7), we assume (B6) or, $\{\varepsilon_{i,j} : i, j \geq 1\}, \{\xi_{i,j} : i, j \geq 1\} \in \mathcal{L}_4 \cap U(\delta)$ for some $\delta > 0$.

9.3   LSD when $p/n \to 0$

9.3.1 $MA(q), q < \infty$

Now we shall discuss the LSD of symmetric polynomials in $\{\hat{\Gamma}_{uX}, \hat{\Gamma}_{uY}\}$ for the case $p, n(p) \to \infty$ such that $p/n \to 0$. Consider the free collections of non-commutative variables $\{w_{X,u,j_1,j_2}\}$ and $\{w_{Y,u,j_1,j_2}\}$, whose free cumulants are as follows:

For all $j_1, j_2, \ldots \geq 1$ and $u_1, u_2, \ldots = 0, \pm 1, \pm 2, \ldots,$

$$w^*_{X,u,j_1,j_2} = w_{X,-u,j_2,j_1}, \quad w^*_{Y,u,j_1,j_2} = w_{Y,-u,j_2,j_1}, \qquad (9.18)$$

$$
\begin{aligned}
\kappa_2(w_{X,u_1,j_1,j_2}, w_{X,u_2,j_3,j_4}) &= \kappa_2(w_{Y,u_1,j_1,j_2}, w_{Y,u_2,j_3,j_4}) \\
&= \frac{1}{2\pi} \int_0^{2\pi} e^{i(j_1-j_2+u_1)\theta} e^{i(j_3-j_4+u_2)\theta} \, d\theta \\
&= \begin{cases} 1, & \text{if } j_2 - j_1 - j_3 + j_4 = u_1 + u_2 \\ 0, & \text{otherwise,} \end{cases} \quad (9.19)
\end{aligned}
$$

$$\kappa_r(w_{X,u_i,j_{2i-1},j_{2i}} : i \leq r) = \kappa_r(w_{Y,u_i,j_{2i-1},j_{2i}} : i \leq r) = 0, \ \forall r \neq 2. \quad (9.20)$$

As mentioned in Section 7.2, the above sequence of free cumulants naturally define a state, say φ_w, on $\mathrm{Span}\{w_{X,u,l,i}, w_{Y,u,l,i} : u, l, i \geq 1\}$. This is because, the moments and free cumulants are in one-to-one correspondence (see (5.22) and (5.24) in Chapter 4).

Recall $(\mathcal{A}_{\mathrm{odd}}, \varphi_{\mathrm{odd}})$ and $\{\eta_j, \eta_j^*, \theta_j, \theta_j^*\}$ respectively in (6.13) and (9.3). Also recall the free product $(\mathcal{B}, \varphi_0)$ of $(\mathcal{A}_{\mathrm{odd}}, \varphi_{\mathrm{odd}})$ and $(\mathrm{Span}\{w_{X,u,l,i}, w_{Y,u,l,i} : u, l, i \geq 1\}, \varphi_w)$ in (7.41). Therefore, $\{w_{X,u,j_1,j_2}\}$, $\{w_{Y,u,l,i}\}$ and $\{\eta_j, \eta_j^*, \theta_j, \theta_j^*\}$ are free in $(\mathcal{B}, \varphi_0)$.

For all $u, q \geq 0$, consider the following polynomial in $\mathcal{B}$

$$S_{X,uq} = \sum_{j_1,j_2=0}^{q} \eta_{j_1} w_{X,u,j_1,j_2} \eta_{j_2}^* \quad \text{and} \quad S_{Y,uq} = \sum_{j_1,j_2=0}^{q} \theta_{j_1} w_{Y,u,j_1,j_2} \theta_{j_2}^*. \quad (9.21)$$

The population autocovariance matrices $\{\Gamma_{uX}\}$ and $\{\Gamma_{uY}\}$ respectively for $\{X_t\}$ and $\{Y_t\}$ are given by

$$\Gamma_{uX} = \sum_{j=0}^{q-u} \psi_j \psi_{j+u}^* \quad \text{and} \quad \Gamma_{uY} = \sum_{j=0}^{q-u} \phi_j \phi_{j+u}^* \ \forall u \geq 0.$$

Note that, by (9.3),

$$
\begin{aligned}
(\mathrm{Span}\{\Gamma_{uX}, &\Gamma_{uX}^*, \Gamma_{uY}, \Gamma_{uY}^* : u \geq 0\}, p^{-1}\mathrm{Tr}) \\
&\to (\mathrm{Span}\{G_{X,uq}, G_{X,uq}^*, G_{Y,uq}, G_{Y,uq}^* : u \geq 0\}, \varphi_{\mathrm{odd}}), \quad (9.22)
\end{aligned}
$$

where

$$G_{X,uq} = \sum_{j=0}^{q-u} \eta_j \eta_{j+u}^* \quad \text{and} \quad G_{Y,uq} = \sum_{j=0}^{q-u} \theta_j \theta_{j+u}^* \quad \forall u, q \geq 0. \qquad (9.23)$$

Now we will state the LSD result for general self-adjoint polynomials $\Pi(\hat{\Gamma}_{uX}, \hat{\Gamma}_{uX}^*, \hat{\Gamma}_{uY}, \hat{\Gamma}_{uY}^* : u \geq 0)$. To describe the limit write it in the form

$$\Pi(\hat{\Gamma}_{uX}, \hat{\Gamma}_{uX}^*, \hat{\Gamma}_{uY}, \hat{\Gamma}_{uY}^* : u \geq 0) = \sum_{l=1}^{T} \beta_l \left(\prod_{i=1}^{k_l} \hat{\Gamma}_{u_{X,l,i}X}^{\epsilon_{X,l,i}} \hat{\Gamma}_{u_{Y,l,i}Y}^{\epsilon_{Y,l,i}} \right), \qquad (9.24)$$

where $\epsilon_{X,l,i}, \epsilon_{Y,l,i} \in \{1, *\}$ and $u_{X,l,i}, u_{Y,l,i} \in \{0, 1, 2, \ldots\}$.

Then we have the following Theorem. Its proof is similar to the proof of Theorem 8.3.1. We leave it as an exercise.

Theorem 9.3.1. *(Bhattacharjee and Bose [2016b]) Consider the model (9.1). Suppose (B5), (B7), and (D) hold and $p, n(p) \to \infty$, $p/n \to 0$. Then the LSD of the self-adjoint polynomial*

$$\sqrt{np^{-1}} \left(\Pi(\hat{\Gamma}_{uX}, \hat{\Gamma}_{uX}^*, \hat{\Gamma}_{uY}, \hat{\Gamma}_{uY}^* : u \geq 0) - \Pi(\Gamma_{uX}, \Gamma_{ux}^*, \Gamma_{uY}, \Gamma_{uY}^* : u \geq 0) \right) \qquad (9.25)$$

exists almost surely and is distributed as

$$\sum_{l=1}^{T} \beta_l \Big(\sum_{\substack{\nu_{X,l,i} \in \{0,1\} \\ \sum_i \nu_{X,l,i}=1}} \sum_{\substack{\nu_{Y,l,i} \in \{0,1\} \\ \sum_i \nu_{Y,l,i}=1}} \prod_{i=1}^{k_l} G_{X,u_{X,l,i}q}^{\epsilon_{X,l,i}(1-\nu_{X,l,i})} S_{X,u_{l,i}q}^{\epsilon_{X,l,i}\nu_{X,l,i}} \qquad (9.26)$$

$$G_{Y,u_{Y,l,i}q}^{\epsilon_{Y,l,i}(1-\nu_{Y,l,i})} S_{Y,u_{l,i}q}^{\epsilon_{Y,l,i}\nu_{Y,l,i}} \Big)$$

where $\{S_{X,uq}, S_{Y,uq}\}$ and $\{G_{X,uq}, G_{Y,uq}\}$ are as in (9.21) and (9.23).

9.3.2 MA(∞)

Proof of the next Theorem uses the same arguments as in the proof of Theorem 8.2.2 and hence we omit it.

Theorem 9.3.2. *(Bhattacharjee and Bose [2016b]) Consider the model (8.16). Suppose (B5), (B7), (D), and (B8) hold and $p, n(p) \to \infty$, $p/n \to 0$. Let $\Pi(\hat{\Gamma}_{uX}, \hat{\Gamma}_{uX}^*, \hat{\Gamma}_{uY}, \hat{\Gamma}_{uY}^* : u \geq 0)$ be decomposed as in (9.24). Then the LSD of the self-adjoint polynomial given in (9.25) exists almost surely and is distributed as*

$$\sum_{l=1}^{T} \beta_l \Big(\sum_{\substack{\nu_{X,l,i} \in \{0,1\} \\ \sum_i \nu_{X,l,i}=1}} \sum_{\substack{\nu_{Y,l,i} \in \{0,1\} \\ \sum_i \nu_{Y,l,i}=1}} \prod_{i=1}^{k_l} G_{X,u_{X,l,i}\infty}^{\epsilon_{X,l,i}(1-\nu_{X,l,i})} S_{X,u_{X,l,i}\infty}^{\epsilon_{X,l,i}\nu_{X,l,i}}$$

$$G_{Y,u_{Y,l,i}\infty}^{\epsilon_{Y,l,i}(1-\nu_{Y,l,i})} S_{Y,u_{Y,l,i}\infty}^{\epsilon_{Y,l,i}\nu_{Y,l,i}} \Big). \qquad (9.27)$$

Recall the classes $\mathcal{L}_r$ and $U(\delta)$. Justification of the following example is similar to the proof of Corollary 8.3.1. We leave it as an exercise.

Example 9.3.1. (a) Consider the model (9.1) with $q < \infty$. Suppose (B5), (B7) and (D) hold and $p, n(p) \to \infty$, $p/n \to 0$. Then the LSD of $\sqrt{np^{-1}}\left(\left((\hat{\Gamma}_{uX} + \hat{\Gamma}_{uX}^*) \pm (\hat{\Gamma}_{uY} + \hat{\Gamma}_{uY}^*)\right) - \left((\Gamma_{uX} + \Gamma_{uX}^*) \pm (\Gamma_{uY} + \Gamma_{uY}^*)\right)\right)$ exists almost surely and it is same as the distribution of

$$\sum_{j_1,j_2=0}^{q}\left[\eta_{j_1}w_{X,u,j_1,j_2}\eta_{j_2}^* + \eta_{j_2}w_{X,-u,j_2,j_1}\eta_{j_1}^* \pm \left\{\theta_{j_1}w_{Y,u,j_1,j_2}\theta_{j_2}^* + \theta_{j_2}w_{Y,-u,j_2,j_1}\theta_{j_1}^*\right\}\right].$$

(b) Under the additional assumption (B8), the result in (a) holds for the model (9.1) once we replace q by ∞.

(c) The results in (a) and (b) hold, if instead of $(B7)$, we assume $(B6)$ or, $\{\varepsilon_{i,j} : i, j \geq 1\}, \{\xi_{i,j} : i, j \geq 1\} \in \mathcal{L}_{2+\delta} \cap U(\delta)$ for some $\delta > 0$.

Exercises

1. Suppose $X_t = \varepsilon_t + A_1\varepsilon_{t-1}$, $Y_t = \xi_t + A_2\xi_{t-1}$ for all t, $A_1 = ((I(|i - j| = 1)))_{p \times p}$, $A_1 = ((I(|i - j| = 2)))_{p \times p}$ and $p/n \to y > 0$. Under appropriate assumptions, show that the LSD of the following matrices exists.

   (a) $\hat{\Gamma}_{1X}\hat{\Gamma}_{1Y} + \hat{\Gamma}_{1Y}^*\hat{\Gamma}_{1X}^*$

   (b) $\hat{\Gamma}_{0X}\hat{\Gamma}_{0Y} + \hat{\Gamma}_{0Y}\hat{\Gamma}_{0X}$

   (c) $\hat{\Gamma}_{0X} + \hat{\Gamma}_{1Y} + \hat{\Gamma}_{1Y}^*$

   Also describe these LSD in terms of polynomials in freely independent variables. Do the appropriate modification when $p/n \to 0$.

2. Give a proof of Lemma 9.2.1.

3. Establish Theorem 9.2.2.

4. Justify Examples 9.2.1 and 9.2.2.

5. Prove Theorems 9.3.1 and 9.3.2.

6. Justify Example 9.3.1.

Chapter 10

GRAPHICAL INFERENCE

Statistical inference for high-dimensional time series is a difficult area and not many methods are known. Some of the methods and results developed in the previous chapters can be used for this purpose.

For example, in Chapter 3 we have seen how banding and tapering methods can be used to estimate the autocovariance matrices of a high-dimensional linear time series. These ideas can also be used for parameter estimation in an IVAR process. For instance, we show how Theorem 8.2.2 can be used to estimate the spectral distribution of the coefficient matrices $\{\psi_j\}$ of the infinite-dimensional MA(q) process defined in (8.6).

Similarly, in Chapters 8 and 9 we have established the LSD for symmetric polynomials of sample autocovariance matrices respectively for one and two independent infinite-dimensional MA processes. These spectral distribution results can be used for graphical inference in high-dimensional time series. For example, we can determine the *unknown order* of infinite-dimensional moving average (MA) and autoregressive (IVAR) processes easily by plotting the CDF of ESD of these polynomials. These results also provide graphical tests for hypotheses on parameter matrices of MA processes.

10.1 MA order determination

There is a very well-known method of order determination when the process is univariate. It is based on autocovariances. Suppose we wish to determine the order q of a univariate MA(q) process. For this process, the population autocovariances of order greater than q are all zero. On the other hand, under reasonable conditions, the sample autocovariances are consistent estimators for the population autocovariances. Hence, a plot of the sequence of sample autocovariances provides a graphical method to identify q; if the plot falls below a small threshold for $u > \hat{q}$, then $\hat{q}$ is an estimate of q.

Now consider the infinite-dimensional MA(q) process defined in (8.6) with unknown parameter matrices and we wish to determine q. Then we know that Γ_u is the null matrix for all $u > q$. Suppose $\{X_{t,p} : 1 \leq t \leq n\}$ is a sample of size n from the process (8.6). Unlike the finite-dimensional case, as we have seen in Chapter 3, the sample autocovariance matrices $\{\hat{\Gamma}_u\}$ are not consistent for the population autocovariance matrices $\{\Gamma_u\}$. Hence, the idea described for univariate MA processes above cannot be extended naively.

Despite that, the LSD results obtained in Chapter 8 for $\{\hat{\Gamma}_u\}$ come to our rescue. The method that we are about to explain is based on Corollaries 8.2.1 (d), 8.2.4 (d), 8.3.1 and 8.3.3. For convenience of the reader, those results along with some of their consequences are collected in the following theorem. Its proof is easy and is omitted.

Consider the following symmetric polynomials,

$$\begin{aligned}
\hat{\Pi}_{1u} &= \hat{\Gamma}_u \hat{\Gamma}_u^*, \quad \Pi_{1u} = \Gamma_u \Gamma_u^*, \\
\hat{\Pi}_{2u} &= \hat{\Gamma}_u \hat{\Gamma}_u^* + \hat{\Gamma}_{u+1} \hat{\Gamma}_{u+1}^*, \quad \Pi_{2u} = \Gamma_u \Gamma_u^* + \Gamma_{u+1} \Gamma_{u+1}^*, \\
\hat{\Pi}_{3u} &= \hat{\Gamma}_u + \hat{\Gamma}_u^*, \quad \Pi_{3u} = \Gamma_u + \Gamma_u^*.
\end{aligned} \tag{10.1}$$

Incidentally, even though the theorem is stated for these above specific polynomials, its conclusions hold for other polynomials if we are ready to make appropriate moment assumptions on $\{\varepsilon_{i,j}\}$. We have restricted to the above polynomials only for simplicity. Existence of all the LSD below are guaranteed by Theorems 8.2.1 and 8.3.1. Recall the classes $\mathcal{L}_r$ and $U(\delta)$ respectively from (8.1) and (8.28).

Theorem 10.1.1. *Consider the model (8.6). Suppose $\{\varepsilon_{i,j}\}$ are independent with $E(\varepsilon_{i,j}) = 0$, $E(\varepsilon_{i,j})^2 = 1$, $\forall i, j$ and $\{\varepsilon_{i,j}\} \in \mathcal{L}_4 \cap U(\delta)$ for some $\delta > 0$. Further $\{\psi_j\}$ are all norm bounded and converge jointly.*

(a) If $p/n \to y > 0$, then the LSD of $\hat{\Pi}_{1u}$ (or of $\hat{\Pi}_{2u}$, $\hat{\Pi}_{3u}$) are identical for $u > q$ and are different for $0 \le u \le q$.

(b) If $p/n \to 0$, then for all $u \ge 0$, the LSD of $\hat{\Pi}_{1u}$, $\hat{\Pi}_{2u}$ and $\hat{\Pi}_{3u}$ are respectively identical with the LSD of Π_{1u}, Π_{2u} and Π_{3u}. Moreover, for $u > q$, these LSD are degenerate at 0.

Suppose we know that the observations are from the time series (8.6) where the unknown q is the parameter of interest and parameter matrices $\{\psi_j\}$ are the nuisance parameters. The above result can be used to identify graphically the value of q as follows. We plot the CDF of the ESD (call it ECDF) of $\hat{\Pi}_{1u}$ (or of $\hat{\Pi}_{2u}$, $\hat{\Pi}_{3u}$) for first few sample autocovariance matrices. When p is moderate compared to n, we say that $\hat{q}$ is an estimate of q if the ECDF of $\hat{\Pi}_{1u}$ (or $\hat{\Pi}_{2u}$, $\hat{\Pi}_{3u}$) with order $u > \hat{q}$ empirically coincide with each other. Similarly, when p is small compared to n, $\hat{q}$ is an estimate of q, if the ECDF of $\hat{\Pi}_{1u}$ (or $\hat{\Pi}_{1u}$, $\hat{\Pi}_{3u}$) with $u > \hat{q}$ is degenerate at 0. By Theorem 10.1.1, $\hat{q}$ is a reasonable estimator of q. An important point to note is that the estimation of the parameter matrices is not required to implement this method.

Models. Let us explore the performance of the above method through simulations. Let I_p and J_p be respectively as in (1.9) and (1.10). Let

$$\varepsilon_t \sim \mathcal{N}_p(0, I_p), \forall t. \tag{10.2}$$

Suppose

$$A_p = 0.5I_p, \quad B_p = 0.5(I_p + J_p),$$
$$C_p = ((I(1 \le i = j \le [p/2]) - I([p/2] < i = j \le p))),$$
$$D_p = ((I(i + j = p + 1))).$$

We consider the following models.

Model 1 $X_t = \varepsilon_t$.

Model 2 $X_t = \varepsilon_t + A_p\varepsilon_{t-1}$.

Model 3 $X_t = \varepsilon_t + B_p\varepsilon_{t-1}$.

Model 4 $X_t = \varepsilon_t + C_p\varepsilon_{t-1} + D_p\varepsilon_{t-2}$.

By Examples 4.1.1-4.1.4, LSD of A_p, B_p, C_p and D_p exist. It is also easy to see that $\{C_p, D_p\}$ converge jointly. Hence, for all the above models, Theorems 8.2.1 and 8.3.1 are applicable.

Suppose we have a time series sample of size n, say S_i, from Model i, $1 \le i \le 4$. Note that no replication is involved.

Case a: $p/n \to y > 0$. We take $p = n = 500$. Pretending that we do not know from which model our sample is from, we wish to determine q. For each $1 \le i \le 4$, we plot the ECDF of $\hat{\Pi}_{1u}$ for $1 \le u \le 4$, based on the sample S_i, in the same graph. See Figure 10.1.

Note that the ECDF of $\hat{\Pi}_{1u}$ coincide in the following cases: (a) for all $u > 0$ in Model 1, (b) for $u > 1$ in Models 2 and 3, and (c) for $u > 2$ in Model 4. Hence, $\hat{q}$ is 0, 1, 1 and 2 respectively for Models 1–4. Thus, the above method identifies q precisely. Incidentally, since LSD of A_p and B_p are identical (see Examples 4.1.1 and 4.1.2—both have mass 1 at 0.5), the ECDF for Models 2 and 3 are almost identical. Similar behavior was observed for the polynomial $\hat{\Pi}_{2u}$ (see Figure 10.2).

For values of n smaller than $n = 300$, convergence often did not occur in our simulations. Some modification may improve the situation for smaller sample sizes. No results in this direction seem to be currently known.

Case b: $p/n \to 0$. For Models 1–4, we draw *one* sample with $n = 500$, $p = n^{0.9} \sim 269$. We plot the ECDF of $\hat{\Pi}_{3u}$ for $1 \le u \le 4$ in Figure 10.3. We have the same conclusions as in Case a.

An alternative approach. It may not be easy to decide if the ESD is close to being degenerate. On the other hand by Corollary 8.3.1(a), the LSD of $0.5\sqrt{np^{-1}}(\hat{\Gamma}_u + \hat{\Gamma}_u^* - \Gamma_u - \Gamma_u^*)$ exits and its description is given in (8.53). However $\{\Gamma_u\}$ are unknown. A way out is to use an appropriate (consistent) estimator of Γ_u. Fortunately this is available from Chapter 3 for certain combinations of n and p.

Recall the appropriate parameter spaces $\Im(\beta, \lambda)$ and $\mathcal{G}(C, \alpha, \eta, \nu)$ from (3.33) and (3.34) respectively, for $\{\psi_j\}$ where the above consistency can be achieved.

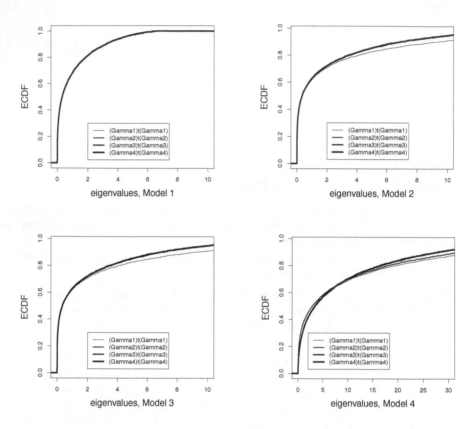

Figure 10.1 $ECDF$ of $\hat{\Pi}_{1u}$, $1 \leq u \leq 4$, $n = p = 500$. In the graphs, Gamma $u = \hat{\Gamma}_u$ and $t(Gamma\ u) = \hat{\Gamma}_u^*$.

Consider the following assumptions.

(C*) $\{\psi_j\} \in \Im(\beta, \lambda) \cap \mathcal{G}(C, \alpha, \eta, \nu)$ for some $0 < \beta, \eta < 1$ and $C, \lambda, \alpha, \nu > 0$.

Let $\varepsilon_{t,j.p}$ be the j-th component of $\varepsilon_{t.p}$.

(C**) For some $\lambda_0 > 0$, $\sup_{j \geq 1} E(e^{\lambda \varepsilon_{t,j.p}}) < \infty$ for all $|\lambda| < \lambda_0$.

Recall the k-banded version of a matrix from (1.15). By Theorem 3.3.3 and Section 3.6, if (C*) and (C**) hold then with $k_n = (n^{-1} \log p)^{-\frac{1}{2(\alpha+1)}}$, we have $\|B_{k_n}(\hat{\Gamma}_u) - \Gamma_u\|_2 = O_p(k_n^{-\alpha})$ for all u. Let

$$\hat{\Pi}_{a,u,B} = \sqrt{np^{-1}}(\hat{\Gamma}_u + \hat{\Gamma}_u^* - B_{k_n}(\hat{\Gamma}_u) - B_{k_n}(\hat{\Gamma}_u^*)).$$

We then have the following theorem.

Theorem 10.1.2. *Consider the model (8.6). Suppose $\{\varepsilon_{i,j}\}$ are independent with $E(\varepsilon_{i,j}) = 0$, $E|\varepsilon_{i,j}|^2 = 1$, $\forall i, j$ and $\{\varepsilon_{i,j}\} \in \mathcal{L}_4 \cap U(\delta)$ for some $\delta > 0$.*

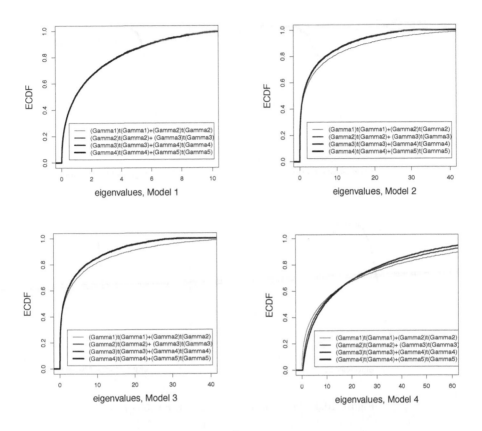

Figure 10.2 *ECDF of $\hat{\Pi}_{2u}$, $1 \leq u \leq 4$, $n = p = 500$.*

Further $\{\psi_j\}$ are all norm bounded and converge jointly, (C*) and (C**) hold and, $n(p), p \to \infty$ such that $p/n \to 0$ and $np^{-(\alpha+1)}(\log p)^\alpha \to 0$. Then for $k_n = \left(n^{-1}\log p\right)^{-\frac{1}{2(\alpha+1)}}$, the LSD (in probability) of $\hat{\Pi}_{a,u,B}$ and $\hat{\Pi}_{a,u}$ are identical. This conclusion also holds for $q = \infty$ if we further assume (B4).

Proof. Observe that by Corollary 8.3.1 and Lemma A.5.1(c) we have

$$\frac{1}{p}\text{Tr}\big(\sqrt{np^{-1}}(\hat{\Gamma}_u + \hat{\Gamma}_u^* - B_{k_n}(\hat{\Gamma}_u) - B_{k_n}(\hat{\Gamma}_u^*))$$

$$-\sqrt{np^{-1}}(\hat{\Gamma}_u + \hat{\Gamma}_u^* - \hat{\Gamma}_u - \hat{\Gamma}_u^*)\big)^2$$

$$= \frac{1}{p}\text{Tr}\big(\hat{\Gamma}_u + \hat{\Gamma}_u^* - B_{k_n}(\hat{\Gamma}_u) - B_{k_n}(\hat{\Gamma}_u^*)\big)^2$$

$$\leq 4np^{-1}\|B_{k_n}(\hat{\Gamma}_u) - \Gamma_u\|_2^2 = o_P\left(\frac{n}{p}\left(\frac{\log p}{n}\right)^{\frac{\alpha}{(\alpha+1)}}\right) \to 0.$$

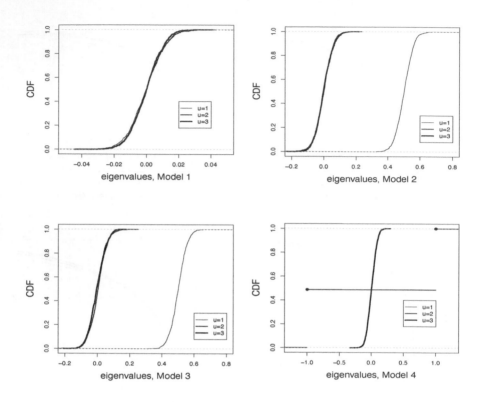

Figure 10.3   *ECDF of $\hat{\Pi}_{3u}$, $1 \leq u \leq 4$, $n = 500, p = n^{0.9}$.*

$\square$

Now by Corollary 8.3.1(a, d) and Theorem 10.1.2, the LSD (in probability) of $\hat{\Pi}_{a,u,B}$ are identical for $u > q$ and are different for $u \leq q$. Thus, we can plot the CDF of the ESD of $\hat{\Pi}_{a,u,B}$ for first few sample autocovariance matrices in the same graph. We say that $\hat{q}$ is an estimate of q, if the ESD of $\hat{\Pi}_{a,u,B}$ with order $u > \hat{q}$ empirically coincide with each other.

Simulations. Consider the $p \times p$ matrix $E = (((|i - j| + 1)^{-1} I(i + j = p + 1)))$ and the following model.

Model 5: $X_t = \varepsilon_t + C\varepsilon_{t-1} + E\varepsilon_{t-2}$.

Let $p = 500, n = p^{1.5}$. For each of the five models, we plot the CDF of the ESD of $\hat{\Pi}_{a,u,B}$ for $1 \leq u \leq 4$ respectively. See Figure 10.4.

The ESD of $\hat{\Pi}_{a,u,B}$ coincide for $u \geq 1$ in Model 1, for $u \geq 2$ in Models 2, 3 and for $u \geq 3$ in Model 5. The matrix D in Model 4 does not belong to the class $\mathcal{G}$ and hence the corresponding ESD are not performing well in the simulation. From the plots, the estimated value of q are $0, 1, 1$ and 2

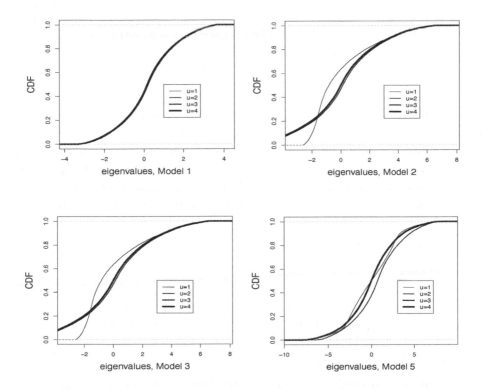

Figure 10.4 *CDF of ESD of* $0.5\sqrt{np^{-1}}(\hat{\Gamma}_u + \hat{\Gamma}_u^* - B_{k_n}(\hat{\Gamma}_u) - B_{k_n}(\hat{\Gamma}_u^*))$, $1 \leq u \leq 4$ *with* $p = 500$, $n = p^{1.5}$ *and* $k_n = \left(n^{-1}\log p\right)^{-\frac{1}{4}}$.

respectively for Models 1–3 and 5. This shows that the method is performing very well.

10.2 AR order determination

Before we discuss order determination of the infinite-dimensional AR process, let us recall the univariate AR(r) process with unknown autoregressive parameters. Unlike the moving average process, the autocovariances do not vanish for higher order. However, the population *partial* autocovariances of order greater than r vanish. Hence, we may plot the sample partial autocovariances to determine r graphically.

We extend this idea to the infinite-dimensional AR(r) process. Suppose

$$X_{t,p} = \sum_{i=1}^{r} A_{i,p} X_{t-i,p} + \varepsilon_t, \ \forall t, \tag{10.3}$$

where $X_{t,p}$ and ε_t are p-dimensional vectors, $\{\varepsilon_t = (\varepsilon_{t,1}, \varepsilon_{t,2}, \ldots, \varepsilon_{t,p})'\}$ are i.i.d. with mean 0 and covariance matrix I_p and, $\{A_{i,p} : 1 \le i \le r\}$ are the $p \times p$ unknown *parameter matrices*. For convenience, we write X_t and A_i respectively for $X_{t,p}$ and $A_{i,p}$. The primary goal is to estimate r.

The idea now is to use the sample partial autocovariance matrices. Since the parameter matrices are assumed to be unknown, we first obtain consistent estimators for these. Then we use these consistent estimators in conjunction with the ideas of the previous section.

Note that we have already obtained such consistent estimators in Chapter 3. Let us denote these consistent estimators by $\{\hat{A}_i^{(r)} : 1 \le i \le r\}$. Now, suppose r is unknown. Consider the residual process $\{\hat{\varepsilon}_t^{(s)}\}$ after fitting the AR(s) process using $\{\hat{A}_i^{(s)} : 1 \le i \le s\}$. In Theorem 10.2.1 and Remark 10.2.1, we argue that the residual process $\{\hat{\varepsilon}_t^{(s)}\}$ behaves like an MA(0) process if and only if $s = r$, the true order of the AR process. We use these results to identify the unknown order of an AR processes.

Consider the following assumptions.

(C1) $\varepsilon_{i,j} \overset{\text{i.i.d.}}{\sim} \mathcal{N}(0,1), \quad \forall i, j \ge 1.$

First recall some notation and results from Chapter 3. For each $1 \le i \le r$, let $A_{i,\infty}$ be the $\infty \times \infty$ extension of the sequence of matrices $\{A_{i,p(n)}\}_{n \ge 1}$. Consider the parameter space $\mathcal{P}(C, \alpha, \epsilon)$ given in (3.59) for $\{A_{i,\infty}\}_{i=1}^r$. We need the following assumption on the parameter matrices $\{A_i\}$.

(C2) $\{A_{i,\infty}\} \in \mathcal{P}(C, \alpha, \epsilon)$ for some $C, \epsilon, \alpha > 0.$

Recall the k-banded version of a matrix from (1.15). Recall $|| \cdot ||_2$ from (1.8). Also recall from (1.14) that by a consistent estimator $\hat{M}_n$ (based on a sample of size n) of M, we mean

$$||\hat{M}_n - M||_2 \overset{P}{\to} 0, \quad \text{as } n \to \infty. \tag{10.4}$$

Let $p/n \to y \in [0, \infty)$. By Theorem 3.5.2, we can say that if (C1) and (C2) hold, then $B_{k_n}(\hat{\Gamma}_u)$ with $k_n = (n^{-1}\log p)^{-1/4}$, is a consistent estimator of Γ_u for each u. Moreover, in Section 3.5, we argued that $(B_{k_n}(\hat{\Gamma}_u))^{-1}$ with $k_n = (n^{-1}\log p)^{-1/4}$, is a consistent estimator of Γ_u^{-1} for each u, provided Γ_u^{-1} exists.

Let

$$\mathcal{Y}_r = (\Gamma_1, \Gamma_2, \ldots, \Gamma_r)^*, \quad \mathcal{A}_r = (A_1^*, A_2^*, \ldots, A_r^*)^*.$$

Let G_r be a block matrix with r^2 many $p \times p$ blocks where the (i,j)-th block is given by

$$G_r(i,j) = \Gamma_{|i-j|} I(i < j) + \Gamma_{|i-j|}^* I(i \ge j), \ 1 \le i, j \le r. \tag{10.5}$$

Consider the Yule–Walker equation,

$$\mathcal{Y}_r = G_r \mathcal{A}_r. \tag{10.6}$$

Consider the following assumption on $\{A_i\}$.

(C3) Γ_0 is non-singular.

By Lemma 3.5.3, if (C1)–(C3) hold, then

$$\mathcal{A}_r = G_r^{-1}\mathcal{Y}_r \qquad (10.7)$$

i.e., each A_i is the finite sum of the finite products of $\{\Gamma_u, \Gamma_u^{-1}, \Gamma_u^*, \Gamma_u^{*-1}\ 1 \leq u \leq r\}$. Moreover, (10.7) provides consistent estimates of A_i, once we replace the population autocovariance matrices $\{\Gamma_u\}$ by their above mentioned consistent estimates $\{B_{k_n}(\hat{\Gamma}_u)\}$ with $k_n = (n^{-1}\log p)^{-1/4}$. Let us denote these estimators of $\{A_i : 1 \leq i \leq r\}$ by $\{\hat{A}_i^{(r)} : 1 \leq i \leq r\}$.

We need the following assumption to guarantee the LSD of symmetric polynomials in $\{\hat{\Gamma}_u, \hat{\Gamma}_u^*\}$.

(C4) $\{A_i\}$ converge jointly.

Define the residuals

$$\hat{\varepsilon}_t^{(s)} = X_t - \sum_{i=1}^{s} \hat{A}_i^{(s)} X_{t-i}, \quad \forall t, s.$$

Now we have the following theorem. It may be noted that even though it will be stated for $\hat{\Pi}_{1u}$, $\hat{\Pi}_{2u}$ and $\hat{\Pi}_{3u}$, the conclusion holds for other polynomials if we are ready to make appropriate moment assumptions on $\{\varepsilon_{i,j}\}$. We have restricted to the above polynomials only for illustrative purposes.

Theorem 10.2.1. *Consider the IVAR(r) process defined in (10.3). Suppose (C1)-(C4) hold and $p/n \to y \in [0, \infty)$. Then for each $u \geq 1$, the LSD (almost sure) of $\hat{\Pi}_{1u}$ (or of $\hat{\Pi}_{2u}$, $\hat{\Pi}_{3u}$) for the process $\{\varepsilon_t\}$ (i.e., for the MA(0) process), coincides with the LSD (in probability) of $\hat{\Pi}_{1u}$ (or of $\hat{\Pi}_{2u}$, $\hat{\Pi}_{3u}$) for $\{\hat{\varepsilon}_t^{(s)}\}$ if $s = r$.*

Proof. For simplicity, we consider only the IVAR(1) process. Let $\{X_t : 1 \leq t \leq n\}$ be a sample of size n from the IVAR(1) process

$$X_{t+1} = \varepsilon_t + AX_{t-1}, \quad \forall t, \qquad (10.8)$$

where ε_t are i.i.d. with mean 0 and variance I_p.

For brevity, we shall prove the theorem only for $\hat{\Pi}_{1u}$ and $\hat{\Pi}_{3u}$ respectively when $p/n \to y > 0$ and $p/n \to 0$.

Recall $|| \cdot ||_2$ in (1.8). By Theorem 3.5.3, under (C2), the consistent estimator $\hat{A}$ of A satisfies

$$||\hat{A} - A|| \xrightarrow{P} 0. \qquad (10.9)$$

Let

$$\hat{\varepsilon}_t^{(1)} = X_t - \hat{A}X_{t-1} = \varepsilon_t + (A - \hat{A})X_{t-1}, \ t \geq 1.$$

Let, for all $k \geq 0$,

$$B_k = n^{-1} \sum_{t=1}^{n-k} \varepsilon_t^{(1)} \varepsilon_{t+k}^{(1)*} \times n^{-1} \sum_{t=1}^{n-k} \varepsilon_{t+k}^{(1)} \varepsilon_t^{(1)*},$$

$$D_k = n^{-1} \sum_{t=1}^{n-k} \varepsilon_t \varepsilon_{t+k}^* \times n^{-1} \sum_{t=1}^{n-k} \varepsilon_{t+k} \varepsilon_t^*,$$

$$E_k = n^{-1} \sum_{t=1}^{n-k} \hat{\varepsilon}_t^{(1)} \hat{\varepsilon}_{t+k}^{(1)*} + n^{-1} \sum_{t=1}^{n-k} \hat{\varepsilon}_{t+k}^{(1)} \hat{\varepsilon}_t^{(1)*},$$

$$F_k = n^{-1} \sum_{t=1}^{n-k} \varepsilon_t \varepsilon_{t+k}^* + n^{-1} \sum_{t=1}^{n-k} \varepsilon_{t+k} \varepsilon_t^*.$$

By Lemma A.5.1(c) in the Appendix, it is enough to show that

$$n^{-1} \text{Tr}(B_k - D_k)^2 \overset{P}{\to} 0, \quad \text{as } p/n \to y > 0 \text{ and} \tag{10.10}$$

$$np^{-2} \text{Tr}(E_k - F_k)^2 \overset{P}{\to} 0, \quad \text{as } p/n \to 0. \tag{10.11}$$

Proof of (10.10). Note that

$$\frac{1}{n} \sum_{t=1}^{n-k} \varepsilon_t^{(1)} \varepsilon_{t+k}^{(1)*}$$

$$= \frac{1}{n} \sum_{t=1}^{n-k} \varepsilon_t \varepsilon_{t+k}^* + (A - \hat{A}) \frac{1}{n} \sum_{t=1}^{n-k} X_{t-1} \varepsilon_{t+k}^* + \frac{1}{n} \sum_{t=1}^{n-k} \varepsilon_t X_{t+k-1}^* (A - \hat{A})^*$$

$$+ (A - \hat{A}) \frac{1}{n} \sum_{t=1}^{n-k} X_{t-1} X_{t+k-1}^* (A - \hat{A})^*$$

$$= \frac{1}{n} \sum_{t=1}^{n-k} \varepsilon_t \varepsilon_{t+k}^* + (A - \hat{A}) \frac{1}{n} \sum_{t=1}^{n-k} X_{t-1} X_{t+k}^* - (A - \hat{A}) \frac{1}{n} \sum_{t=1}^{n-k} X_{t-1} X_{t+k-1}^* A^*$$

$$+ \frac{1}{n} \sum_{t=1}^{n-k} X_t X_{t+k-1}^* (A - \hat{A})^* - A \frac{1}{n} \sum_{t=1}^{n-k} X_{t-1} X_{t+k-1}^* (A - \hat{A})^*$$

$$+ (A - \hat{A}) \frac{1}{n} \sum_{t=1}^{n-k} X_{t-1} X_{t+k-1}^* (A - \hat{A})^*$$

$$= G_1 + G_2 + G_3 + G_4 + G_5 + G_6, \text{ (say)}. \tag{10.12}$$

Therefore,

$$B_k = \sum_{j,l=1}^{6} G_j G_l^* \quad \text{and} \quad D_k = G_1 G_1^*.$$

By Hölder's inequality,

$$n^{-1}\mathrm{Tr}(B_k - D_k)^2 = \sum_{\substack{1 \le j_1, j_2, l_1, l_2 \le 6 \\ (j_1, l_1), (j_2, l_2) \ne (1,1)}} n^{-1}\mathrm{Tr}(G_{j_1} G_{l_1}^* G_{j_2} G_{l_2}^*),$$

$$\le \sum_{\substack{1 \le j_1, j_2, l_1, l_2 \le 6 \\ (j_1, l_1), (j_2, l_2) \ne (1,1)}} \left(\prod^2 \frac{1}{n}\mathrm{Tr}(G_{j_s}^* G_{j_s})^2 \frac{1}{n}\mathrm{Tr}(G_{l_s}^* G_{l_s})^2 \right)^{1/4}. \quad (10.13)$$

Therefore, to show (10.10), it is enough to prove

(i) $\frac{1}{n}\mathrm{Tr}(G_1^* G_1)^2 = O_P(1)$ and

(ii) $\frac{1}{n}\mathrm{Tr}(G_i^* G_i)^2 = o_P(1)$, $\forall 2 \le i \le 6$.

To prove (i) and (ii), we need the following lemma.

Lemma 10.2.1. *Consider MA(q) or MA(∞) processes respectively defined in (8.6) and (8.16). Suppose (C1) and (C4) hold and $p/n \to y > 0$. Then for any symmetric polynomial Π,*

$$\frac{1}{n}\mathrm{Tr}(\Pi(\hat{\Gamma}_u, \hat{\Gamma}_u^* : u \ge 0)) \overset{a.s.}{\to} \lim E\frac{1}{n}\mathrm{Tr}(\Pi(\hat{\Gamma}_u, \hat{\Gamma}_u^* : u \ge 0)), \quad (10.14)$$

and hence

$$\frac{1}{n}\mathrm{Tr}(\Pi(\hat{\Gamma}_u, \hat{\Gamma}_u^* : u \ge 0)) = O_P(1).$$

Proof. By (M1) and (M4) in the proof of Theorems 8.2.1 and 8.2.2, under (C1) and (C4) and as $p/n \to y > 0$, we have

$$\frac{1}{n}E\mathrm{Tr}(\Pi(\hat{\Gamma}_u, \hat{\Gamma}_u^* : u \ge 0)) \to \lim E\frac{1}{n}\mathrm{Tr}(\Pi(\hat{\Gamma}_u, \hat{\Gamma}_u^* : u \ge 0))$$
$$E(n^{-1}\mathrm{Tr}(\Pi(\hat{\Gamma}_u, \hat{\Gamma}_u^* : u \ge 0)) - En^{-1}\mathrm{Tr}(\Pi(\hat{\Gamma}_u, \hat{\Gamma}_u^* : u \ge 0)))^4 = O(n^{-4}).$$

Hence, by Borel-Cantelli lemma, the first part of Lemma 10.2.1 follows. The second part is trivial. $\qquad\square$

Note that G_1 is $\hat{\Gamma}_k$ for the MA(0) process $\{\varepsilon_t\}$. Therefore, (i) follows immediately by Lemma 10.2.1.

We now prove (ii), first for $i = 6$. Recall $\|\cdot\|_2$ in (1.8). Note that for any $n \times n$ matrix A,

$$n^{-1}\mathrm{Tr}(A^* A) \le \|A\|_2^2. \quad (10.15)$$

Note that $G_6 = (A - \hat{A})\hat{\Gamma}_k(A - \hat{A})^*$. Therefore, by Hölder's inequality,

$$\frac{1}{n}\mathrm{Tr}(G_6^* G_6)^2 \le \left(\frac{1}{n}\mathrm{Tr}[((A - \hat{A})(A - \hat{A})^*)^8] \frac{1}{n}\mathrm{Tr}[(\hat{\Gamma}_k^* \hat{\Gamma}_k)^8] \right)^{1/2}$$

$$\le \|A - \hat{A}\|_2^8 \left(\frac{1}{n}\mathrm{Tr}(\hat{\Gamma}_k^* \hat{\Gamma}_k)^8 \right)^{1/2}. \quad (10.16)$$

By (10.9), $||A - \hat{A}||_2 = o_P(1)$ and by Lemma 10.2.1, $\frac{1}{n}\text{Tr}[(\hat{\Gamma}_k^*\hat{\Gamma}_k)^8] = O_P(1)$. Hence, $\frac{1}{n}\text{Tr}[(G_6^*G_6)^2] = o_P(1)$. Therefore, (ii) is proved for $i = 6$.

Similar arguments go through for $i = 2, 3, 4, 5$. Hence, the proof of (10.11) is complete. Therefore, (10.10) is proved.

Proof of (10.11). Note that by (10.12), $E_k = \sum_{j=1}^{6} G_j + \sum_{j=1}^{6} G_j^*$ and $F_k = G_1 + G_1^*$. Hence,

$$\frac{n}{p^2}\text{Tr}(E_k - F_k)^2 = \sum_{j,l=2}^{6} \frac{n}{p^2}\text{Tr}(G_j + G_j^*)(G_l + G_l^*). \tag{10.17}$$

By Hölder's inequality for all $2 \le j, l \le 6$

$$\left|\frac{n}{p^2}\text{Tr}(G_j + G_j^*)(G_l + G_l^*)\right| \le \frac{4n}{p}\left[p^{-1}\text{Tr}(G_j^*G_j)p^{-1}\text{Tr}(G_l^*G_l)\right]^{1/2}. \tag{10.18}$$

Now the proof of (10.11) for IVAR(1) is completed using arguments similar to those used in the proof of (10.10).

Similar argument works for other polynomials. This completes the proof of Theorem 10.2.1 for the IVAR(1) process. The IVAR(r), $r \ge 2$, can be tackled using the above ideas. $\square$

Remark 10.2.1. *Instead of r, if we use any other positive integer $s < r$, then the residual process $\{\hat{\varepsilon}_t^{(s)}\}$ does not behave like the MA(0) process. Instead of giving a rigorous proof, we limit ourselves to the following heuristic idea. Suppose $\{X_t\}$ is an IVAR(2) process i.e.,*

$$X_t = A_1 X_{t-1} + A_2 X_{t-2} + \varepsilon_t, \quad \forall t \tag{10.19}$$

and we fit the IVAR(1) process using $\hat{A}_1^{(1)}$. Let $\hat{B} = A_1 - \hat{A}_1^{(1)}$. Therefore,

$$\hat{\varepsilon}_t^{(1)} = X_t - \hat{A}_1^{(1)}X_{t-1} = \hat{B}X_{t-1} + A_2 X_{t-2} + \varepsilon_t. \tag{10.20}$$

Let $B = A_1 - \Gamma_0^{-1}\Gamma_1$. Using the fact that $||\hat{A}_1^{(1)} - \Gamma_0^{-1}\Gamma_1||_2 \xrightarrow{P} 0$ (by Theorem 3.5.2), it is easy to see that the LSD of Π_{iu} for the process $\{\hat{\varepsilon}_t^{(1)}\}$ coincides with the corresponding LSD (in probability) for

$$\tilde{\varepsilon}_t^{(1)} = BX_{t-1} + A_2 X_{t-2} + \varepsilon_t. \tag{10.21}$$

Note that under (C2), by Theorem 3.5.1, $\{X_t\}$ can be expressed as

$$X_t = \varepsilon_t + \sum_{j=1}^{\infty} \phi_j \varepsilon_{t-j}, \quad \forall t, \tag{10.22}$$

where $\{\phi_j\}$ are functions of A_2 and B. Therefore,

$$\tilde{\varepsilon}_t^{(1)} = \sum_{j=0}^{\infty} \theta_j \varepsilon_{t-j}, \ \ where \ \theta_0 = I_p, \ \theta_1 = B, \ \theta_{j+2} = B\phi_{j+1} + A_2\phi_j, \ j \geq 2.$$

Note that $\{\tilde{\varepsilon}_t^{(1)}\}$ is an MA(∞) process. Then using similar idea as in the proofs of Corollaries 8.2.4(d) and 8.3.1(d), if $\{\theta_j\}$ are norm bounded and converge jointly, it is easy to prove that the LSD of $\hat{\Pi}_{iu}$ for the process $\{\varepsilon_t\}$ do not coincide with the corresponding LSD (in probability) for $\{\tilde{\varepsilon}_t^{(1)}\}$.

Under suitable conditions, $\{A_1, A_2\}$ and $\{\theta_j\}$ do indeed converge jointly. Proof of this requires some work and we omit it.

Therefore, for each $u \geq 0$, the LSD of $\hat{\Pi}_{iu}$ for the process $\{\varepsilon_t\}$ coincides with the LSD (in probability) for $\{\hat{\varepsilon}_t^{(r)}\}$. Instead of r, if we use any other positive integer $s < r$, then the residual process $\{\hat{\varepsilon}_t^{(s)}\}$ does not behave like the MA(0) process. As by Theorem 10.1.1, ECDF of $\hat{\Pi}_{iu}$ for $u = 1, 2$ coincide (almost surely) for the MA(0) process, to determine the order of the IVAR process, it is enough to check whether the ECDF of $\hat{\Pi}_{iu}$ of $\{\hat{\varepsilon}_t^{(r)}\}$ for $u = 1, 2$ coincide or not. Therefore, if we plot the ECDF of $\hat{\Pi}_{iu}$ $u = 1, 2$ for the residual process $\{\hat{\varepsilon}_t^{(s)}\}$ in the same graph, the two distribution functions will coincide only when $s = r$. Hence, we have the following method.

Identification of the unknown order r. Successively fit an IVAR(s) for $s = 0, 1, 2, \ldots$ and for each s, plot the ECDF of $\hat{\Pi}_{1u}$ (or $\hat{\Pi}_{2u}$, $\hat{\Pi}_{3u}$), $u = 1, 2$ for residuals $\{\hat{\varepsilon}_t^{(s)}\}$ in the same graph.

We say that $\hat{r}$ is an estimate of the unknown order r of the IVAR process, if the ECDF of $\hat{\Pi}_{1u}$ (or $\hat{\Pi}_{2u}$, $\hat{\Pi}_{3u}$), $u = 1, 2$ do not coincide for all $s < \hat{r}$ and coincide for $s = \hat{r}$.

Simulations. Consider the following IVAR processes. Suppose $\{\varepsilon_t\}$ is as in (10.2).

Model 5 $X_t = \varepsilon_t + 0.5X_{t-1}$.

Model 6 $X_t = \varepsilon_t + 0.5X_{t-1} + 0.2X_{t-2}$.

Assuming that we do not know the parameter matrices, we use their consistent estimators discussed above.

For Model 5, we plot the ECDF of $\hat{\Pi}_{1u}$ (or $\hat{\Pi}_{3u}$), $u = 1, 2$ for the residual process $\{\hat{\varepsilon}_t^{(1)}\}$ and for $n = p = 500$ (or $n = 500$, $p = n^{0.9} = 269$) in the same graph and observe that they coincide. See Row 1 left panel in Figure 10.5 (or Figure 10.6). Therefore, 1 is an estimate of the order of Model 5.

For Model 6, we do the same but the two ECDF do not coincide (see Row 1 right panel in Figures 10.5 and 10.6). In Row 2 of Figures 10.5 and 10.6, the same two ECDF are plotted for $\{\hat{\varepsilon}_t^{(2)}\}$ and they coincide and hence 2 is an estimate of the order for Model 6.

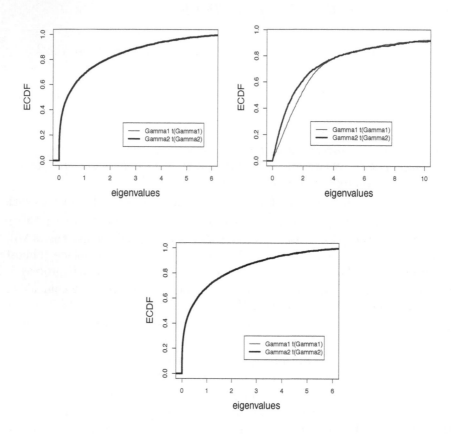

Figure 10.5 $n = p = 500$. *Row 1 left: ECDF of $\hat{\Pi}_{11}$ and $\hat{\Pi}_{12}$ for residuals for an $AR(1)$ fit in Model 5. Row 1 right: same for an $AR(1)$ fit in Model 6. Row 2: same for an $AR(2)$ fit in Model 6.*

10.3   Graphical tests for parameter matrices

In this section we show how certain hypotheses about parameter matrices can be graphically tested using the LSD results. Throughout this section, we assume that all the relevant deterministic (non-random) matrices converge jointly. By $A = B$, we will mean that the (asymptotic) eigenvalue distributions of A and B are identical. In our methods, matrices with asymptotically same eigenvalue distribution are indistinguishable. We shall deal with some specific simple hypothesis, and only for the case $p/n \to 0$. While it will be easy to see how to implement these ideas when $p/n \to y > 0$, further investigations would be needed to generalize our prescription to deal with more general null and alternative hypotheses. For each hypothesis, we provide graphical tests based on the LSD of non-scaled but centered sample autocovariance matrices. These are easy to derive using Theorem 8.3.1 and Corollary 8.3.1.

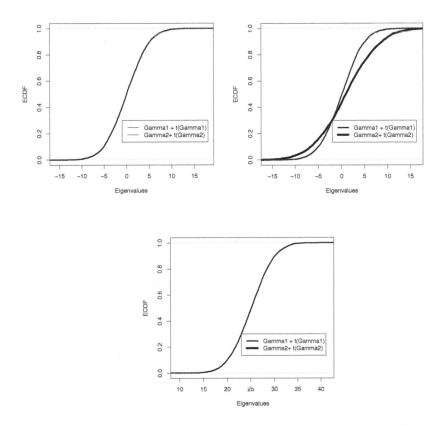

Figure 10.6 $n = 500$, $p = n^{0.9}$. *Row 1 left: ECDF of $\hat{\Pi}_{31}$ and $\hat{\Pi}_{32}$ for residuals after fitting AR(1) in Model 5. Row 1 right: same after fitting AR(1) in Model 6. Row 2: same after fitting AR(2) in Model 6.*

One sample.

Example 10.3.1. Consider the model given in (8.6). Let A be a $p \times p$ matrix with a non-degenerate LSD. Suppose we wish to test

$$H_0 : \psi_q = A \quad \text{against} \quad H_1 : \psi_q \neq A. \tag{10.23}$$

As $\Gamma_q = \psi_q^*$, this is equivalent to testing

$$H_0 : \Gamma_q = A^* \quad \text{against} \quad H_1 : \Gamma_q \neq A^*. \tag{10.24}$$

Using Corollary 8.3.1, it is easy to establish that, under H_0 the LSD of $(\hat{\Gamma}_q + \hat{\Gamma}_q^* - A - A^*)$ is degenerate at 0 and, under H_1 this LSD is identical with the LSD of $(\Gamma_q + \Gamma_q^* - A - A^*)$ which is non-degenerate. Thus, to test (10.23) graphically, we plot the eigenvalue distribution of $(\hat{\Gamma}_q + \hat{\Gamma}_q^* - A - A^*)$. If it appears degenerate at 0, then H_0 is accepted. Else we reject H_0.

Example 10.3.2. Consider the model (8.6). In general, we do not have any ready-made answer for testing $H_0 : \psi_j = A$ against $H_1 : \psi_j \neq A$ when $0 \leq j \leq (q-1)$. To explain why, consider the MA(3) process with coefficient matrices $\{\psi_0 = I, \psi_1, \psi_2, \psi_3\}$. In this case,

$$\Gamma_1 = \psi_1^* + \psi_1\psi_2^* + \psi_2\psi_3^*, \ \Gamma_2 = \psi_2^* + \psi_1\psi_3^*, \ \Gamma_3 = \psi_3^*, \ \Gamma_u = 0 \ \forall u \geq 4. \tag{10.25}$$

Suppose we wish to test

$$H_0 : \ \psi_2 = A \quad \text{against} \quad H_1 : \ \psi_2 \neq A. \tag{10.26}$$

Γ_1, Γ_2 and Γ_3 under H_0 are given by

$$\Gamma_{1H_0} = \psi_1^* + \psi_1 A^* + A\psi_3^*,, \ \Gamma_2 = A^* + \psi_1\psi_3^*, \ \Gamma_3 = \psi_3^*. \tag{10.27}$$

If ψ_1 and ψ_3 are known, then by Corollary 8.3.1, it is easy to see that under H_0, the LSD of $(\hat{\Gamma}_1 + \hat{\Gamma}_1^* - \Gamma_{1H_0} - \Gamma_{1H_0}^*)$ is degenerate at 0, whereas under H_1, this LSD is identical with the LSD of $(\Gamma_1 + \Gamma_1^* - \Gamma_{1H_0} - \Gamma_{1H_0}^*)$. Thus, we may plot the ESD of $(\hat{\Gamma}_1 + \hat{\Gamma}_1^* - \Gamma_{1H_0} - \Gamma_{1H_0}^*)$ and accept H_0 if it appears degenerate at 0.

Clearly, the above method does not work when ψ_1 and ψ_3 are unknown. Moreover, it is hard to estimate these coefficient matrices. If we use the method of moments, we get the system of equations given in (10.25) after replacing the population autocovariance matrices by the sample autocovariance matrices. These equations are not easy to solve. Moreover appropriate regularization could be needed before consistency is achieved. There is scope for further research here.

Example 10.3.3. Next, consider the hypotheses

$$H_0 : \ \psi_j = A_j \ \forall j \quad \text{against} \quad H_1 : \ \psi_j \neq A_j \ \text{for at least one } j, \tag{10.28}$$

for some known $p \times p$ matrices $\{A_j\}$. As H_0 specifies all the coefficient matrices $\{\psi_j : 0 \leq j \leq q\}$, a natural testing method should be based on $\hat{\Gamma}_u$ for all $1 \leq u \leq q$. Let

$$\hat{G}_q = \sum_{u=1}^{q} \hat{\Gamma}_u, \quad G_q = \sum_{u=1}^{q} \Gamma_u = \sum_{u=0}^{q}\sum_{j=0}^{q-u} \psi_j\psi_{j+u}^* \quad \text{and} \tag{10.29}$$

$$G_{qH_0} = \sum_{u=0}^{q}\sum_{j=0}^{q-u} A_j A_{j+u}^* \ (\text{i.e., } G_q \text{ under } H_0). \tag{10.30}$$

By Theorem 8.3.1, it can easily be proved that under H_0, the LSD of $(\hat{G}_q +$

$\hat{G}_q^* - G_{qH_0} - G_{qH_0}^*$) is degenerate at 0 whereas under H_1, this LSD is identical with the LSD of $(G_q + G_q^* - G_{qH_0} - G_{qH_0}^*)$. Thus, to test (10.28) graphically, we plot the eigenvalue distribution of $(\hat{G}_q + \hat{G}_q^* - G_{qH_0} - G_{qH_0}^*)$. We accept H_0 if this ESD appears degenerate at 0.

Two samples.

Now suppose we have samples from two independent infinite-dimensional MA processes given in (9.1):

$$X_t = \sum_{j=0}^{q} \psi_j \varepsilon_{t-j} \ \text{ and } \ Y_t = \sum_{j=0}^{q} \phi_j \xi_{t-j} \ \ \forall t \geq 1 \tag{10.31}$$

where $\varepsilon_t = (\varepsilon_{t,1}, \varepsilon_{t,2}, \dots, \varepsilon_{t,p})'$ and $\xi_t = (\xi_{t,1}, \xi_{t,2}, \dots, \xi_{t,p})'$ are all $p \times 1$ independent vectors and $p, n \to \infty$. Let $\hat{\Gamma}_{uX}$ (Γ_{uX}) and $\hat{\Gamma}_{uY}$ (Γ_{uY}) be the sample (population) autocovariance matrices of order u respectively for $\{X_t\}$ and $\{Y_t\}$.

Example 10.3.4. Suppose we wish to test

$$H_0 : \psi_q = \phi_q \ \text{ against } \ H_1 : \psi_q \neq \phi_q. \tag{10.32}$$

Recall $\hat{\Pi}_{3u}$ from (10.1). Let $\hat{\Pi}_{3q,X}$ and $\hat{\Pi}_{3q,Y}$ be $\hat{\Pi}_{3u}$ with $u = q$ respectively for the processes $\{X_t\}$ and $\{Y_t\}$. It is immediate from Example 9.3.1 that under H_0, the LSD of $(\hat{\Pi}_{3q,X} - \hat{\Pi}_{3q,Y})$ is degenerate at 0 and under H_1, this LSD is identical with the LSD of $(\psi_q + \psi_q^* - \phi_q - \phi_q^*)$, which is non-degenerate. Thus, a graphical method to test (10.32) is to plot the eigenvalue distribution of $(\hat{\Pi}_{3q,X} - \hat{\Pi}_{3q,Y})$. If it appears to be degenerate at 0, we accept H_0.

Example 10.3.5. Consider the hypotheses

$$H_0 : \ \psi_j = \phi_j \ \forall j \ \text{ against } \ H_1 : \ \psi_j \neq \phi_j \ \text{ for at least one } j. \tag{10.33}$$

Recall the definition of $\{G_q, \hat{G}_q\}$ in (10.29). $\{G_{q,X}, \hat{G}_{q,X}\}$ and $\{G_{q,Y}, \hat{G}_{q,Y}\}$ have obvious meaning. By Example 9.3.1, under H_0 the LSD of $(\hat{G}_{q,X} + \hat{G}_{q,X}^* - \hat{G}_{q,Y} - \hat{G}_{q,Y}^*)$ is degenerate at 0 whereas under H_1, this LSD is identical with the LSD of $(G_{q,X} + G_{q,X}^* - G_{q,Y} - G_{q,Y}^*)$.

Therefore, to test (10.33), we can plot the eigenvalue distribution of $(\hat{G}_{q,X} + \hat{G}_{q,X}^* - \hat{G}_{q,Y} - \hat{G}_{q,Y}^*)$ and accept H_0 if the distribution appears to be degenerate at 0.

Hypothesis testing for q: a simple model.

Consider a random sample $\{X_t : 1 \leq t \leq n\}$ from the process

$$X_t = \sum_{j=0}^{q} \varepsilon_{t-j} \ \ \forall t \geq 0. \tag{10.34}$$

Based on the above sample, suppose we wish to test

$$H_0 : q = q_0 \quad \text{against} \quad H_1 : q = q_1. \tag{10.35}$$

Let

$$\hat{\Pi}_{a,u} = \sqrt{np^{-1}}(\hat{\Pi}_{3u} - \Pi_{3u}). \tag{10.36}$$

The LSD of $\hat{\Pi}_{a,u}$ and its free probability description, given in Corollary 8.3.1, are useful in this context. Recall the semi-circle distribution described in Definition 4.2.2. In Theorem 10.3.1 below we show that the LSD of $\hat{\Pi}_{a,u}$, for $u \geq q$, are all semi-circle distributions with some variance $\sigma_{q,u}^2$. Expression for $\sigma_{q,u}^2$ for any arbitrary q and u, is cumbersome and is given in (10.37). By Corollary 10.3.1, whenever $u > q$, $\sigma_{q,u}^2$ depends on q only. For example, $\sigma_{1,u}^2 = 3 \; \forall u > 1$ and $\sigma_{2,u}^2 = 9.5 \; \forall u > 2$. For more details, see Examples 10.3.6 and 10.3.7.

The following lemma provides the shape of the LSD of $\hat{\Pi}_{a,u}$ for the model (10.34). This is useful to test the hypotheses given in (10.35). Let,

$$\sigma_{q,u}^2 = 0.5 \sum_{\delta=1}^{q+u} (C_{q,\delta,u} + C_{q,-\delta,u})^2 + C_{q,0,u}^2, \quad \text{where} \tag{10.37}$$

$$C_{q,\delta,u} = \#\{(j_1, j_2) : j_1 - j_2 + u = \delta, \; 0 \leq j_1, j_2 \leq q\}. \tag{10.38}$$

Theorem 10.3.1. *Consider the process given in (10.34). Suppose (B1) holds and $\{\varepsilon_{i,j} : i, j \geq 1\} \in \mathcal{L}_4 \cap U(\delta)$ for some $\delta > 0$. Further suppose $p/n \to 0$. Then for all $u \geq 1$, the almost sure LSD of $\hat{\Pi}_{a,u}$ is the semi-circle variable with variance $\sigma_{q,u}^2$.*

Proof. By (8.43), (8.45) and Corollary 8.3.1(a), for the model (10.34), $\eta_j = 1 \; \forall \; 0 \leq j \leq q$ and the LSD of $\hat{\Pi}_{a,u}$ is given by

$$g_{a,u} = \frac{1}{2} \sum_{j_1,j_2=0}^{q} (\omega_{u,j_1,j_2} + \omega_{-u,j_2,j_1}), \tag{10.39}$$

where $\{\omega_{u,j_1,j_2} : \omega_{u,j_1,j_2}^* = \omega_{-u,j_2,j_1}, \; 0 \leq j_1, j_2 \leq q\}$. Its free cumulants of order greater than 2 are 0 and those of order 2 satisfy (8.41).

Suppose $\{\omega_u : \omega_{-u} = \omega_u^*, \; u = 0, \pm 1, \pm 2, \ldots\}$ are non-commutative variables with free cumulants

$$\kappa_r(\omega_{u_i} : 1 \leq i \leq r) = 0.25I(u_1 = -u_2, r = 2) \; \forall r \geq 1. \tag{10.40}$$

Therefore, $g_{a,u}$ has the same distribution as

$$\sum_{j_1,j_2=0}^{q} (\omega_{|j_1-j_2+u|} + \omega_{-|j_1-j_2+u|}). \tag{10.41}$$

Let $s_\delta = \omega_\delta + \omega_{-\delta}$. Then (10.41) equals

$$\sum_{\delta=0}^{q+u} s_\delta(\#\{(j_1, j_2) : |j_1 - j_2 + u| = \delta, \ 0 \le j_1, j_2 \le q\})$$

$$= \sum_{\delta=1}^{q+u} (C_{q,\delta,u} + C_{q,-\delta,u}) s_\delta + C_{q,0,u} s_0 \qquad (10.42)$$

Note that, by (10.40), for all $r \ge 1$,

$$\kappa_r(s_{\delta_i} : 1 \le i \le r) = 0.5I(\delta_1 = \delta_2 \ne 0, r = 2) + I(\delta_1 = \delta_2 = 0, r = 2). \quad (10.43)$$

Note that by (10.41)–(10.43), $\mathrm{Var}(g_{a,u}) = \sigma_{q,u}^2$. Now it remains to prove that $g_{a,u}$ is a semi-circle variable. However, a variable is semi-circle if it is self-adjoint and all its free cumulants vanish except of order 2. Now, by (10.42) and (10.43), $g_{a,u}$ has all free cumulants zero except $\kappa_2(g_{a,u}, g_{a,u}) = \sigma_{q,u}^2$. This completes the proof of Lemma 10.3.1. $\qquad \square$

Corollary 10.3.1. *Consider the model (10.34). Then the LSD of $\hat{\Pi}_{a,u}$, for $u > q$, are identical and is the semi-circle distribution with variance σ_q^2 given by*

$$\sigma_q^2 = \frac{1}{2} \sum_{u=-q}^{q} (q + 1 - |u|)^2. \qquad (10.44)$$

Proof. To verify this, note that, for a fix $u > q$,

$$C_{q,\delta,u} = ((q + 1) - |\delta - u|)I(-q + u \le \delta \le q + u).$$

Therefore, by (10.37),

$$\sigma_{q,1,u}^2 = \frac{1}{2} \sum_{\delta=-q+u}^{q+u} (q + 1 - |\delta - u|)^2 = \sigma_q^2, \quad \forall u > q. \qquad (10.45)$$

Hence, Corollary 10.3.1 is established. $\qquad \square$

Example 10.3.6. Consider the model (10.34) with $q = 1$. Then the LSD of $\hat{\Pi}_{a,1}$ is the semi-circle distribution with variance 3.5. Also, the LSD of $\hat{\Pi}_{a,u}$ are identical for $u > 1$ and it is the semi-circle distribution with variance 3.

To see this first suppose $u = 1$. Then by (10.38), $C_{1,0,1} = 1, C_{1,1,1} = 2$, $C_{1,2,1} = 1$ and $C_{1,\delta,1} = 0$ for other δ. Therefore, by (10.37), $\sigma_{1,1}^2 = 3.5$. Now consider $u > 1$. Then we have $C_{0,u-1,1} = 1, C_{1,u,1} = 2, C_{1,u+1,1} = 1$ and $C_{1,\delta,1} = 0$ for other δ. This gives $\sigma_{1,u}^2 = 3 \ \forall u > 1$.

Example 10.3.7. Consider the model (10.34) with $q = 2$. Then the LSD of $\hat{\Pi}_{a,1}$ and $\hat{\Pi}_{a,2}$ are semi-circle distributions with variances 14.5 and 10 respectively. Also, the LSD of $\hat{\Pi}_{a,u}$ are identical for $u > 2$ and it is the semi-circle distribution with variance 9.5.

To verify this note that $C_{2,-1,1} = C_{2,0,1} = 1, C_{2,0,1} = C_{2,1,1} = 2, C_{2,1,1} = C_{2,2,1} = 3, C_{2,2,1} = C_{2,3,1} = 2, C_{2,3,1} = C_{2,4,1} = 1$ and $C_{2,\delta,u} = 0$ for other δ and $u = 1, 2$. Therefore, $\sigma_{2,1}^2 = 14.5$ and $\sigma_{2,2}^2 = 10$. Moreover, for $u > 2$, $C_{2,u-2,u} = 1, C_{2,u-1,u} = 2, C_{2,u,u} = 3, C_{2,u+1,u} = 2, C_{2,u+2,u} = 1, C_{2,\delta,u} = 0$ for all other δ and hence $\sigma_{2,u}^2 = 9.5$, for all $u > 2$.

For $u > q$, let σ_q^2 denote the common value of $\sigma_{q,u}^2$, which can be computed using (10.44). An easy graphical way to test the hypotheses in (10.35) is to plot the ESD of $\hat{\Pi}_{a,u}$ for $u = \max(q_0, q_1) + 1$. If the support of the distribution is $[-2\sigma_{q_i}, 2\sigma_{q_i}]$, then we accept H_i.

It may be noted that all methods described here are based on the symmetric sum of $\{\hat{\Gamma}_u, \hat{\Gamma}_u^*\}$, but other symmetric polynomials can also be used. We have restricted to the symmetric sum only for illustration.

Exercises

1. Run the simulation given in Figures 10.1, 10.2, and 10.3 for $n = 200, 300, 500, 700$. Comment on the convergence of the ESD.

2. Let $X_t = \sum_{i=1}^{10} A^i \varepsilon_{t-i}$ and $p/n \to 0$. Suggest two graphical tests for

$$H_0 : A = I_p \quad \text{against} \quad H_1 : A \neq I_p$$

based on the discussion in Examples 10.3.1 and 10.3.3. Simulate and compute empirical power of both the tests. Hence, compare their performances.

3. Let $X_t = \varepsilon_t + A\varepsilon_{t-1} + \varepsilon_{t-2}$, $A = \frac{1}{p}((1-\rho)I_p + \rho J_p) + I_p$ and $p/n \to 0$. Suggest a graphical test for

$$H_0 : \rho = 0.5 \quad \text{against} \quad H_1 : \rho \neq 0.5.$$

Modify the test when $p/n \to y > 0$.

4. Let $X_t = \varepsilon_t + A\varepsilon_{t-1} + B\varepsilon_{t-2}$, $A = ((\theta^{|i-j|}))_{p \times p}$, $B = ((\rho^{|i-j|}))_{p \times p}$ and $p/n \to 0$. Suggest a graphical test for

$$H_0 : \theta = 0.7, \rho = 0.5 \quad \text{against} \quad H_1 : \theta = 0.9, \rho = 0.3.$$

Modify the test when $p/n \to y > 0$.

5. Let $X_t = \sum_{i=1}^{q} \lambda_i \varepsilon_{t-i}$ where $\lambda_i \in \mathbb{R} \ \forall i$ and $p/n \to 0$. Discuss appropriate assumptions on $\{\lambda_i : 1 \leq i \leq q\}$ so that LSD of $\hat{\Gamma}_u + \hat{\Gamma}_u^*$ follows the semi-circle law. Hence, test $H_0 : q = 10$ against $H_1 : q = 15$.

6. Let $X_t = AX_{t-1} + \varepsilon_t$ and $p/n \to 0$. Suggest graphical tests for

$$H_0 : A = O \quad \text{against} \quad H_1 : A \neq O \quad \text{and}$$
$$H_0 : A = I_p \quad \text{against} \quad H_1 : A \neq I_p. \tag{10.46}$$

Modify the tests for $p/n \to y > 0$.

7. Let $X_t = AX_{t-1} + B\varepsilon_{t-1} + \varepsilon_t$ and $p/n \to 0$. Suggest graphical tests for the following hypotheses:

$$H_0 : A = O \quad \text{against} \quad H_1 : A \neq O,$$
$$H_0 : B = O \quad \text{against} \quad H_1 : B \neq O,$$
$$H_0 : A = O, B = I_p \quad \text{against} \quad H_1 : A = I_p, B = O \quad \text{and}$$
$$H_0 : A = O, B = O \quad \text{against} \quad H_1 : A_p \neq O, B = O.$$

Modify the tests for $p/n \to y > 0$.

8. Simulate and obtain empirical power of the tests given in Examples 10.3.4 and 10.3.5. Hence, comment on the consistency of the tests.

9. Give a proof of Theorem 10.1.1.

10. Show that under suitable conditions, $\{A_1, A_2\}$ and $\{\theta_j\}$ in Remark 10.2.1 converge jointly.

Chapter 11

TESTING WITH TRACE

The most commonly used test statistic in high-dimensional models is the linear spectral statistic. While there are a few asymptotic results for general spectral statistics for the sample covariance matrix, there does not seem to be any such results known for sample autocovariance matrices. It may be noted that the trace is a particular linear spectral statistics. Due to its special structure, this statistic is easier to handle than a general linear spectral statistics. The asymptotic normality of the trace has been proved in the literature for several high-dimensional matrices. In Chapters 6 and 7, we have seen trace results for polynomials in generalized covariance matrices.

We now extend those ideas and discuss asymptotic normality of the trace of any polynomial in sample autocovariance matrices for one and two independent infinite-dimensional MA processes. These results will then be used in *significance testing* of some hypotheses in one and two sample high-dimensional time series. The subject is in its infancy and currently only specific simple hypotheses can be tested. These ideas appear to have significant potential.

11.1   One sample trace

First suppose that we have observations on one infinite-dimensional MA(q) or MA(∞) process defined in (8.6) and (8.16). For convenience of the reader, we recall Assumptions (B1), (B3), (B), and (B4) from Chapter 8. Recall also the classes $\mathcal{L}$ and $C(\delta, p)$ respectively from (8.1) and (8.3).

(B1) $\{\varepsilon_{i,j}\}$ are independently distributed with mean 0 and variance 1.

(B3) $\{\varepsilon_{i,j} : 1 \le i \le p,\ 1 \le j \le n\} \in \mathcal{L} \cap C(\delta, p)$, for all $p > 1$.

(B) $\{\psi_j, \psi_j^*\}$ are norm bounded and jointly converge.

(B4) $\sum_{j=0}^{\infty} \sup_p ||\psi_j||_2 < \infty$.

Let $\pi(\cdot)$ be a polynomial defined on a set of matrices closed under the transpose operation. Suppose

$$\hat{\Pi} = \pi(\hat{\Gamma}_u, \hat{\Gamma}_u^* : u \ge 0) \quad \text{and} \quad \Pi = \pi(\Gamma_u, \Gamma_u^* : u \ge 0). \tag{11.1}$$

Denote

$$R_\Pi = \sqrt{np^{-1}}(\hat{\Pi} - \Pi).$$

Let
$$\sigma_\Pi^2 = \lim E[(\mathrm{Tr}(\hat{\Pi}) - E\mathrm{Tr}(\hat{\Pi}))^2], \quad \sigma_R^2 = E(\mathrm{Tr}(R_\Pi))^2.$$

Note that σ_Π^2 and σ_R^2 are finite under our assumptions. Then we have the following theorem

Theorem 11.1.1. *Consider the MA(q) process (8.6) with $q < \infty$. Suppose (B1), (B3), and (B) hold.*

(a) If $p/n \to y > 0$ then $\mathrm{Tr}(\hat{\Pi}) - E\mathrm{Tr}(\hat{\Pi}) \xrightarrow{D} \mathcal{N}(0, \sigma_\Pi^2)$.

(b) If $p/n \to 0$ then $\mathrm{Tr}(R_\Pi) \xrightarrow{D} \mathcal{N}(0, \sigma_R^2)$. Results in (a) and (b) also hold for the MA(∞) process (8.16) if in addition we assume (B4).

Proof. (a) We use Lemmas 6.3.1 and 8.2.1. In these lemmas put
$$\begin{aligned}
\mathcal{P}_i &= \mathrm{Tr}(\Pi(\Delta_u, \Delta_u^* : u \geq 0)), \ \forall i \geq 1 \text{ and} && (11.2) \\
\tilde{\mathcal{P}}_i &= \mathrm{Tr}(\Pi(\hat{\Gamma}_u, \hat{\Gamma}_u^* : u \geq 0)), \ \forall i \geq 1. && (11.3)
\end{aligned}$$

Therefore,
$$\begin{aligned}
\mathcal{P}_i^0 &= E\mathrm{Tr}(\Pi(\Delta_u, \Delta_u^* : u \geq 0)), \ \forall i \geq 1 \text{ and} && (11.4) \\
\tilde{\mathcal{P}}_i^0 &= E\mathrm{Tr}(\Pi(\hat{\Gamma}_u, \hat{\Gamma}_u^* : u \geq 0)), \ \forall i \geq 1. && (11.5)
\end{aligned}$$

Also note that, by (11.3) and (11.5), we have
$$\lim E\big[(\tilde{\mathcal{P}}_i - \tilde{\mathcal{P}}_i^0)(\tilde{\mathcal{P}}_j - \tilde{\mathcal{P}}_j^0)\big] = \sigma_\Pi^2, \ \forall i, j \geq 1. \qquad (11.6)$$

Therefore,
$$\lim E(\mathrm{Tr}(\Pi(\hat{\Gamma}_u, \hat{\Gamma}_u^* : u \geq 0)) - E\mathrm{Tr}(\Pi(\hat{\Gamma}_u, \hat{\Gamma}_u^* : u \geq 0)))^T$$
$$= \lim E\Big(\prod_{i=1}^{T}(\tilde{\mathcal{P}}_i - \tilde{\mathcal{P}}_i^0)\Big), \ \text{(by (11.3) and (11.5))}$$
$$= \lim E\Big(\prod_{i=1}^{T}(\mathcal{P}_i - \mathcal{P}_i^0)\Big), \ \text{(by Lemma 8.2.1(b))}$$
$$= \begin{cases} 0 \text{ if } T = 2d - 1, \\ \sum_{\mathcal{S}_d} \prod_{k=1}^{d} \lim E\big[(\mathcal{P}_{i_{2k-1}} - \mathcal{P}_{i_{2k-1}}^0)(\mathcal{P}_{i_{2k}} - \mathcal{P}_{i_{2k}}^0)\big], \text{ if } T = 2d. \end{cases}$$
$$\text{(by Lemma 6.3.1)}$$
$$= \begin{cases} 0 \text{ if } T = 2d - 1, \\ \sum_{\mathcal{S}_d} \prod_{k=1}^{d} \lim E\big[(\tilde{\mathcal{P}}_{i_{2k-1}} - \tilde{\mathcal{P}}_{i_{2k-1}}^0)(\tilde{\mathcal{P}}_{i_{2k}} - \tilde{\mathcal{P}}_{i_{2k}}^0)\big], \text{ if } T = 2d. \end{cases}$$
$$\text{(by Lemma 8.2.1 (b))}$$
$$= \begin{cases} 0, & \text{if } T = 2d - 1 \\ (\#\mathcal{S}_d)\sigma_\Pi^{2d}, & \text{if } T = 2d \end{cases} \quad \text{(by (11.6))}$$
$$= \begin{cases} 0, \text{if } T = 2d - 1 \\ (\text{number of pair partitions of } \{1, \dots, 2d\})\sigma_\Pi^{2d}, \text{if } T = 2d. \end{cases} \qquad (11.7)$$

which is nothing but the T-th order raw moment of $\mathcal{N}(0, \sigma_\Pi^2)$. This completes the proof of (a).

(b) Similar arguments as in (a), works for (b) also. Hence, we omit it.   □

Recall $\hat{\Pi}_{iu}$ from (10.1). The following example relaxes (B3) and provides asymptotic normality of $\hat{\Pi}_{iu}$ for $i = 1, 2, 3$ under weaker assumptions. This claim follows when we use Theorem 11.1.1 in conjunction with the truncation technique used in the proof of Corollaries 8.3.1(c) and 8.3.3(c). We leave the details as an exercise.

Recall the classes $\mathcal{L}_r$ and $U(\delta)$ defined respectively in (8.1) and (8.28). Let

$$\delta_{i,u}^2 = \lim E(\mathrm{Tr}(\hat{\Pi}_{iu}) - E(\mathrm{Tr}(\hat{\Pi}_{iu})))^2, \tag{11.8}$$

$$\sigma_{i,u}^2 = \lim E(\mathrm{Tr}(\sqrt{np^{-1}}(\hat{\Pi}_{iu} - \Pi_{iu})))^2.$$

Example 11.1.1. Consider the process (8.6) with $q < \infty$. Suppose (B1), (B) hold and $\{\varepsilon_{i,j} : i, j \geq 1\} \in \mathcal{L}_4 \cap U(\delta)$ for some $\delta > 0$.

(a) If $p/n \to y > 0$ then for all $u \geq 0$,

$$(\mathrm{Tr}(\hat{\Pi}_{iu}) - E\mathrm{Tr}(\hat{\Pi}_{iu})) \xrightarrow{\mathcal{D}} \mathcal{N}(0, \delta_{i,u}^2), \quad \forall i = 1, 2, 3.$$

(b) If $p/n \to 0$ then for all $u \geq 0$,

$$\mathrm{Tr}(\sqrt{np^{-1}}(\hat{\Pi}_{iu} - \Pi_{iu})) \xrightarrow{\mathcal{D}} \mathcal{N}(0, \sigma_{i,u}^2), \quad \forall i = 1, 2, 3.$$

Results in (a) and (b) continue to hold for the MA(∞) process (8.16) if in addition we assume ($B4$).

Let us now consider some special cases and investigate the finite sample accuracy of the normality via simulations.

Example 11.1.2. Let $X_t = \varepsilon_t, \forall t$. We consider $\varepsilon_t \sim \mathcal{N}_p(0, I_p)$, where ε_t's are independent. Then using Theorem 11.1.1(a), it is easy to see that, when $p/n \to 1$,

$$\mathrm{Tr}\hat{\Gamma}_0 - n \xrightarrow{\mathcal{D}} \mathcal{N}(0, 2),$$

$$\mathrm{Tr}(\hat{\Gamma}_1 \hat{\Gamma}_1^*) - n + 1 \xrightarrow{\mathcal{D}} \mathcal{N}(0, 10),$$

$$\mathrm{Tr}(\hat{\Gamma}_1 + \hat{\Gamma}_1^*) \xrightarrow{\mathcal{D}} \mathcal{N}(0, 4). \tag{11.9}$$

Moreover, using Theorem 11.1.1(b), when $p/n \to 0$, we have

$$\mathrm{Tr}\sqrt{np^{-1}}(\hat{\Gamma}_0 - I) \xrightarrow{\mathcal{D}} \mathcal{N}(0, 1),$$

$$\mathrm{Tr}\sqrt{np^{-1}}(\hat{\Gamma}_1 + \hat{\Gamma}_1^*) \xrightarrow{\mathcal{D}} \mathcal{N}(0, 2).$$

Simulation results given in Row 1 (left and right panels, Figure 11.1), Row 2 (left panel Figure 11.1) and Row 1 (left and right panels, Figure 11.2) show that the distributions are indeed asymptotic normal.

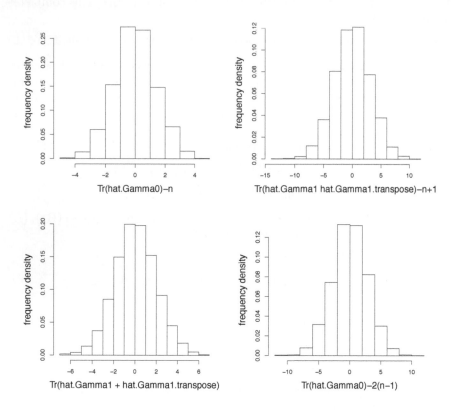

Figure 11.1 $n = p = 700$ and 700 replications. Row (1) left, Row (1) right and Row (2) left represent respectively the histogram of $\mathrm{Tr}(\hat{\Gamma}_0) - n$, $\mathrm{Tr}(\hat{\Gamma}_1\hat{\Gamma}_1^*) - n + 1$ and $\mathrm{Tr}(\hat{\Gamma}_1 + \hat{\Gamma}_1^*)$, when $X_t = \varepsilon_t$. Row (2) right represents the histogram of $\mathrm{Tr}(\hat{\Gamma}_0) - 2(n-1)$, when $X_t = \varepsilon_t + \varepsilon_{t-1}$.

Example 11.1.3. Let $X_t = \varepsilon_t + \varepsilon_{t-1}$. Then using Theorem 11.1.1 (a), it is easy to see that, when $p/n \to 1$, we have

$$\mathrm{Tr}\hat{\Gamma}_0 - 2(n-1) \xrightarrow{\mathcal{D}} \mathcal{N}(0,8).$$

Moreover, using Theorem 11.1.1(b), when $p/n \to 0$, we have

$$\mathrm{Tr}\sqrt{np^{-1}}(\hat{\Gamma}_0 - 2I_p) \xrightarrow{\mathcal{D}} \mathcal{N}(0,6),$$
$$\mathrm{Tr}\sqrt{np^{-1}}(\hat{\Gamma}_1 + \hat{\Gamma}_1^* - 2I_p) \xrightarrow{\mathcal{D}} \mathcal{N}(0,12).$$

Again, simulation result given in Row 2 (right panel, Figure 11.1) and Row 2 (left and right panels, Figure 11.2) support the above convergences.

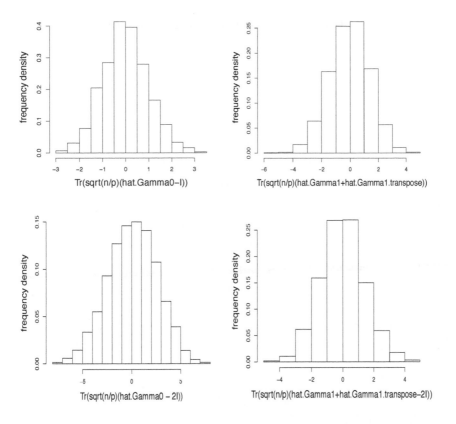

Figure 11.2 $p = n^{0.9}$ *and 700 replications. Row (1) left and Row (1) right represent respectively the histogram of* $(\mathrm{Tr}\sqrt{np^{-1}}(\hat{\Gamma}_0 - I))$ *and* $\mathrm{Tr}\sqrt{np^{-1}}(\hat{\Gamma}_1 + \hat{\Gamma}_1^*)$, *when* $X_t = \varepsilon_t$. *Row (2) left and Row (2) right respectively represent the histogram of* $(\mathrm{Tr}\sqrt{np^{-1}}(\hat{\Gamma}_0 - 2I))$ *and* $\mathrm{Tr}\sqrt{np^{-1}}(\hat{\Gamma}_1 + \hat{\Gamma}_1^* - 2I)$, *when* $X_t = \varepsilon_t + \varepsilon_{t-1}$.

11.2   Two sample trace

Now suppose we have samples from two independent infinite-dimensional MA processes given in (9.1)

$$X_t = \sum_{j=0}^{q} \psi_j \varepsilon_{t-j} \ \text{ and } \ Y_t = \sum_{j=0}^{q} \phi_j \xi_{t-j} \ \ \forall t \geq 1 \tag{11.10}$$

where $\varepsilon_t = (\varepsilon_{t,1}, \varepsilon_{t,2}, \ldots, \varepsilon_{t,p})'$ and $\xi_t = (\xi_{t,1}, \xi_{t,2}, \ldots, \xi_{t,p})'$ are all $p \times 1$ independent vectors and $p, n \to \infty$. Let $\hat{\Gamma}_{uX}$ (Γ_{uX}) and $\hat{\Gamma}_{uY}$ (Γ_{uY}) be the sample (population) autocovariance matrices of order u respectively for the processes $\{X_t\}$ and $\{Y_t\}$.

We now state the asymptotic normality result for the trace of polynomials in $\{\hat{\Gamma}_{uX}, \hat{\Gamma}_{uX}^*, \hat{\Gamma}_{uY}, \hat{\Gamma}_{uY}^* : u \geq 0\}$. Let,

$$\begin{aligned}
\hat{\Pi}_{XY} &= \pi(\hat{\Gamma}_{uX}, \hat{\Gamma}_{uX}^*, \hat{\Gamma}_{uY}, \hat{\Gamma}_{uY}^* : u \geq 0), \\
\Pi_{XY} &= \pi(\Gamma_{uX}, \Gamma_{uX}^*, \Gamma_{uY}, \Gamma_{uY}^* : u \geq 0).
\end{aligned}$$

and

$$\begin{aligned}
\delta_{\Pi_{XY}}^2 &= \lim E(\mathrm{Tr}(\hat{\Pi}_{XY}) - E(\mathrm{Tr}(\hat{\Pi}_{XY})))^2, \qquad (11.11) \\
\sigma_{\Pi_{XY}}^2 &= \lim E(\sqrt{np^{-1}}\mathrm{Tr}(\hat{\Pi}_{XY} - \Pi_{XY}))^2.
\end{aligned}$$

Recall the classes of independent variables $\mathcal{L}$ and $C(\delta, p)$ from (8.2) and (8.3). Consider the following assumptions from Chapter 9.

(B5) $\{\varepsilon_{i,j}, \xi_{i,j}\}$ are independently distributed with mean 0 and variance 1.

(B7) $\{\varepsilon_{i,j}, i \leq p, \ j \leq n\}$, $\{\xi_{i,j}, i \leq p, \ j \leq n\} \in \mathcal{L} \cup C(\delta, p)$ for all $p \geq 1$ and for some $\delta > 0$.

(D) $\{\psi_j, \phi_j : \ j \geq 0\}$ are norm bounded and converge jointly.

(B8) $\sum_{j=0}^{\infty} \left[\sup_p ||\psi_j||_2 + \sup_p ||\phi_j||_2 \right] < \infty$.

Proof of the following two sample version of Theorem 11.1.1 is left as an exercise.

Theorem 11.2.1. *Consider the model (11.10) with $q < \infty$. Suppose (B5), (B7), and (D) hold.*

(a) If $p/n \to y > 0$ then

$$\mathrm{Tr}(\hat{\Pi}_{XY}) - E(\mathrm{Tr}(\hat{\Pi}_{XY})) \xrightarrow{\mathcal{D}} \mathcal{N}(0, \delta_{\Pi_{XY}}^2).$$

(b) If $p/n \to 0$ then

$$\mathrm{Tr}(\sqrt{np^{-1}}(\hat{\Pi}_{XY} - \Pi_{XY})) \xrightarrow{\mathcal{D}} \mathcal{N}(0, \sigma_{\Pi_{XY}}^2).$$

Also these continues to hold for the model (11.10) with $q = \infty$ if in addition we assume (B8).

Recall $\hat{\Pi}_{iu}$, Π_{iu}, $\delta_{i,u}$ and $\sigma_{i,u}$ from (10.1) and (11.8). Let $\hat{\Pi}_{iu,X}$ ($\Pi_{iu,X}$) and $\hat{\Pi}_{iu,Y}$ ($\Pi_{iu,Y}$) be the values of $\hat{\Pi}_{iu}$ (Π_{iu}) respectively for the processes $\{X_t\}$ and $\{Y_t\}$. Similar interpretation is meant for $\delta_{i,u,X}$, $\sigma_{i,u,X}$, $\delta_{i,u,Y}$ and $\sigma_{i,u,Y}$. Let,

$$\begin{aligned}
\hat{\Pi}_{m,(u,v),(X,Y)} &= \hat{\Gamma}_{u,X} \hat{\Gamma}_{v,Y}^*, \quad \Pi_{m,(u,v),(X,Y)} = \Gamma_{u,X} \Gamma_{v,Y}^*, \\
\delta_{m,(u,v),(X,Y)}^2 &= \lim E(\mathrm{Tr}(\hat{\Pi}_{m,(u,v),(X,Y)}) - E(\mathrm{Tr}(\hat{\Pi}_{m,(u,v),(X,Y)})))^2 \\
\sigma_{m,(u,v),(X,Y)}^2 &= \lim E(\sqrt{np^{-1}}(\mathrm{Tr}(\hat{\Pi}_{m,(u,v),(X,Y)} - \Pi_{m,(u,v),(X,Y)})))^2.
\end{aligned}$$

Recall the classes $\mathcal{L}_r$ and $U(\delta)$ defined respectively in (8.1) and (8.28). The next example is the two sample version of Example 11.1.1. The proof of the statement is left as an exercise.

Example 11.2.1. Consider the model (11.10) with $q < \infty$. Suppose (B5), (D) hold and $\{\varepsilon_{i,j} : i, j \geq 1\}, \{\xi_{i,j} : i, j \geq 1\} \in \mathcal{L}_4 \cap U(\delta)$ for some $\delta > 0$.
(a) Suppose $p/n \to y > 0$. Then for $i = 1, 2, 3$ and $u, v \geq 0$,

$$\mathrm{Tr}(\hat{\Pi}_{iu,X} \pm \hat{\Pi}_{iu,Y}) - E(\mathrm{Tr}(\hat{\Pi}_{iu,X} \pm \hat{\Pi}_{iu,Y})) \overset{\mathcal{D}}{\to} \mathcal{N}(0, \delta^2_{i,u,X} + \delta^2_{i,u,Y})$$

$$\mathrm{Tr}(\hat{\Pi}_{m,(u,v),(X,Y)}) - E(\mathrm{Tr}(\hat{\Pi}_{m,(u,v),(X,Y)})) \overset{\mathcal{D}}{\to} \mathcal{N}(0, \delta^2_{m,(u,v),(X,Y)}).$$

(b) Suppose $p/n \to 0$. Then for $i = 1, 2, 3$ and $u, v \geq 0$,

$$\mathrm{Tr}(\sqrt{np^{-1}}((\hat{\Pi}_{iu,X} \pm \hat{\Pi}_{iu,Y}) - (\Pi_{iu,X} \pm \Pi_{iu,Y})) \overset{\mathcal{D}}{\to} \mathcal{N}(0, \sigma^2_{i,u,X} + \sigma^2_{i,u,Y})$$

$$\mathrm{Tr}(\sqrt{np^{-1}}(\hat{\Pi}_{m,(u,v),(X,Y)} - \Pi_{m,(u,v),(X,Y)})) \overset{\mathcal{D}}{\to} \mathcal{N}(0, \sigma^2_{m,(u,v),(X,Y)}).$$

These results continue to hold for (11.10) with $q = \infty$ if we also assume $(B8)$.

11.3   Testing

In Chapter 10, we dealt with some specific simple hypothesis and provided graphical tests for the case $p/n \to 0$. In this section we show how we can perform significance tests for these hypotheses. These tests are all based on the asymptotic normality of appropriately centered and scaled trace of sample autocovarance matrices. These normality results can be derived from Theorem 11.1.1 and Corollary 11.1.1. As before all the relevant deterministic (non-random) matrices are assumed to converge jointly. By $A = B$, we mean that the (asymptotic) eigenvalue distributions of A and B are identical.

For any $-q \leq u_1, u_2, \dots, u_r \leq q$, let us define

$$T_{(u_1,u_2,\dots,u_r),p} = \sqrt{np^{-1}}\left[\mathrm{Tr}\left(\sum_{j=1}^{r} \hat{\Gamma}_{u_j}\right) - \mathrm{Tr}\left(\sum_{j=1}^{r} \Gamma_{u_j}\right)\right], \qquad (11.12)$$

$$H_{(u_1,u_2,\dots,u_r),p} = \sum_{j=1}^{r}(\Gamma_{v-u_j} + \Gamma_{v+u_j})\sum_{j=1}^{r}(\Gamma_{v-u_j} + \Gamma_{v+u_j})^*, \quad (11.13)$$

$$\hat{H}_{(u_1,u_2,\dots,u_r),p} = \sum_{j=1}^{r}(\hat{\Gamma}_{v-u_j} + \hat{\Gamma}_{v+u_j})\sum_{j=1}^{r}(\hat{\Gamma}_{v-u_j} + \hat{\Gamma}_{v+u_j})^*, \quad (11.14)$$

$$\sigma^2_{(u_1,u_2,\dots,u_r)} = 0.5 \sum_{v=-\infty}^{\infty} \lim p^{-1}\mathrm{Tr}(H_{(u_1,u_2,\dots,u_r),p}), \qquad (11.15)$$

$$\hat{\sigma}^2_{(u_1,u_2,\dots,u_r),p} = 0.5 \sum_{v=-\infty}^{\infty} p^{-1}\mathrm{Tr}(\hat{H}_{(u_1,u_2,\dots,u_r),p}). \qquad (11.16)$$

Recall the classes $\mathcal{L}_r$ and $U(\delta)$ defined respectively in (8.1) and (8.28).

Theorem 11.3.1. *Consider the model (8.6) and let $0 \leq q < \infty$. Suppose*

(B1), (B) hold and $\{\varepsilon_{i,j} : i, j \geq 1\} \in \mathcal{L}_4 \cap U(\delta)$ *for some* $\delta > 0$. *Further suppose* $p/n \to 0$. *Then for any* $-q \leq u_1, u_2, \ldots, u_r \leq q$,

(a) $T_{(u_1,u_2,\ldots,u_r),p} \xrightarrow{\mathcal{D}} \mathcal{N}(0, \sigma^2_{(u_1,u_2,\ldots,u_r)})$ *and*

(b) $T_{(u_1,u_2,\ldots,u_r),p}/\hat{\sigma}_{(u_1,u_2,\ldots,u_r),p} \xrightarrow{\mathcal{D}} \mathcal{N}(0,1)$.

Proof. (a) Asymptotic normality of $T_{(u_1,u_2,\ldots,u_r),p}$ follows from Theorem 11.1.1. To complete the proof, we only need to establish

$$\lim V(T_{(u_1,u_2,\ldots,u_r),p}) = \lim E(T_{(u_1,u_2,\ldots,u_r),p})^2 = \sigma^2_{(u_1,u_2,\ldots,u_r)}. \qquad (11.17)$$

Recall $\{\Delta_u\}$ from (6.3). Let

$$\tilde{T}_{(u_1,u_2,\ldots,u_r),p} = \sqrt{np^{-1}}\Big[\text{Tr}\Big(\sum_{j=1}^{r} \Delta_{u_j}\Big) - \text{Tr}\Big(\sum_{j=1}^{r} \Gamma_{u_j}\Big)\Big]. \qquad (11.18)$$

As discussed in the proof of Theorem 7.3.1, conclusions of Lemma 6.3.1 hold for the appropriately normalized autocovariances when $p/n \to 0$. Therefore,

$$\lim E(T_{(u_1,u_2,\ldots,u_r),p})^2 = \lim E(\tilde{T}_{(u_1,u_2,\ldots,u_r),p})^2. \qquad (11.19)$$

Note that

$$\lim E(\tilde{T}_{(u_1,u_2,\ldots,u_r),p})^2 \qquad (11.20)$$

$$= \lim \frac{n}{p}\Big[E\Big(\text{Tr}\Big(\sum_{j=1}^{r}\Delta_{u_j}\Big)\Big)^2 - 2E\Big(\text{Tr}\Big(\sum_{j=1}^{r}\Delta_{u_j}\Big)\Big)\text{Tr}\Big(\sum_{j=1}^{r}\Gamma_{u_j}\Big)$$

$$+ \Big(\text{Tr}\Big(\sum_{j=1}^{r}\Gamma_{u_j}\Big)\Big)^2\Big]$$

$$= \lim(T_1 - 2T_2\sqrt{T_3} + T_3), \quad \text{(say)}.$$

Now, by simple matrix algebra and using the fact that $\text{Tr}(P_u) = I(u = 0)$, it is easy to see that

$$\begin{aligned}
T_2 &= \sqrt{\frac{n}{p}}E\Big(\text{Tr}\Big(\sum_{j=1}^{r}\Delta_{u_j}\Big)\Big) \\
&= \sqrt{\frac{n}{p}}E\Big(\text{Tr}\Big(\sum_{j=1}^{r}\frac{1}{n}\sum_{i,k=0}^{q}\psi_i Z P_{i-k+u_j} Z^*\psi_k^*\Big)\Big) \\
&= \sqrt{\frac{n}{p}}\sum_{j=1}^{r}\sum_{i,k=0}^{q}E\Big(\frac{1}{n}\text{Tr}(\psi_i Z P_{i-k+u_j} Z^*\psi_k^*)\Big) \\
&= \sqrt{\frac{n}{p}}\sum_{j=1}^{r}\sum_{i=0}^{q-u_j}\text{Tr}(\psi_i\psi_{i+u_j}^*) \\
&= \sqrt{\frac{n}{p}}\sum_{j=1}^{r}\text{Tr}(\Gamma_{u_j}) = \sqrt{\frac{n}{p}}\text{Tr}\Big(\sum_{j=1}^{r}\Gamma_{u_j}\Big) = \sqrt{T_3} \qquad (11.21)
\end{aligned}$$

and

$$\lim T_1 \;=\; \lim T_3 + \sum_{\substack{j,l=1}}^{r} \sum_{\substack{i_1,k_1,i_2,k_2=0 \\ i_1+i_2+u_j+u_l=k_1+k_2}}^{q} \lim \frac{1}{p}\mathrm{Tr}(\psi_{i_1}\psi_{k_2}^*\psi_{i_2}\psi_{k_1}^*)$$

$$+ \sum_{\substack{j,l=1}}^{r} \sum_{\substack{i_1,k_1,i_2,k_2=0 \\ i_1+k_2+u_j=i_2+k_1+u_l}}^{q} \lim \frac{1}{p}\mathrm{Tr}(\psi_{i_1}\psi_{k_2}^*\psi_{i_2}\psi_{k_1}^*). \qquad (11.22)$$

Therefore, by (11.20)–(11.22),

$$\lim E(\tilde{T}_{(u_1,u_2,\ldots,u_r),p})^2 \;=\; \sum_{\substack{j,l=1}}^{r} \sum_{\substack{i_1,k_1,i_2,k_2=0 \\ i_1+i_2+u_j+u_l=k_1+k_2}}^{q} \lim \frac{1}{p}\mathrm{Tr}(\psi_{i_1}\psi_{k_2}^*\psi_{i_2}\psi_{k_1}^*)$$

$$+ \sum_{\substack{j,l=1}}^{r} \sum_{\substack{i_1,k_1,i_2,k_2=0 \\ i_1+k_2+u_j=i_2+k_1+u_l}}^{q} \lim \frac{1}{p}\mathrm{Tr}(\psi_{i_1}\psi_{k_2}^*\psi_{i_2}\psi_{k_1}^*)$$

$$=\; T_{11} + T_{12}, \;(\text{say}). \qquad (11.23)$$

Now,

$$T_{11} \;=\; \sum_{\substack{j,l=1}}^{r} \sum_{\substack{i_1,k_1,i_2,k_2=0 \\ i_1+i_2+u_j+u_l=k_1+k_2}}^{q} \lim \frac{1}{p}\mathrm{Tr}(\psi_{i_1}\psi_{i_1+(k_2-i_1-u_j)+u_j}^*\psi_{k_1+(i_2-k_1+u_l)-u_l}\psi_{k_1}^*)$$

$$=\; \sum_{\substack{j,l=1}}^{r} \sum_{v_1=-\infty}^{\infty} \lim \frac{1}{p}\mathrm{Tr}\Big(\sum_{i_1=0}^{q-v_1} \psi_{i_1}\psi_{i_1+v_1+u_j}^*\Big)\Big(\sum_{i_2=0}^{q-v_2} \psi_{i_2}\psi_{i_2+v_1-u_l}^*\Big)$$

$$(\text{put } \psi_j = 0 \text{ for } j > q)$$

$$=\; \sum_{\substack{j,l=1}}^{r} \sum_{v_1=-\infty}^{\infty} \lim \frac{1}{p}\mathrm{Tr}(\Gamma_{v_1+u_j}\Gamma_{v_1-u_l}^*)$$

$$=\; \frac{1}{2}\sum_{\substack{j,l=1}}^{r} \sum_{v_1=-\infty}^{\infty} \Big[\lim \frac{1}{p}\mathrm{Tr}(\Gamma_{v_1+u_j}\Gamma_{v_1-u_l}^*) + \lim \frac{1}{p}\mathrm{Tr}(\Gamma_{v_1+u_l}\Gamma_{v_1-u_j}^*)\Big].$$

Similarly,

$$T_{12} \;=\; 0.5\sum_{\substack{j,l=1}}^{r} \sum_{v_1=-\infty}^{\infty} \lim \frac{1}{p}\mathrm{Tr}(\Gamma_{v_1+u_j}\Gamma_{v_1+u_j}^*)$$

$$+ 0.5\sum_{\substack{j,l=1}}^{r} \sum_{v_1=-\infty}^{\infty} \lim \frac{1}{p}\mathrm{Tr}(\Gamma_{v_1-u_l}\Gamma_{v_1-u_j}^*).$$

Hence, by (11.23),

$$\lim E(\tilde{T}_{(u_1,u_2,\ldots,u_r),p})^2 = \sigma_{(u_1,u_2,\ldots,u_r)}^2. \qquad (11.24)$$

Now by (11.19), (11.17) is established. This completes the proof of (a).

(b) By the analogous result of Lemma 6.3.1 for $p/n \to 0$, it is easy to see that $\hat{\sigma}^2_{(u_1,u_2,\ldots,u_r),p} \xrightarrow{P} \sigma^2_{(u_1,u_2,\ldots,u_r)}$. Hence, the proof of (b) is complete by Slutsky's theorem.   □

One sample tests.

Example 11.3.1. Consider the model (8.6). Let A be a square matrix of order p with a non-degenerate LSD. Suppose we wish to test

$$H_0 : \psi_q = A \quad \text{against} \quad H_1 : \psi_q \neq A. \tag{11.25}$$

As $\Gamma_q = \psi_q^*$, this is equivalent to testing

$$H_0 : \Gamma_q = A^* \quad \text{against} \quad H_1 : \Gamma_q \neq A^*. \tag{11.26}$$

The following corollary of Theorem 11.3.1(b) which is immediate once we put $r = 1$, $u_1 = q$ and note that under H_0, $\Gamma_q = A^*$, leads to a test of the hypotheses (11.25).

Corollary 11.3.1. *Consider the model (8.6) and the hypotheses given in (11.25). Suppose (B1), (B) hold and $\{\varepsilon_{i,j} : i,j \geq 1\} \in \mathcal{L}_4 \cap U(\delta)$ for some $\delta > 0$. Further suppose $p/n \to 0$. Then under H_0,*

$$\sqrt{np^{-1}}(\text{Tr}(\hat{\Gamma}_q) - \text{Tr}(A))/\hat{\lambda}_q \xrightarrow{D} \mathcal{N}(0,1), \quad \text{where} \tag{11.27}$$

$$\hat{\lambda}_q^2 = 0.5 \sum_{v=-2q}^{2q} p^{-1}\text{Tr}(\hat{\Gamma}_{v-q} + \hat{\Gamma}_{v+q})(\hat{\Gamma}_{v-q} + \hat{\Gamma}_{v+q})^*. \tag{11.28}$$

Moreover, in Corollary 11.3.2 we show that under H_1, for some $0 < \lambda < \infty$,

$$\sqrt{np^{-1}}(\text{Tr}(\hat{\Gamma}_q) - \text{Tr}(\Gamma_q))/\hat{\lambda}_q \xrightarrow{D} \mathcal{N}(0,\lambda^2). \tag{11.29}$$

Therefore, we can use the statistic on the left side of (11.27) and reject H_0 if it is large.

Example 11.3.2. Consider the model (8.6). In general, we do not have any ready-made answer for testing $H_0 : \psi_j = A$ against $H_1 : \psi_j \neq A$ for some fixed $0 \leq j \leq (q-1)$ and $p \times p$ matrix A. To explain why, consider the MA(3) process with coefficient matrices $\{\psi_0 = I, \psi_1, \psi_2, \psi_3\}$. In this case,

$$\Gamma_1 = \psi_1^* + \psi_1\psi_2^* + \psi_2\psi_3^*, \ \Gamma_2 = \psi_2^* + \psi_1\psi_3^*,$$
$$\Gamma_3 = \psi_3^*, \ \Gamma_u = 0 \ \forall u \geq 4. \tag{11.30}$$

Suppose we wish to test

$$H_0 : \ \psi_2 = A \quad \text{against} \quad H_1 : \ \psi_2 \neq A. \tag{11.31}$$

Under H_0, Γ_1, Γ_2, and Γ_3 are given by

$$\Gamma_{1H_0} = \psi_1^* + \psi_1 A^* + A\psi_3^*,, \quad \Gamma_2 = A^* + \psi_1\psi_3^*, \quad \Gamma_3 = \psi_3^*. \tag{11.32}$$

By Theorem 11.1.1, for some $0 < a, b < \infty$,

$$\text{under } H_0, \quad \sqrt{np^{-1}}(\text{Tr}(\hat{\Gamma}_1) - \text{Tr}(\Gamma_{1H_0})) \xrightarrow{\mathcal{D}} \mathcal{N}(0, a^2) \text{ and}$$

$$\text{under } H_1, \quad \sqrt{np^{-1}}(\text{Tr}(\hat{\Gamma}_1) - \text{Tr}(\Gamma_1)) \xrightarrow{\mathcal{D}} \mathcal{N}(0, b^2).$$

Here a and b are respectively functions of $\{\lim p^{-1}\text{Tr}(\Gamma_{uH_0}\Gamma_{uH_0}) : -3 \le u, v \le 3\}$ and $\{\lim p^{-1}\text{Tr}(\Gamma_u\Gamma_v) : -3 < u, v < 3\}$. Just as in Example 11.3.1, we can make use of these results to test (11.31). One can use autocovariance of order 2 also. $\hat{\Gamma}_3$ cannot be used to test (11.31) as Γ_3 is not a function of ψ_2 and hence it makes no distinction between H_0 and H_1.

Clearly, the above method does not work when ψ_1 and ψ_3 are unknown. Moreover, it is hard to estimate these coefficient matrices. If we consider the method of moments, we get the system of equations given in (11.30) after replacing the population autocovariance matrices by the sample autocovariance matrices. These equations are not easy to solve. Moreover appropriate regularization could be needed before consistency is achieved. This needs further investigation.

Example 11.3.3. Next, consider the hypotheses

$$H_0 : \psi_j = A_j \; \forall j \quad \text{against} \quad H_1 : \psi_j \ne A_j \; \text{for at least one } j, \quad (11.33)$$

for some known $p \times p$ matrices $\{A_j\}$. As H_0 specifies all the coefficient matrices $\{\psi_j : 0 \le j \le q\}$, a natural testing method should be based on $\hat{\Gamma}_u$ for all $1 \le u \le q$. Let

$$\hat{G}_q = \sum_{u=1}^{q} \hat{\Gamma}_u, \quad G_q = \sum_{u=1}^{q} \Gamma_u = \sum_{u=0}^{q}\sum_{j=0}^{q-u} \psi_j\psi_{j+u}^* \quad \text{and} \quad (11.34)$$

$$G_{qH_0} = \sum_{u=0}^{q}\sum_{j=0}^{q-u} A_j A_{j+u}^* \quad \text{(i.e., } G_q \text{ under } H_0\text{).} \tag{11.35}$$

A test statistic can be proposed for (11.33) using $\text{Tr}(\hat{G}_q)$. This is based on the the following corollary which follows immediately from Theorem 11.3.1 once we put $r = q$, $u_i = i$, $\forall 1 \le i \le q$ and noting that under H_0, $G_q = G_{qH_0}$.

Corollary 11.3.2. *Consider the model (8.6) and the hypotheses given in (11.33). Suppose (B1), (B) hold and $\{\varepsilon_{i,j} : i, j \ge 1\} \in \mathcal{L}_4 \cap U(\delta)$ for some $\delta > 0$. Further suppose $p/n \to 0$. Then under H_0,*

$$\sqrt{np^{-1}}(\text{Tr}(\hat{G}_q) - \text{Tr}(G_{qH_0}))/\hat{\delta}_q \xrightarrow{\mathcal{D}} \mathcal{N}(0, 1), \quad where$$

$$\hat{\delta}_q^2 = \sum_{v=-\infty}^{\infty} p^{-1}\text{Tr}\Big(\sum_{\substack{-q \le i \le q \\ i \ne 0}} \hat{\Gamma}_{v-i} \Big)\Big(\sum_{\substack{-q \le i \le q \\ i \ne 0}} \hat{\Gamma}_{v-i} \Big)^*. \tag{11.36}$$

Moreover, under H_1, for some $0 < \delta < \infty$

$$\sqrt{np^{-1}}(\mathrm{Tr}(\hat{G}_q) - \mathrm{Tr}(G_q))/\hat{\delta}_q \xrightarrow{\mathcal{D}} \mathcal{N}(0, \delta^2). \tag{11.37}$$

Note that centering in (11.36) and (11.37) are different. Therefore, we can use $\sqrt{np^{-1}}(\mathrm{Tr}(\hat{G}_q) - \mathrm{Tr}(G_{qH_0}))/\hat{\delta}_q$ as a test statistic and reject H_0 if it is large in absolute value.

Two samples tests.

Consider the two independent MA(q) processes in (11.10).

Example 11.3.4. Suppose we wish to test

$$H_0 : \psi_q = \phi_q \quad \text{against} \quad H_1 : \psi_q \neq \phi_q. \tag{11.38}$$

Similar to the one sample cases, we can devise a test statistic. Recall $\hat{\lambda}_q$ from (11.28). Let $\hat{\lambda}_{q,X}$ and $\hat{\lambda}_{q,Y}$ be $\hat{\lambda}_q$ respectively for the processes $\{X_t\}$ and $\{Y_t\}$. Define $\hat{\lambda}_{q,XY}^2 = \hat{\lambda}_{q,X}^2 + \hat{\lambda}_{q,Y}^2$. By the independence of these processes and Corollary 11.3.1, it is easy to see that under H_0,

$$\sqrt{np^{-1}}(\mathrm{Tr}(\hat{\Gamma}_{q,X}) - \mathrm{Tr}(\hat{\Gamma}_{q,Y}))/\hat{\lambda}_{q,XY} \xrightarrow{\mathcal{D}} \mathcal{N}(0, 1)$$

and under H_1, for some $0 < \tilde{\lambda} < \infty$,

$$\sqrt{np^{-1}}((\mathrm{Tr}(\hat{\Gamma}_{q,X}) - \mathrm{Tr}(\hat{\Gamma}_{q,Y})) - (\mathrm{Tr}(\psi_q) - \mathrm{Tr}(\phi_q)))/\hat{\lambda}_{q,XY} \xrightarrow{\mathcal{D}} \mathcal{N}(0, \tilde{\lambda}^2).$$

Therefore, we can use $\sqrt{np^{-1}}(\mathrm{Tr}(\hat{\Gamma}_{q,X}) - \mathrm{Tr}(\hat{\Gamma}_{q,Y}))/\hat{\lambda}_{q,XY}$ as our test statistic and reject H_0 if it is large in absolute value.

Example 11.3.5. Consider the hypotheses

$$H_0 : \psi_j = \phi_j \; \forall j \quad \text{against} \quad H_1 : \psi_j \neq \phi_j \; \text{for at least one } j. \tag{11.39}$$

Recall the definition of $\{G_q, \hat{G}_q\}$ in (11.34). Let $\{G_{q,X}, \hat{G}_{q,X}\}$, $\{G_{q,Y}, \hat{G}_{q,Y}\}$ and $\hat{\delta}_{q,X}$ and $\hat{\delta}_{q,Y}$ have their obvious meaning.

Define $\hat{\delta}_{q,XY}^2 = \hat{\delta}_{q,X}^2 + \hat{\delta}_{q,Y}^2$. By Corollary 11.3.2 and the independence of $\{X_t\}$ and $\{Y_t\}$, it is easy to see that under H_0,

$$\sqrt{np^{-1}}(\mathrm{Tr}(\hat{G}_{q,X}) - \mathrm{Tr}(\hat{G}_{q,Y}))/\hat{\delta}_{q,XY} \xrightarrow{\mathcal{D}} \mathcal{N}(0, 1)$$

and under H_1, for some $0 < \tilde{\delta} < \infty$

$$\sqrt{np^{-1}}((\mathrm{Tr}(\hat{G}_{q,X}) - \mathrm{Tr}(\hat{G}_{q,Y})) - (\mathrm{Tr}(G_{q,X}) - \mathrm{Tr}(G_{q,Y})))/\hat{\delta}_{q,XY} \xrightarrow{\mathcal{D}} \mathcal{N}(0, \tilde{\delta}^2).$$

Thus, we can use $\sqrt{np^{-1}}(\mathrm{Tr}(\hat{G}_{q,X}) - \mathrm{Tr}(\hat{G}_{q,Y}))/\hat{\delta}_{q,XY}$ as a test statistic and reject H_0 if it is large in absolute value. If $\mathrm{Tr}(G_{q,X}) = \mathrm{Tr}(G_{q,Y})$, then we can consider $r > 1$ such that $\mathrm{Tr}(G_{q,X}^r) \neq \mathrm{Tr}(G_{q,Y}^r)$ and use weak convergence of appropriately centered and scaled $(\mathrm{Tr}(\hat{G}_{q,X}^r) - \mathrm{Tr}(\hat{G}_{q,Y}^r))$ for testing.

Testing for q: a simple model.

Consider a random sample $\{X_t : 1 \leq t \leq n\}$ from the process

$$X_t = \sum_{j=0}^{q} \varepsilon_{t-j} \quad \forall t \geq 0. \tag{11.40}$$

Based on the above sample, suppose we wish to test

$$H_0 : q = q_0 \quad \text{against} \quad H_1 : q = q_1. \tag{11.41}$$

Recall Π_{3u} and $\hat{\Pi}_{3u}$ from (10.1). Let

$$\hat{\Pi}_{a,u} = \sqrt{np^{-1}}(\hat{\Pi}_{3u} - \Pi_{3u}). \tag{11.42}$$

Let us assume $q_0 < q_1$. Similar method would work for the reverse case. The following two corollaries are useful to design a test statistic for the hypothesis (11.41).

Corollary 11.3.3. *Consider the model (11.40) and the hypotheses in (11.41) with $q_0 < q_1$. Suppose (B1) holds and $\{\varepsilon_{i,j} : i, j \geq 1\} \in \mathcal{L}_4 \cap U(\delta)$ for some $\delta > 0$. Further suppose $p/n \to 0$. Then*

$$\sqrt{np^{-1}}(\mathrm{Tr}(\hat{\Gamma}_u) - (q + 1 - |u|)I(|u| \leq q)) \xrightarrow{\mathcal{D}} \mathcal{N}(0, \tau_{u,q}^2), \quad \text{where}$$

$$\tau_{u,q}^2 = \frac{1}{2} \sum_{v=-\infty}^{\infty} (a_{v-u} + a_{v+u})^2, a_j = ((q+1) - |j|)I(|j| \leq q). \tag{11.43}$$

Proof. First observe that, for the model (11.40)

$$\Gamma_u = ((q+1) - |u|)I(|u| \leq q)I_p, \tag{11.44}$$

where I_p is the identity matrix of order p.

Now put $r = 1$ and $u_1 = u$ in (11.12)–(11.16). As $p^{-1}\mathrm{Tr}(I_p) = 1$, this yields

$$T_{(u),p} = \sqrt{np^{-1}}(\mathrm{Tr}(\hat{\Gamma}_u) - ((q+1) - |u|)I(|u| \leq q)),$$

$$\sigma_{(u)}^2 = \frac{1}{2} \sum_{v=-\infty}^{\infty} (a_{v-u} + a_{v+u})^2, a_j = ((q+1) - |j|)I(|j| \leq q).$$

This completes the proof of Corollary 11.3.3. $\qquad\qquad\qquad\square$

Corollary 11.3.4. *Consider the model (11.40) and the hypotheses in (11.41) with $q_0 < q_1$. Suppose (B1) holds and $\{\varepsilon_{i,j} : i, j \geq 1\} \in \mathcal{L}_4 \cap U(\delta)$ for some $\delta > 0$. Further suppose $p/n \to 0$. Then under H_0,*

$$\sqrt{np^{-1}}\mathrm{Tr}(\hat{\Gamma}_{q_0+1}) \xrightarrow{\mathcal{D}} \mathcal{N}(0, 2\sigma_{q_0}^2), \tag{11.45}$$

where $\sigma_{q_0}^2$ is given by

$$\sigma_{q_0}^2 = \frac{1}{2} \sum_{u=-q_0}^{q_0} (q_0 + 1 - |u|)^2. \tag{11.46}$$

Moreover, under H_1,

$$\sqrt{np^{-1}} \mathrm{Tr}(\hat{\Gamma}_{q_0+1} - (q_1 - q_0)) \xrightarrow{D} \mathcal{N}(0, \tau^2), \quad where \tag{11.47}$$

$$\tau^2 = \frac{1}{2} \sum_{v=-\infty}^{\infty} (a_{v-(q_0+1)} + a_{v+(q_0+1)})^2, \tag{11.48}$$

$$a_u = ((q_1 + 1) - |u|)I(|u| \le q_1). \tag{11.49}$$

Proof. To verify (11.45), put $q = q_0$, $u = q_0 + 1$ in Corollary 11.3.3. Then

$$\sqrt{np^{-1}} \mathrm{Tr}(\hat{\Gamma}_{q_0+1}) \xrightarrow{D} \mathcal{N}(0, \tau_{q_0+1,q_0}^2), \quad where$$

$$\tau_{q_0+1,q_0}^2 = \frac{1}{2} \sum_{v=-\infty}^{\infty} (a_{v-q_0-1} + a_{v+q_0+1})^2, a_j = ((q_0 + 1) - |j|)I(|j| \le q_0).$$

Now,

$$a_{v-q_0-1} = (q_0 + 1) - |(q_0 + 1) - v|, \; a_{v+q_0+1} = 0 \;\text{ when } 1 \le v \le q_0 + 1,$$
$$a_{v-q_0-1} = (q_0 + 1) - |v - (q_0 + 1)|, \; a_{v+q_0+1} = 0 \text{ when } q_0 + 2 \le v \le 2q_0 + 1,$$
$$a_{v+q_0+1} = (q_0 + 1) - |(q_0 + 1) + v|, \; a_{v-q_0-1} = 0 \text{ when } -2q_0 - 1 \le v \le -1,$$
$$a_{v+q_0+1} = 0, \; a_{v-q_0-1} = 0 \;\text{ when } |v| > 2q_0 + 1 \text{ and } v = 0.$$

Therefore,

$$2\tau_{q_0+1,q_0}^2 = \sum_{v=1}^{q_0+1} (q_0 + 1 - |(q_0 + 1) - v|)^2 + \sum_{v=q_0+2}^{2q_0+1} (q_0 + 1 - |v - (q_0 + 1)|)^2$$

$$+ \sum_{v=-2q_0-1}^{-1} (q_0 + 1 - |(q_0 + 1) + v|)^2$$

$$= \sum_{v=0}^{q_0} (q_0 + 1 - |v|)^2 + \sum_{v=-q_0}^{-1} (q_0 + 1 - |v|)^2 + \sum_{v=-q_0}^{q_0} (q_0 + 1 - |v|)^2$$

$$= 2 \sum_{v=-q_0}^{q_0} (q_0 + 1 - |v|)^2 = 4\sigma_{q_0}.$$

Thus, (11.45) is established.

(11.47) is immediate from Corollary 11.3.3 by putting $q = q_1$, $u = q_0 + 1$ and observing that $q_0 < q_1$.

This completes the proof of Corollary 11.3.4. $\square$

Hence, we can use $\sqrt{np^{-1}}\mathrm{Tr}(\hat{\Gamma}_{q_0+1})/\sqrt{2}\sigma_{q_0}$ as the test statistic and reject H_0 at $100\alpha\%$ level of significance if $|T| > z_\alpha$, where z_α is the upper α point of the standard normal distribution.

It may be noted that even though all methods in Section 11.3 are based on the symmetric sum of $\{\hat{\Gamma}_u, \hat{\Gamma}_u^*\}$, other polynomials can also be used. We have restricted to the symmetric sum only for illustration. Calculation of asymptotic variances become increasingly difficult with higher order polynomials.

Exercises

1. Let $X_t = AX_{t-1} + \varepsilon_t$ where $\varepsilon_t \overset{\text{i.i.d.}}{\sim} \mathcal{N}(0,1)$ and $p/n \to y > 0$. Find the asymptotic distribution of the appropriately centered and scaled version of the following statistics: (a) $\mathrm{Tr}(\hat{\Gamma}_1)$, (b) $\mathrm{Tr}(\hat{\Gamma}_2\hat{\Gamma}_2^*)$, (c) $\mathrm{Tr}(\hat{\Gamma}_1\hat{\Gamma}_1^*) + \mathrm{Tr}(\hat{\Gamma}_2\hat{\Gamma}_2^*)$, and (d) $\mathrm{Tr}(\hat{\Gamma}_0) + \mathrm{Tr}(\hat{\Gamma}_1\hat{\Gamma}_1^*) + \mathrm{Tr}(\hat{\Gamma}_2\hat{\Gamma}_2^*)$. What will happen when $p/n \to 0$? Also state sufficient conditions on A that are required to establish the above asymptotic results.

2. Let $X_t = \sum_{i=1}^{10} A^i \varepsilon_{t-i}$ and $p/n \to 0$. Suggest two test statistics for

$$H_0 : A = I_p \quad \text{against} \quad H_1 : A \neq I_p$$

based on the discussion in Examples 11.3.1 and 11.3.3. Establish the null and alternative asymptotic distribution of these test statistics and state appropriate assumptions required for this. Hence, mention the critical regions. Simulate and compute empirical power of both the tests and compare them. How can one modify these tests when $p/n \to y > 0$?

3. Let $X_t = \varepsilon_t + A_1\varepsilon_{t-1} + B_1\varepsilon_{t-2}$ and $Y_t = \xi_t + A_2\xi_{t-1} + B_2\xi_{t-2}$ where $\{\varepsilon_t\}$ and $\{\xi_t\}$ are independently distributed and $p/n \to 0$. Suggest tests for the following hypotheses:

$$H_0 : B_1 = B_2 \quad \text{against} \quad H_1 : B_1 \neq B_2,$$
$$H_0 : A_1 = A_2, B_1 = B_2 \quad \text{against} \quad H_1 : A_1 \neq A_2 \text{ or } B_1 \neq B_2,$$
$$H_0 : A_1 = A_2, B_1 = I_p, B_2 = 0 \quad \text{against} \quad H_1 : A_1 \neq A_2, B_1 = I_p, B_2 = 0.$$

Modify the tests appropriately when $p/n \to y > 0$.

4. Suggest significance tests for the hypotheses given in Exercises 2-7 of Chapter 10.

5. Establish Theorem 11.1.1(b).

6. Justify Examples 11.1.1, 11.1.2, and 11.1.3.

7. Give a proof of Theorem 11.2.1.

8. Justify Example 11.2.1.

Appendix

SUPPLEMENTARY PROOFS

A.1 Proof of Lemma 6.3.1

We prove the result when only one Z matrix is involved. This proof applies mutas mutandis when more than one independent $\{Z_u\}$ matrices are involved. This is done in two steps. In the first step we show that it is enough to prove the lemma for monomials $\{m_i\}$ in $\{\mathbb{P}_u, \mathbb{P}_u^*\}$. In the second step we verify that result for monomials.

Step 1. Consider arbitrary $p \times p$ matrices $\{A_{ik}, C_{ik} : 1 \le k \le r_i, 1 \le i \le T\}$ $\subset \text{Span}\{B_{2i-1}, B_{2i-1}^*\}$ and $n \times n$ matrices $\{B_{ik} : 1 \le k \le r_i, 1 \le i \le T\} \subset \text{Span}\{B_{2i}, B_{2i}^*\}$. Define

$$\pi_i = n^{-r_i} \text{Tr}(\prod_{k-1}^{r_i} A_{ik} Z B_{ik} Z^* C_{ik}) \text{ and } \pi_i^0 = E\pi_i, \ 1 \le i \le T. \quad \text{(A.1)}$$

For all $d \ge 1$, consider the equations

$$\lim E \left[\Pi_{i=1}^T \left(\pi_i - \pi_i^0 \right) \right] \quad \text{(A.2)}$$

$$= \begin{cases} 0 \text{ if } T = 2d - 1, \\ \sum_{\mathcal{S}_d} \prod_{k=1}^d \lim E\left[\left(\pi_{i_{2k-1}} - \pi_{i_{2k-1}}^0 \right) \left(\pi_{i_{2k}} - \pi_{i_{2k}}^0 \right) \right], \text{ if } T = 2d. \end{cases}$$

We now prove that if (A.2) holds then (6.22) holds. That is, it is enough to prove the lemma for monomials only.

Note that for some matrices $\{A_{iks}, C_{iks}\} \in \text{Span}\{B_{2i-1}, B_{2i-1}^*\}$ and $\{B_{iks}\} \in \text{Span}\{B_{2i}, B_{2i}^*\}$, we can write

$$\mathcal{P}_i = \sum_{k=1}^{t_i} \text{Tr}\left(n^{-r_k} \prod_{s=1}^{r_k} A_{iks} Z B_{iks} Z^* C_{iks}\right) = \sum_{k=1}^{t_i} S_{ik}, \text{ (say)}.$$

$$\mathcal{P}_i^0 = \sum_{i=1}^{t_i} E(S_{ik}) = \sum_{i=1}^{t_i} S_{ik}^0, \text{ say}.$$

Therefore,

$$\lim E(\prod_{i=1}^{T}(\mathcal{P}_i - \mathcal{P}_i^0))$$

$$= \lim E(\prod_{i=1}^{T}\sum_{k=1}^{t_i}(S_{ik} - S_{ik}^0)) = \lim E(\prod_{i=1}^{T}\sum_{k=1}^{t_i}T_{ik}), \text{ say}$$

$$= \lim E(\sum_{1\le k_i \le t_i}\prod_{i=1}^{T}T_{ik_i}) = \sum_{1\le k_i \le t_i}\lim E(\prod_{i=1}^{T}T_{ik_i})$$

$$= \begin{cases} 0, & \text{if } T = 2d-1 \\ \sum_{k_i \le t_i}\sum_{\mathcal{S}_d}\prod_{s=1}^{d}\lim E(T_{i_{2s-1}k_{i_{2s-1}}}T_{i_{2s}k_{i_{2s}}}), & \text{if } T = 2d. \end{cases}$$

The last equality holds by (A.2). Therefore, (6.22) follows from (A.2) when T is odd.

When T is even, we have

$$\prod_{s=1}^{d}\sum_{k_{2s-1},k_{2s}}\lim E(T_{i_{2s-1}k_{i_{2s-1}}}T_{i_{2s}k_{i_{2s}}}) = \prod_{k=1}^{d}\lim E\big[(\mathcal{P}_{i_{2k-1}}-\mathcal{P}_{i_{2k-1}}^0)(\mathcal{P}_{i_{2k}}-\mathcal{P}_{i_{2k}}^0)\big].$$

Therefore, summing the above over $\mathcal{S}_d$, (6.22) follows from (A.2) for all $T \ge 1$.

Step 2. Proof of (A.2). Let $A(i,j)$ be the (i,j)-th element of the matrix A. Note that, for all $1 \le i \le T$,

$$n^{r_i}\pi_i = \text{Tr}(\prod_{k=1}^{r_i}A_{ik}ZB_{ik}Z^*C_{ik}) \qquad (A.3)$$

$$= \sum_{\substack{1\le u_t^{(i)}\le p,\ 1\le v_s^{(i)}\le P\ (i) \\ 1\le t\le 3r_i,\ 1\le s\le 2r_i u_{3r_i+1}^{(i)}=u_1^{(i)}}}\prod_{k=1}^{r_i}A_{ik}(u_{3k-2}^{(i)},u_{3k-1}^{(i)})\varepsilon_{u_{3k-1}^{(i)},v_{2k-1}^{(i)}}B_{ik}(v_{2k-1}^{(i)},v_{2k}^{(i)})$$

$$\times\varepsilon_{u_{3k}^{(i)},v_{2k}^{(i)}}C_{ik}(u_{3k}^{(i)},u_{3k+1}^{(i)}).$$

For fixed $1 \le i \le T$, we define

$$\mathcal{U}_i = \{(u_{3k+\delta}^{(i)},v_{2k+\delta}^{(i)}) : 1 \le k \le r_i, \delta = -1,0, u_{3r_i+1}^{(i)} = u_1^{(i)}\}. \qquad (A.4)$$

Note that $\mathcal{U}_i$ is the set of all indices attached with ε's in the expansion of π_i given in (A.3). An index $(u_{3k+\delta}^{(i)},v_{2k+\delta}^{(i)})$ is said to be matched if there is at least one $(k',\delta',i') \ne (k,\delta,i)$ with $(u_{3k+\delta}^{(i)},v_{2k+\delta}^{(i)}) = (u_{3k'+\delta'}^{(i')},v_{2k'+\delta'}^{(i')})$. Now note that $E\big[\Pi_{i=1}^{T}(\pi_i - \pi_i^0)\big]$ involves all indices in $\cup_{i=1}^{T}\mathcal{U}_i$ (if we expand $\{\pi_i\}$ using the last equality of (A.3)). As $\{\varepsilon_{i,j}\}$ are independent and have mean 0, all indices in $\cup_{i=1}^{T}\mathcal{U}_i$ need to be matched to guarantee a non-zero contribution. For each $1 \leq i \leq T$, consider the following sets of matched indices.

B_i = set of all matches where for each (k, δ), there is at least one

$\quad\quad (k', \delta') \neq (k, \delta)$ with $(u_{3k+\delta}^{(i)}, v_{2k+\delta}^{(i)}) = (u_{3k'+\delta'}^{(i)}, v_{2k'+\delta'}^{(i)})$ and for $i \neq i'$,

$\quad\quad$ there is no (k', δ', i') such that $(u_{3k+\delta}^{(i)}, v_{2k+\delta}^{(i)}) = (u_{3k'+\delta'}^{(i')}, v_{2k'+\delta'}^{(i')})$.

Consider the disjoint decomposition $\cup_{i=1}^{T+1} C_i$ of all possible matches of indices in $\cup_{i=1}^{T}\mathcal{U}_i$, where

$$C_1 = B_1, \; C_i = (\cap_{j=1}^{i-1} B_j^c) \cap B_i \; \forall 2 \leq i \leq T, \; C_{T+1} = \cap_{i=1}^{T} B_i^c. \quad\quad (A.5)$$

Let for any set A, E_A be the usual expectation restricting on the set A. We shall first show that

$$E\left[\Pi_{i=1}^{T}\left(\pi_i - \pi_i^0\right)\right] = E_{C_{T+1}}(\Pi_{i=1}^{T}\pi_i). \quad\quad (A.6)$$

For this purpose, we need more analysis for the set C_i. Define

S_i = set of all matches of indices in $\mathcal{U}_i$, and

S_{-i} = set of all matches of indices in $\cup_{j\neq i}\mathcal{U}_j$ such that for each

$\quad\quad 1 \leq j < i$, there is at least one index in $\mathcal{U}_j$ which matches

$\quad\quad$ with some index in $\mathcal{U}_k, k \neq j, i$.

Note that

$$C_i = (\cap_{j=1}^{i-1} B_j^c) \cap B_i = \{(\sigma_1 \cup \sigma_2) : \sigma_1 \in S_i, \; \sigma_2 \in S_{-i}\}.$$

Also note that

$$E(\Pi_{i=1}^{T}(\pi_i - \pi_i^0)) = E_{C_i}((\Pi_{j=1}^{i}\pi_j)\Pi_{j=i+1}^{T}(\pi_j - \pi_j^0)) + \text{ other terms.}$$

Then for all $2 \leq i \leq T$, we have

$$E_{C_i}(\Pi_{j=1}^{i}\pi_j\Pi_{j=i+1}^{T}(\pi_j - \pi_j^0)) = \sum_{\sigma \in C_i} E_\sigma(\Pi_{j=1}^{i}\pi_j\Pi_{j=i+1}^{T}(\pi_j - \pi_j^0))$$

$$= \sum_{\sigma_1 \in S_i, \; \sigma_2 \in S_{-i}} E_{\sigma_1}(\pi_i) E_{\sigma_2}(\Pi_{j=1}^{i-1}\pi_j\Pi_{j=i+1}^{T}(\pi_j - \pi_j^0)) \quad\quad (A.7)$$

$$\text{[as } C_i \subset B_i \text{ and under } B_i, \{\varepsilon_{u,v} : (u,v) \in \mathcal{U}_i\} \text{ are}$$

$$\text{independent of } \{\varepsilon_{u,v} : (u,v) \in \cup_{j\neq i}\mathcal{U}_j\}]$$

$$= \sum_{\sigma_1 \in S_i} E_{\sigma_1}(\pi_i) \sum_{\sigma_2 \in S_{-i}} E_{\sigma_2}(\Pi_{j=1}^{i-1}\pi_j\Pi_{j=i+1}^{T}(\pi_j - \pi_j^0))$$

$$= \pi_i^0 (E_{\cap_{j=1}^{i-1} B_j^c}(\Pi_{j=1}^{i-1}\pi_j\Pi_{j=i+1}^{T}(\pi_j - \pi_j^0))).$$

Similarly,

$$E_{B_1}(\pi_1 \Pi_{i=2}^T (\pi_i - \pi_i^0)) \;=\; \pi_1^0 E(\Pi_{i=2}^T (\pi_i - \pi_i^0)). \qquad (A.8)$$

Now, the left side of (A.6) equals,

$$E_{B_1}(\pi_1 \Pi_{i=2}^T(\pi_i - \pi_i^0)) + E_{B_1^c}(\pi_1 \Pi_{i=2}^T(\pi_i - \pi_i^0)) - \pi_1^0 E(\Pi_{i=2}^T(\pi_i - \pi_i^0))$$

$$= E_{B_1^c}(\pi_1 \Pi_{i=2}^T(\pi_i - \pi_i^0))$$

$$= E_{B_1^c \cap B_2}(\pi_1 \pi_2 \Pi_{i=3}^T(\pi_i - \pi_i^0)) + E_{B_1^c \cap B_2^c}(\pi_1 \pi_2 \Pi_{i=3}^T(\pi_i - \pi_i^0))$$

$$\qquad\qquad - \pi_2^0 E_{B_1^c}(\pi_1 \Pi_{i=3}^T(\pi_i - \pi_i^0))$$

$$= E_{B_1^c \cap B_2^c}(\pi_1 \pi_2 \Pi_{i=3}^T(\pi_i - \pi_i^0)), \quad \text{(by (A.7), for } T = 2)$$

$$\vdots$$

$$= E_{B_1^c \cap B_2^c \cap \cdots \cap B_T^c}(\Pi_{i=1}^T \pi_i), \text{ (by repeated use of (A.7), } 3 \le i \le T$$

$$= E_{C_{T+1}}(\Pi_{i=1}^T \pi_i).$$

Therefore, (A.6) is established.

Next we shall analyze the set C_{T+1} and identify the set of matches which contribute in the limit.

Two index sets $\mathcal{U}_i$ and $\mathcal{U}_{i'}$ are said to be connected if there is (k, δ) and (k', δ') with $(u_{3k+\delta}^{(i)}, v_{2k+\delta}^{(i)}) = (u_{3k+\delta}^{(i')}, v_{2k'+\delta'}^{(i')})$. Also a collection of index sets $\{\mathcal{U}_{i_1}, \mathcal{U}_{i_2}, \ldots, \mathcal{U}_{i_s}\}$, $s \ge 2$, is said to form a connected group if for each $1 \le k \le s-1$, $\mathcal{U}_{i_k}$ and $\mathcal{U}_{i_{k+1}}$ is connected. Note that, in a typical match in C_{T+1}, for each i, $\mathcal{U}_i$ is connected with some other $\mathcal{U}_{i'}$, $i' \ne i$. Therefore, each match in C_{T+1} corresponds to some disjoint connected groups each of length at least 2. Consider the following disjoint decomposition of C_{T+1}.

$$C_{T+1} \;=\; \bigcup_{\substack{2 \le g_1, \ldots, g_R \le T \\ \sum_{j=1}^R g_j = T, \; R \ge 1}} G(g_1, \ldots, g_R), \text{ where}$$

$$G(g_1, \ldots, g_R) \;=\; \text{set of all such matches in } C_{T+1} \text{ which form exactly}$$
$$R \text{ connected groups of length } g_1, \ldots, g_R.$$

Hence, by (A.6), we have

$$E\left[\Pi_{i=1}^T (\pi_i - \pi_i^0) \right] \;=\; E_{C_{T+1}}(\Pi_{i=1}^T \pi_i) \qquad\qquad (A.9)$$

$$= \sum_{\substack{2 \le g_1, \ldots, g_R \le T \\ \sum_{j=1}^R g_j = T, \; R \ge 1}} E_{G(g_1, \ldots, g_R)}(\Pi_{i=1}^T \pi_i).$$

Consider a typical match σ in $G(g_1, \ldots, g_R)$ with connected groups $G_{\sigma 1}, \ldots, G_{\sigma R}$ respectively of lengths $g_1, \ldots, g_R$. Note that, for a fixed σ, $\{G_{\sigma k} : 1 \le k \le R\}$ forms a partition of $\{\mathcal{U}_k : 1 \le k \le T\}$. Further if $i \ne j$, then no index in $G_{\sigma i}$ matches with any index in $G_{\sigma j}$. Hence, by independence of $\{\varepsilon_{i,j}\}$,

$$E_{G(g_1,\ldots,g_R)}(\Pi_{i=1}^T \pi_i) = \sum_\sigma \prod_{k=1}^R E_{G_{\sigma k}}(\pi_1,\ldots,\pi_T), \text{ where} \quad \text{(A.10)}$$

$$E_{G_{\sigma k}}(\pi_1,\ldots,\pi_T) = E_{G_{\sigma k}}\Big(\prod_{\substack{i_j: \; u_{i_j} \in G_{\sigma k} \\ 1 \le j \le g_j}} \pi_{i_j} \Big), \; \forall 1 \le k \le R,$$

and $E_{G_{\sigma k}}$ is the usual expectation restricting on the matches in $G_{\sigma k}$. For the moment assume that the following claim is true. We shall prove the claim later.

Claim. $E_{G_{\sigma k}}(\pi_1,\ldots,\pi_T) = O(n^{-g_k+2})$, $\forall \sigma, k$.

Therefore, for all $\sigma \in G(g_1,\ldots,g_R)$,

$$\prod_{k=1}^R E_{G_{\sigma k}}(\pi_1,\ldots,\pi_T) = O(n^{-\Sigma(g_j-2)}) = O(n^{-T+2R}). \quad \text{(A.11)}$$

As $G(g_1,\ldots,g_R)$ is a finite set, by (A.10), we have

$$E_{G(g_1,\ldots,g_R)}(\Pi_{i=1}^T \pi_i) = O(n^{-T+2R}). \quad \text{(A.12)}$$

Note that as $g_1,\ldots,g_R \ge 2$, the maximum possible value of R is $[T/2]$, the greatest integer $\le T/2$.

First suppose T is odd. Then we always have $T - 2R > 0$ and hence, using (A.12), $\lim E_{G(g_1,\ldots,g_R)}(\Pi_{i=1}^T \pi_i) = 0$. As a consequence, using (A.9), we have

$$\lim E\left[\Pi_{i=1}^T \left(\pi_i - \pi_i^0\right)\right] = 0, \text{ if } T \text{ is odd}, \quad \text{(A.13)}$$

proving (A.2) when T is odd.

Now suppose T is even, say $T = 2d$. Then note that

$$T - 2R \begin{cases} = 0, & \text{for } G(2,\ldots,2), \; R = d \\ > 0, & \text{otherwise.} \end{cases}$$

Therefore, by (A.12),

$$\lim E_{G(g_1,\ldots,g_R)}(\Pi_{i=1}^T \pi_i) = 0, \text{ if } G(g_1,\ldots,g_R) \ne G(2,\ldots,2), \quad \text{(A.14)}$$

and hence by (A.9), we have

$$\lim E\left[\Pi_{i=1}^T \left(\pi_i - \pi_i^0\right)\right] = \lim E_{G(2,\ldots,2)}(\Pi_{i=1}^T \pi_i). \quad \text{(A.15)}$$

It remains to identify the right side of (A.15) as the right side of (A.2). Note that a typical match in $G(2,2,\ldots,2)$ involves d groups each with length 2. Hence, there is a one-to-one correspondence of $G(2,2,\ldots,2)$ and $\mathcal{S}_d$, set of all pair partitions of $\{1, 2, \ldots, 2d\}$. The one-to-one correspondence is as follows. Consider $\sigma = \{(i_1, i_2), (i_3, i_4), \ldots, (i_{2d-1}, i_{2d})\} \in \mathcal{S}_d$, then for every $1 \leq k \leq d$, $\{\mathcal{U}_{i_{2k-1}}, \mathcal{U}_{i_{2k}}\}$ forms a connected group. Therefore, by (A.10), we have

$$E_{G(2,\ldots,2)}(\Pi_{i=1}^T \pi_i) = \sum_{\sigma \in \mathcal{S}_d} \prod_{k=1}^d E_{\{\mathcal{U}_{i_{2k-1}}, \mathcal{U}_{i_{2k}}\}}(\pi_1, \ldots, \pi_T). \qquad (A.16)$$

Let D be the set of all such matches of indices in $\mathcal{U}_{i_{2k-1}} \cup \mathcal{U}_{i_{2k}}$ such that $\{\mathcal{U}_{i_{2k-1}}, \mathcal{U}_{i_{2k}}\}$ are connected. Note that

$$
\begin{aligned}
E_{\{\mathcal{U}_{i_{2k-1}}, \mathcal{U}_{i_{2k}}\}}(\pi_1, \ldots, \pi_T) &= \sum_{\sigma \in D} E_\sigma(\pi_{i_{2k-1}} \pi_{i_{2k}}) \\
&= E\big((\pi_{i_{2k-1}} - \pi_{i_{2k-1}}^0)(\pi_{i_{2k}} - \pi_{i_{2k}}^0)\big), \quad (A.17) \\
&\qquad \text{by (A.6) for } T = 2.
\end{aligned}
$$

Therefore, by (A.16) and (A.17), we have

$$E_{G(2,\ldots,2)}(\Pi_{i=1}^T \pi_i) = \sum_{\sigma \in \mathcal{S}_d} \prod_{k=1}^d E\big((\pi_{i_{2k-1}} - \pi_{i_{2k-1}}^0)(\pi_{i_{2k}} - \pi_{i_{2k}}^0)\big). \qquad (A.18)$$

Now substituting (A.18) in (A.15), (A.2) is established for $T = 2d$. Therefore, by Steps 1 and 2, proof of (6.22) and hence of Lemma 6.3.1 is complete when one Z matrix is involved, provided the claim is true.

Proof of claim. As $\{\pi_i\}$ are commutative, it is enough to show that

$$E_C(\pi_1 \pi_2 \cdots \pi_g) = O(n^{-g+2}), \qquad (A.19)$$

where C is the set of all matches of indices in $\cup_{i=1}^g \mathcal{U}_i$ such that $\{\mathcal{U}_i : 1 \leq i \leq g\}$ forms a connected group. Recall $\{\pi_i\}$ from (A.3). Consider the following decomposition of C.

$$C = \bigcup_{1 \leq t_j \leq k_j \leq r_j} C(k_j, t_j : 1 \leq j \leq g), \text{ where}$$

$$
\begin{aligned}
C(k_j, t_j : 1 \leq j \leq g) = \ & \text{set of matches (pair, non-pair,} \qquad (A.20) \\
& \text{crossing, non-crossing) in } C \text{ such that} \\
& (u_{3k_i}^{(i)}, v_{2k_i}^{(i)}) = (u_{3t_{i+1}-1}^{(i+1)}, v_{2t_{i+1}-1}^{(i+1)}), \forall 1 \leq i \leq g-1.
\end{aligned}
$$

Therefore,

$$E_C(\pi_1 \pi_2 \cdots \pi_g) = \sum_{1 \leq t_j \leq k_j \leq r_j} E_{C(k_j, t_j : 1 \leq j \leq g)}(\pi_1 \pi_2 \cdots \pi_g). \qquad (A.21)$$

Now for convenience of writing, let us denote, for all $1 \leq i \leq g$,

$$
\begin{aligned}
D_i &= (\Pi_{k=1}^{t_i-1} A_{ik}(Z/\sqrt{n}) B_{ik}(Z^*/\sqrt{n}) C_{ik}) A_{it_i}, \\
E_i &= B_{it_i}(Z^*/\sqrt{n}) C_{it_i}(\Pi_{k=t_i+1}^{k_i-1} A_{ik}(Z/\sqrt{n}) B_{ik}(Z^*/\sqrt{n}) C_{ik}) A_{ik_i} Z B_{ik_i}, \\
F_i &= C_{ik_i}(\Pi_{k=k_i+1}^{r_i} A_{ik}(Z/\sqrt{n}) B_{ik}(Z^*/\sqrt{n}) C_{ik}), \text{ and, hence,} \\
n\pi_i &= \text{Tr}(D_i Z E_i Z^* F_i), \forall 1 \leq i \leq g. \qquad (A.22)
\end{aligned}
$$

Therefore,

$$n^g E_{\mathcal{C}(k_j,t_j:1<j\leq g)}(\pi_1\pi_2\cdots\pi_g)$$
$$= E_{\mathcal{C}(k_j,t_j:1\leq j\leq g)}\big(\Pi_{i=1}^g \mathrm{Tr}(D_i Z E_i Z^* F_i)\big),\ \text{(by (A.22))}$$
$$= \sum_{\substack{\{u_{ij},v_{ik},\ 1\leq i\leq g\}\\ j=1,2,3,\ k=1,2}} E_{\mathcal{C}(k_j,t_j:1\leq j\leq g)}\big(\Pi_{i=1}^g D_i(u_{i1},u_{i2})\varepsilon_{u_{i2},v_{i1}} E_i(v_{i1},v_{i2})$$
$$\times \varepsilon_{u_{i3},v_{i2}} F_i(u_{i3},u_{i1})\big)$$
$$= \sum_{\sigma\in\mathcal{C}(k_j,t_j:1\leq j\leq g)}\ \sum_{\substack{\{u_{ij},v_{ik},\ 1\leq i\leq g\}\\ j=1,2,3,\ k=1,2}} E_\sigma\big(\Pi_{i=1}^g D_i(u_{i1},u_{i2})\varepsilon_{u_{i2},v_{i1}}$$
$$\times E_i(v_{i1},v_{i2})\varepsilon_{u_{i3},v_{i2}} F_i(u_{i3},u_{i1})\big).$$

Now by (A.20), we have $(u_{i3},v_{i2}) = (u_{(i+1)2},v_{(i+1)1})$, $\forall 1\leq i\leq g-1$ and therefore,

$$\sum_{\substack{\{u_{ij},v_{ik},\ 1\leq i\leq g\}\\ j=1,2,3,\ k=1,2}} E_\sigma\big(\Pi_{i=1}^g D_i(u_{i1},u_{i2})\varepsilon_{u_{i2},v_{i1}} E_i(v_{i1},v_{i2})\varepsilon_{u_{i3},v_{i2}} F_i(u_{i3},u_{i1})\big)$$
$$= E_\sigma\big(\mathrm{Tr}(Z(\Pi_{i=1}^g E_i Z^* Z)Z^*(\Pi_{i=1}^g F_{g+1-i}D_{g+1-i}))\big).$$

Hence,

$$n^g E_{\mathcal{C}(k_j,t_j:1\leq j\leq g)}(\pi_1\pi_2\cdots\pi_g) \tag{A.23}$$
$$= \sum_{\sigma\in\mathcal{C}(k_j,t_j:1\leq j\leq g)} E_\sigma\big(\mathrm{Tr}(Z(\Pi_{i=1}^g E_i Z^* Z)Z^*(\Pi_{i=1}^g F_{g+1-i}D_{g+1-i}))\big).$$

Using the same idea as in the proof of the (M1) condition for Theorem 3.1, one can show that

$$\lim n^{-2} E_\sigma \mathrm{Tr}(Z(\Pi_{i=1}^g E_i Z^* Z)Z^*(\Pi_{i=1}^g F_{g+1-i}D_{g+1-i}))$$
$$= \begin{cases} O(1), & \text{if } \sigma \text{ is a non-crossing pair match} \\ o(1), & \text{if } \sigma \text{ is not a non-crossing or pair match}. \end{cases}$$

Hence, by (A.23), $E_{\mathcal{C}(k_j,t_j:1\leq j\leq g)}(\pi_1\pi_2\cdots\pi_g) = O(n^{-g+2})$. Therefore, by (A.21), (A.19) follows and the claim is established.

This completes the proof of Lemma 6.3.1 for one independent matrix Z. Note that if we have more than one independent matrix $\{Z_u\}$, then also the above proof will remain unchanged except that $\varepsilon_{i,j}$ (the (i,j)-th element of Z) will be replaced by $\varepsilon_{u,i,j}$ (the (i,j)-th element of Z_u). $\qquad\square$

A.2   Proof of Theorem 6.4.1(a)

Let $\Pi = \Pi(d,e,z)$ be any polynomial in d,e where $z \in \mathbb{C}^+$. Let $\bar{\Pi} = \Pi(\bar{d},\bar{e},z)$ be its embedded version. To prove Theorem 6.4.1(a), we need a lemma that provides expression for $\bar{\varphi}(\bar{\Pi}\bar{\delta}^r)$. Let us use the shorthand notation for simplicity.

$$\sum_1 = \sum_{l_0,l_1,\ldots,l_{t-1}=1}^q , \quad \sum_2 = \sum_{1=k_0<k_1<k_2<\ldots<k_{t-1}\leq r} ,$$

$$\sum_3 = \sum_{l_1,l_2,\ldots,l_r=1}^q , \quad \bar{d}_{l_t} = \bar{\Pi}\bar{d}_{l_0}, \quad k_t = r+1.$$

Lemma A.2.1. *Suppose (A1)-(A3) hold and $p,n \to \infty$, $p/n \to y > 0$. Then*

$$\bar{\varphi}(\bar{\Pi}\bar{\delta}^r) = \sum_{t=1}^r \sum_1 \sum_2 (1+y)^t \bar{\varphi}\left(\prod_{u=0}^{t-1} \bar{\varphi}(\bar{e}_{l_u}\bar{\delta}^{k_{u+1}-k_u-1}\bar{d}_{l_{u+1}})\underline{f}_{l_u} \right).$$

Proof. Recall that $K(\sigma)$ is the Kreweras complement of σ defined in Section 5.2. Therefore, by (5.34), we have

$$
\begin{aligned}
\bar{\varphi}(\bar{\Pi}\bar{\delta}^r) &= (1+y)^r \bar{\varphi}(\bar{\Pi}(\sum_{i=1}^q \bar{d}_i s \underline{f}_i s \bar{e}_i)^r) \\
&= (1+y)^r \sum_3 \bar{\varphi}(\bar{\Pi} \prod_{u=1}^r \bar{d}_{l_u} s \underline{f}_{l_u} s \bar{e}_{l_u})) \\
&= (1+y)^r \sum_3 \sum_{\sigma \in NC_2(2r)} \bar{\varphi}_{K(\sigma)}[\underline{f}_{l_1}, \bar{e}_{l_1}\bar{d}_{l_2}, \underline{f}_{l_2}, \bar{e}_{l_2}\bar{d}_{l_3}, \ldots, \bar{e}_{l_r}\bar{\Pi}\bar{d}_{l_1}] \\
&= \sum_{\sigma \in NC_2(2r)} \tau_\sigma
\end{aligned}
\tag{A.24}
$$

where

$$\tau_\sigma = (1+y)^r \sum_3 \bar{\varphi}_{K(\sigma)}[\underline{f}_{l_1}, \bar{e}_{l_1}\bar{d}_{l_2}, \underline{f}_{l_2}, \bar{e}_{l_2}\bar{d}_{l_3}, \ldots, \bar{e}_{l_r}\bar{\Pi}\bar{d}_{l_1}].$$

Now to compute (A.24), we consider the decomposition of $NC_2(2r) = \cup_{t=1}^r \mathcal{P}_t^{2r}$, where

$$\mathcal{P}_1^{2r} = \{\sigma \in NC_2(2r) : \{1,2\} \in \sigma\}$$

and for all $2 \leq t \leq r$,

$$
\begin{aligned}
\mathcal{P}_t^{2r} = \{&\sigma \in NC_2(2r) : \sigma = \{2k_0 - 1, 2k_{t-1}\}, \{2k_0, 2k_1 - 1\}, \{2k_1, 2k_2 - 1\}, \\
&\ldots, \{2k_{t-2}, 2k_{t-1} - 1\}, 1 = k_0 < k_1 < \cdots < k_{t-1} \leq r\}.
\end{aligned}
$$

Hence, (A.24) is equivalent to

$$\bar{\varphi}(\bar{\Pi}\bar{\delta}^r) = \sum_{t=1}^r \mathcal{T}_t, \tag{A.25}$$

where for all $1 \le t \le r$,

$$
\begin{aligned}
\mathcal{T}_t &= \sum_{\sigma \in \mathcal{P}_t^{2r}} \mathcal{T}_\sigma = (1+y)^r \sum_{\sigma \in \mathcal{P}_t^{2r}} \sum_3 \bar{\varphi}_{K(\sigma)}[\underline{f}_{l_1}, \bar{c}_{l_1} \bar{d}_{l_2}, \underline{f}_{l_2}, \bar{e}_{l_2} \bar{d}_{l_3}, \dots, \bar{e}_{l_r} \bar{\Pi} \bar{d}_{l_1}] \\
&= (1+y)^t \sum_1 \sum_2 \bar{\varphi}\Big(\prod_{u=0}^{t-1} \underline{f}_{l_u} \Big) \prod_{u=0}^{t-1} \bar{\varphi}(\bar{e}_{l_u} \bar{\delta}^{k_{u+1}-k_u-1} \bar{d}_{l_{u+1}}) \\
&= (1+y)^t \sum_1 \sum_2 \bar{\varphi}\Big(\prod_{u=0}^{t-1} \bar{\varphi}(\bar{e}_{l_u} \bar{\delta}^{k_{u+1}-k_u-1} \bar{d}_{l_{u+1}}) \underline{f}_{l_u} \Big).
\end{aligned}
\tag{A.26}
$$

Note that $\sigma \in \mathcal{P}_t^{2r}$ implies that one block of $K(\sigma)$ is $\{2k_0 - 1, 2k_1 - 1, \dots, 2k_{t-1} - 1\}$ and its other blocks are subsets of $\{2k_u, 2k_u + 1, \dots, 2k_{u+1} - 2, \}: 0 \le u \le t - 2$. Thus, the last equality above follows by using (5.34). In addition, recall (6.34) and note that each $\bar{\delta}$ involves a factor of $(1+y)$. In the last equality we have $\sum_{u=0}^{t-1}(k_{u+1} - k_u - 1) = r - t$ many $\bar{\delta}$ and therefore they absorb a factor $(1+y)^{r-t}$. This explains the presence of the factor $(1+y)^r$ in the final expression. Hence, Lemma A.2.1 follows from (A.25) and (A.26). $\square$

Proof of (6.42).

$$
\begin{aligned}
\bar{A}_{j_1,j_2,j_3}(z, \underline{f}, \bar{\Pi}) &= \sum_{r=1}^{\infty} z^{-r}(1+y)\bar{\varphi}(\bar{\Pi} \bar{d}_{j_1} \bar{e}_{j_3} \bar{\delta}^{r-1}) \underline{f}_{j_2} \\
&= T_1 + T_2, \quad (\text{say})
\end{aligned}
\tag{A.27}
$$

where

$$
\begin{aligned}
T_1 &= \frac{1+y}{z} \bar{\varphi}(\bar{\Pi} \bar{d}_{j_1} \bar{e}_{j_3}) \underline{f}_{j_2}, \\
T_2 &= \sum_{r=1}^{\infty} z^{-(r+1)}(1+y)\bar{\varphi}(\bar{\Pi} \bar{d}_{j_1} \bar{e}_{j_3} \bar{\delta}^r) \underline{f}_{j_2}.
\end{aligned}
\tag{A.28}
$$

Let us use the shorthand notation $\sum_4 = \sum_{i_1,i_2,\dots,i_t=1}^{\infty}$. Using Lemma A.2.1 with $\bar{\Pi}$ replaced by $\bar{\Pi} \bar{d}_{j_1} \bar{e}_{j_3}$ (so that now $\bar{d}_{l_t} = \bar{\Pi} \bar{d}_{j_1} \bar{e}_{j_3} \bar{d}_{l_0}$),

$$
\begin{aligned}
T_2 &= \frac{1+y}{z} \sum_{r=1}^{\infty} \sum_{t=1}^{r} \sum_1 \sum_2 z^{-r}(1+y)^t \bar{\varphi}\Big(\prod_{u=0}^{t-1} \bar{\varphi}(\bar{e}_{l_u} \bar{\delta}^{k_{u+1}-k_u-1} \bar{d}_{l_{u+1}}) \underline{f}_{l_u} \Big) \underline{f}_{j_2} \\
&= \frac{1+y}{z} \sum_{t=1}^{\infty} \sum_{r=t}^{\infty} \sum_1 \sum_2 z^{-r}(1+y)^t \bar{\varphi}\Big(\prod_{u=0}^{t-1} \bar{\varphi}(\bar{e}_{l_u} \bar{\delta}^{k_{u+1}-k_u-1} \bar{d}_{l_{u+1}}) \underline{f}_{l_u} \Big) \underline{f}_{j_2} \\
&= \frac{1+y}{z} \sum_{t=1}^{\infty} \sum_1 \sum_4 z^{-(i_1+i_2+\cdots+i_t)}(1+y)^t \bar{\varphi}\Big(\prod_{u=0}^{t-1} \bar{\varphi}(\bar{e}_{l_u} \bar{\delta}^{i_{u+1}-1} \bar{d}_{l_{u+1}}) \underline{f}_{l_u} \Big) \underline{f}_{j_2} \\
&= \frac{1+y}{z} \sum_{t=1}^{\infty} \sum_1 \bar{\varphi}\Big(\prod_{u=0}^{t-1} \sum_{i_{u+1}=1}^{\infty} z^{-i_{u+1}}(1+y)\bar{\varphi}(\bar{e}_{l_u} \bar{\delta}^{i_{u+1}-1} \bar{d}_{l_{u+1}}) \underline{f}_{l_u} \Big) \underline{f}_{j_2}
\end{aligned}
$$

$$= \frac{1+y}{z} \sum_{t=1}^{\infty} \sum_{1} \bar{\varphi}\Big(\sum_{i_t=1}^{\infty} z^{-i_t} \bar{R}_{i_t,l_0,l_{t-1},l_{t-1}}(\underline{f}, \bar{\Pi}\bar{d}_{j_1}\bar{e}_{j_3})$$

$$\times \prod_{u=0}^{t-2} \sum_{i_{u+1}=1}^{\infty} z^{-i_{u+1}} \bar{R}_{i_{u+1},l_{u+1},l_u,l_u}(\underline{f},1) \Big)$$

$$= \frac{1+y}{z} \sum_{t=1}^{\infty} \sum_{1} \bar{\varphi}\big(\bar{A}_{l_0,l_{t-1},l_{t-1}}(z,\underline{f},\bar{\Pi}\bar{d}_{j_1}\bar{e}_{j_3})$$

$$\prod_{u=0}^{t-2} \bar{A}_{l_{u+1},l_u,l_u}(z,\underline{f},1)\big). \tag{A.29}$$

Hence, (6.42) is established by (A.27)–(A.29).

Proof of (6.43). Note that

$$m_{\bar{\mu}}(z) = \bar{\varphi}((\bar{\delta}-z)^{-1}) = -\sum_{i=1}^{\infty} z^{-i}\bar{\varphi}(\bar{\delta}^{i-1}) = -(1+y)^{-1}\bar{A}_{0,0,0}(z,\underline{f},1).$$

Thus, (6.43) follows by (6.42) and observing that $\bar{d}_0 = 1$, $\bar{e}_0 = 1$ and $\underline{f}_0 = 1$. This completes the proof of Theorem 6.4.1.

A.3   Proof of Theorem 7.2

In the proof, we will frequently need to calculate expressions of the form

$$\frac{1}{p}E\mathrm{Tr}\Big(A_1\frac{Z}{\sqrt{n}}A_2\frac{Z^*}{\sqrt{n}}A_3\frac{Z}{\sqrt{n}}\cdots\frac{Z^*}{\sqrt{n}}A_{2k+1}\Big). \tag{A.30}$$

Now, the above equals

$$\frac{1}{n^k p}E\sum_{u,v}\Big(\prod_{i=1}^{k} A_{2i-1}(u_{2i-1},u_{2i})\varepsilon_{u_{2i},v_{2i-1}}A_{2i}(v_{2i-1},v_{2i})\varepsilon_{u_{2i+1},v_{2i}}\Big)$$

$$\times A_{2k+1}(u_{2k+1},u_1). \tag{A.31}$$

Let

$$\mathcal{T} = \{(u_2,v_1),(u_3,v_2),\ldots,(u_{2k},v_{2k-1}),(u_{2k+1},v_{2k})\}. \tag{A.32}$$

Note that $\mathcal{T}$ is the set of all indices attached with ε's. In (A.30), (Z,Z^*) appear alternately and there are k such (Z,Z^*). When $\delta = 0$, $(u_{2i+\delta},v_{2i+\delta-1})$ is attached with the i-th Z. Similarly, when $\delta = -1$, $(u_{2i+\delta},v_{2i+\delta-1})$ is attached with the i-th Z^*. Pairs of indices $(u_{2i+\delta},v_{2i+\delta-1})$ and $(u_{2i'+\delta'},v_{2i'+\delta'-1})$ are said to be matched if $(i',\delta') \neq (i,\delta)$ and $(u_{2i+\delta},v_{2i+\delta-1}) = (u_{2i'+\delta'},v_{2i'+\delta'-1})$.

As $\{\varepsilon_{i,j}\}$ are independent with mean 0, only matched indices in $\mathcal{T}$ need to be considered.

There is a natural bijection between $\mathcal{T}$ in (A.32) and $\{1, 2, \ldots, 2k\}$: a match in $\mathcal{T}$ forms a partition of $\mathcal{T}$, where the matched indices form blocks. Since we have ruled out singleton blocks, the relevant set of matches of $\mathcal{T}$ is in bijection (induced by the above bijection) with

$$\mathcal{P}_{2k} = \{\pi : \pi \text{ is a partition of } \{1, \ldots, 2k\} \text{ with no singleton block}\}. \quad (A.33)$$

Let for any set A, E_A be the usual expectation restricting on the set A. Then from the above discussions,

$$
\begin{aligned}
E\mathrm{Tr}&\left(A_1 \frac{Z}{\sqrt{n}} A_2 \frac{Z^*}{\sqrt{n}} A_3 \cdots \frac{Z^*}{\sqrt{n}} A_{2k+1}\right) \\
&= E_{\mathcal{P}_{2k}} \mathrm{Tr}\left(A_1 \frac{Z}{\sqrt{n}} A_2 \frac{Z^*}{\sqrt{n}} A_3 \cdots \frac{Z^*}{\sqrt{n}} A_{2k+1}\right) \\
&= \sum_{\sigma \in \mathcal{P}_{2k}} E_\sigma \mathrm{Tr}\left(A_1 \frac{Z}{\sqrt{n}} A_2 \frac{Z^*}{\sqrt{n}} A_3 \cdots \frac{Z^*}{\sqrt{n}} A_{2k+1}\right). \quad (A.34)
\end{aligned}
$$

Now to compute (A.34), let us first concentrate on $\sigma \in \mathcal{P}_{2k} \cap NC(2k)$, where $NC(2k)$ is as in (5.13). Lemma A.3.1 provides an upper bound for the terms in (A.34). This will be useful in the proof of Theorem 7.2.1.

Lemma A.3.1. *Suppose* (A1) *holds and the matrices* $\{A_i\}$ *are all norm bounded. Suppose* $\sigma \in NC(2k) \cap \mathcal{P}_{2k}$ *has* $K_{i,\sigma}$ *blocks of size* $i \geq 2$ *of which* $K_{i,1,\sigma}$ *and* $K_{i,2,\sigma}$ *start with odd and even indices respectively. Then for some* $C > 0$,

(a)

$$\left|\frac{1}{p} E_\sigma \mathrm{Tr}(A_1 \frac{Z}{\sqrt{n}} A_2 \frac{Z^*}{\sqrt{n}} \cdots \frac{Z^*}{\sqrt{n}} A_{2k+1})\right| \leq C y_n^{K_{2,2,\sigma}} (y_n p^{-1})^K. \quad (A.35)$$

where

$$K_\sigma = \sum_{i \geq 2} (0.5 K_{2i-1,\sigma} + K_{2i,\sigma}).$$

(b) The same upper bound in (A.35) also holds for

$$|E_\sigma(A_1 \frac{Z}{\sqrt{n}} A_2 \frac{Z^*}{\sqrt{n}} A_3 \frac{Z}{\sqrt{n}} \cdots \frac{Z^*}{\sqrt{n}} A_{2k+1})(u, u')|, \ \forall u, u' \leq p. \quad (A.36)$$

Proof. Recall the classes $\mathcal{L}$ and $C(\delta, p)$ respectively in (4.23) and (4.25). Note that in (A1) we assume **one** of the following.

(A1a) $((\varepsilon_{i,j})) \in C(\delta, n)$ for some $\delta \in (0, 2]$.

(A1b) $((\varepsilon_{i,j})) \in \mathcal{L}$.

Under (A1a) for some $C > 0$,

$$E|\varepsilon_{a,b}|^r \leq (\sqrt{n})^{r-4} E|\varepsilon_{a,b}|^4 \leq C n^{r/2-2}, \ \forall r \geq 4. \quad (A.37)$$

Under (A1b) too, for some $C > 0$,

$$E|\varepsilon_{a,b}|^r \leq C \leq Cn^{r/2-2}, \; \forall r \geq 4. \tag{A.38}$$

Recall $||\cdot||_2$ in (1.8). Let $A(i,j)$ be the (i,j)-th element of the matrix A. Note that, as $\{A_i\}$ are norm bounded matrices (by (A2) and (A3a)), for some $C > 0$ and for all a, b,

$$
\begin{aligned}
|A_i(a,b)| &\leq \sqrt{(A_iA_i^*)(a,a)}, \quad \text{by Cauchy-Schwartz inequality} \\
&\leq ||A_i||_2 < C, \; \forall 1 \leq i \leq 2k+1. \tag{A.39}
\end{aligned}
$$

We shall use (A.37), (A.38), and (A.39) to prove the lemma. The proof will proceed for (a) and (b) simultaneously by induction on k. Let $k = 1$. Note that $NC(2) \cap \mathcal{P}_2 = \{\{1,2\}\}$. By (A.31), for some $C > 0$, we have

$$
\begin{aligned}
&\left|\frac{1}{p}E\mathrm{Tr}\left(A_1\frac{Z}{\sqrt{n}}A_2\frac{Z^*}{\sqrt{n}}A_3\right)\right| \\
&= \left|\frac{1}{np}E_{\{1,2\}}\sum_{u_1,a,b,c,d}A_1(u_1,a)\varepsilon_{a,b}A_2(b,d)\varepsilon_{c,d}A_3(c,u_1)\right| \\
&= \left|\frac{1}{np}E_{\{a=c,b=d\}}\sum_{u_1,a,b}A_1(u_1,a)\varepsilon_{a,b}A_2(b,d)\varepsilon_{c,d}A_3(c,u_1)\right| \\
&= \left|\frac{1}{np}\sum_{u_1,a,b}A_1(u_1,a)A_2(b,b)A_3(a,u_1)E(\varepsilon_{a,b}^2)\right| \\
&= \left|\frac{1}{p}(\mathrm{Tr}(A_1A_3))||\frac{1}{n}\mathrm{Tr}(A_2)|, \quad \text{as } E(\varepsilon_{a,b}^2) = 1 \\
&\leq \sqrt{\frac{1}{p}(\mathrm{Tr}(A_1^*A_1))\frac{1}{p}(\mathrm{Tr}(A_3^*A_3))}\;|\frac{1}{n}\mathrm{Tr}(A_2)| \\
&\leq ||A_1||_2\,||A_2||_2\,||A_3||_2 \leq C,
\end{aligned}
$$

as $\{A_i\}$ are norm bounded. Therefore, as $K_{2,1,\{1,2\}} = 1$, $K_{2,2,\{1,2\}} = 0$ and $K_{i,\{1,2\}} = 0 \; \forall i \geq 3$, (a) is proved for $k = 1$.

Next, for some $C > 0$,

$$
\begin{aligned}
&\left|E\left(A_1\frac{Z}{\sqrt{n}}A_2\frac{Z^*}{\sqrt{n}}A_3\right)(u_1,u_2)\right| \\
&= \left|\frac{1}{n}E_{\{1,2\}}\sum_{a,b,c,d}A_1(u_1,a)\varepsilon_{a,b}A_2(b,d)\varepsilon_{c,d}A_3(c,u_2)\right| \\
&= \left|\frac{1}{n}E_{\{a=c,b=d\}}\sum_{a,b}A_1(u_1,a)\varepsilon_{a,b}A_2(b,d)\varepsilon_{c,d}A_3(c,u_2)\right| \\
&= \left|\frac{1}{n}\sum_{a,b}A_1(u_1,a)A_2(b,b)A_3(a,u_2)E(\varepsilon_{a,b}^2)\right| \\
&= |((A_1A_3)(u_1,u_2))|\,|\frac{1}{n}\mathrm{Tr}(A_2)|, \quad \text{as } E(\varepsilon_{a,b}^2) = 1 \\
&\leq C, \quad \text{by applying (A.39) on } A_1A_3 \text{ and } A_2.
\end{aligned}
$$

Hence, (b) is proved for $k = 1$. Suppose (a) and (b) hold for all $k \leq m - 1$. Now we shall show that they are true for $k = m$ also.

Since σ is non-crossing, it always has at least one block B with adjacent indices. If we drop any one of those blocks, then again we have a non-crossing partition, say σ^*, of the remaining indices. Note that $\sigma^* \in NC(2k) \cap \mathcal{P}_{2k}$ for some $k \leq m - 1$. Therefore, (A.35) holds for σ^*. Then four situations can arise depending on the length of B (even/odd) and the index of the starting element of B (even/odd). Here we shall show the details for the case where B is of even length and starts with an odd index. Similar argument works for other cases.

Since σ has a block $B = \{2j - 1, 2j, \ldots, 2s\}$, there exists (a, b) such that

$$(u_{2i+\delta}, v_{2i+\delta-1}) \begin{cases} = (a, b), \forall j \leq i \leq s, \delta = 0, 1 \\ \neq (a, b), \forall i < j \text{ or } i > s, \ \delta = 0, 1. \end{cases}$$

Moreover, note that $\sigma^* \in NC(2(m - s + j - 1)) \cap \mathcal{P}_{2(m-s+j-1)}$ and

$$
\begin{aligned}
K_{2i-1,1,\sigma^*} &= K_{2i-1,1,\sigma}, \ K_{2i-1,2,\sigma^*} = K_{2i-1,2,\sigma}, \ i \geq 1, \\
K_{2i,2,\sigma^*} &= K_{2i,2,\sigma}, \ i \geq 1, \\
K_{2i,1,\sigma^*} &= K_{2i,1,\sigma}, \ i \neq s - j + 1, \\
K_{2(s-j+1),1,\sigma^*} &= K_{2(s-j+1),1,\sigma} - 1.
\end{aligned}
\tag{A.40}
$$

Let

$$
\begin{aligned}
D_1 &= A_1 \frac{Z}{\sqrt{n}} A_2 \frac{Z^*}{\sqrt{n}} A_3 \cdots \frac{Z^*}{\sqrt{n}} A_{2j-1} \\
D_2 &= A_{2s+1} \frac{Z}{\sqrt{n}} A_{2s+2} \frac{Z^*}{\sqrt{n}} A_{2s+3} \cdots A_{2k} \frac{Z^*}{\sqrt{n}} A_{2k+1}.
\end{aligned}
$$

Case I. Let $s - j = 0$. Then by (A.31), for some $C_1, C_2 > 0$, we have

$$
\begin{aligned}
&\left| \frac{1}{p} E_\sigma \mathrm{Tr}\left(A_1 \frac{Z}{\sqrt{n}} A_2 \frac{Z^*}{\sqrt{n}} A_3 \frac{Z}{\sqrt{n}} \cdots \frac{Z^*}{\sqrt{n}} A_{2k+1} \right) \right| \\
&= \left| \frac{1}{np} E_\sigma \sum_{u_1, a, b} D_1(u_1, a) \varepsilon_{a,b} A_{2s}(b, b) \varepsilon_{a,b} D_2(a, u_1) \right| \\
&= \left| \frac{1}{np} \sum_{u_1, u, b} E_{\sigma^*}(D_1(u_1, a) A_{2s}(b, b) D_2(a, u_1)) E(\varepsilon_{a,b}^2) \right| \\
&= \left| \frac{1}{p} E_{\sigma^*}(\mathrm{Tr}(D_1 D_2)) \right| \left| \frac{1}{n} \mathrm{Tr}(A_{2s}) \right|, \text{ as } E(\varepsilon_{a,b}^2) = 1 \\
&\leq C_1 \left| \frac{1}{p} E_{\sigma^*}(\mathrm{Tr}(D_1 D_2)) \right|, \text{ by applying (A.39) on } A_{2s} \\
&\leq C_2 y_n^{K_{2,2,\sigma^*}} (y_n p^{-1})^{\sum_{i \geq 2}(0.5K_{2i-1,\sigma^*} + K_{2i,\sigma^*})} \\
&\qquad \qquad \text{by applying } (a) \text{ on } k = m - 1 \\
&= C_2 y_n^{K_{2,2,\sigma}} (y_n p^{-1})^{\sum_{i \geq 2}(0.5K_{2i-1,\sigma} + K_{2i,\sigma})}, \text{ by (A.40).}
\end{aligned}
$$

Hence, (a) is proved for $k = m$ and $s - j = 0$.

Case II. Let $s - j > 0$. Then by (A.31), for some $C_1, C_2, C_3, C_4 > 0$, we have

$$|\frac{1}{p}E_\sigma \mathrm{Tr}\big(A_1 \frac{Z}{\sqrt{n}} A_2 \frac{Z^*}{\sqrt{n}} A_3 \frac{Z}{\sqrt{n}} \cdots \frac{Z^*}{\sqrt{n}} A_{2k+1}\big)|$$

$$= |\frac{1}{n^{s-j+1}p}E_\sigma \sum_{u_1,a,b} D_1(u_1,a) \prod_{i=j}^{s-1} \varepsilon_{a,b} A_{2i}(b,b)\varepsilon_{a,b}A_{2i+1}(a,a)$$

$$\times \varepsilon_{a,b}A_{2s}(b,b)\varepsilon_{a,b}D_2(a,u_1)|$$

$$= |\frac{1}{n^{s-j+1}p}\sum_{a,b} \prod_{i=j}^{s-1} A_{2i+1}(a,a) \prod_{i=j}^{s} A_{2i}(b,b)$$

$$\times E_{\sigma^*}\big(\sum_{u_1} D_1^*(a,u_1)D_2^*(u_1,a)\big)E(\varepsilon_{a,b}^{2s-2j+2})\big)|$$

$$\leq [\frac{1}{np}\sum_{a,b}\big(\prod_{i=j}^{s-1} A_{2i+1}^2(a,a) \prod_{i=j}^{s} A_{2i}^2(b,b)\big)$$

$$\times \big(E_{\sigma^*}\big(\sum_{u_1} D_1^*(a,u_1)D_2^*(u_1,a)\big)\big)^2]^{1/2}$$

$$\times \frac{1}{n^{s-j}}[\frac{1}{np}\sum_{a,b}(E(\varepsilon_{a,b}^{2s-2j+2}))^2]^{1/2}, \quad (\text{``Cauchy-Schwartz'' on } \Sigma_{a,b})$$

$$\leq [\frac{1}{np}\sum_{a,b}\prod_{i=j}^{s-1} A_{2i+1}^2(a,a) \prod_{i=j}^{s} A_{2i}^2(b,b)$$

$$\times \big(E_{\sigma^*}\big(\sum_{u_1} D_1^*(a,u_1)D_2^*(u_1,a)\big)\big)^2]^{1/2}$$

$$\times \frac{1}{n^{s-j}}[\frac{1}{np}\sum_{a,b}(C_1 n^{s-j-1} \sup_{a,b} E(\varepsilon_{a,b}^4))^2]^{1/2},$$

$$\text{(by (A.37) and (A.38))}$$

$$\leq [\frac{1}{np}\sum_{a,b}\big(\prod_{i=j}^{s-1} A_{2i+1}^2(a,a) \prod_{i=j}^{s} A_{2i}^2(b,b)\big)$$

$$\times \big(E_{\sigma^*}\big(\sum_{u_1} D_1^*(a,u_1)D_2^*(u_1,a)\big)\big)^2]^{1/2} C_2 \frac{n^{s-j-1}}{n^{s-j}},$$

$$\text{(as } \sup_{a,b} E(\varepsilon_{a,b}^4) < \infty)$$

$$\leq \frac{C_3}{n}[\frac{1}{np}\sum_{a,b}\big(E_{\sigma^*}\big(\sum_{u_1} D_1^*(a,u_1)D_2^*(u_1,a)\big)\big)^2]^{1/2},$$

$$\text{by applying (A.39) on } \big(\prod_{i=j}^{s-1} A_{2i+1}^2(a,a) \prod_{i=j}^{s} A_{2i}^2(b,b)\big)$$

$$\leq \frac{C_3}{n}\Big[\frac{1}{p}\sum_a \big(E_{\sigma^*}(D_1^* D_2^*(a,a))\big)^2\Big]^{1/2}$$

$$= \frac{C_3}{p}y_n\Big[\frac{1}{p}\sum_a \big(E_{\sigma^*}(D_1^* D_2^*(a,a))\big)^2\Big]^{1/2}$$

$$\leq C_4\frac{y_n}{p}y_n^{K_{2,2,\sigma^*}}(y_n p^{-1})^{\sum_{i\geq 2}(0.5K_{2i-1,\sigma^*}+K_{2i,\sigma^*})}$$

by applying (b) on σ^* for $k = m - s + j - 1$

$$= C_4 y_n^{K_{2,2,\sigma}}(y_n p^{-1})^{\sum_{i\geq 2}(0.5K_{2i-1,\sigma}+K_{2i,\sigma})}, \quad \text{(by (A.40))}.$$

Therefore, (a) is proved for $k = m$ and $s - j > 0$ and hence proof of (a) is complete. One can similarly prove (b). This completes the proof of Lemma A.3.1. $\qquad\square$

(a) We prove the theorem only for $U = 1$. Similar argument works for $U > 1$. Now for $U = 1$, we have only $\{\mathbb{P}_{l,(1,1,\dots,1)}\}$, $\{\mathcal{R}_{l,(1,1,\dots,1)}\}$, $\{w_{1,l,j}\}$ and $\{\alpha_{l,(1,1,\dots,1)}\}$. Let us denote them respectively by $\{\mathbb{P}_l\}$, $\{\mathcal{R}_l\}$, $\{w_{l,j}\}$ and $\{\alpha_l\}$. Let us also write $\{\mathbb{G}_l\}$ for $\{\mathbb{G}_{l,k_l}\}$.

Let π be any polynomial. Then by Definition 5.1.4, it is enough to prove

$$\lim p^{-1} E \mathrm{Tr}(\pi(\mathcal{R}_l : l \geq 1)) = \varphi(\pi(\alpha_l : l \geq 1)). \tag{A.41}$$

Note that, for any $\{\mathcal{R}_{l_{ti}}\}$ from $\{\mathcal{R}_l\}$ and constants $\{c_i\}$, we can write $\pi(\mathcal{R}_l : l \geq 1) = \sum_{i=1}^{T} c_i \prod_{t=1}^{T_i} \mathcal{R}_{l_{ti}}$. Therefore, it is enough to establish, for each $1 \leq i \leq T$

$$\lim p^{-1} E \mathrm{Tr}\Big(\prod_{t=1}^{T_i} \mathcal{R}_{l_{ti}}\Big) = \begin{cases} 0, & \text{if } T_i \text{ is odd} \\ \varphi(\prod_{t=1}^{T_i} \alpha_{l_{ti}}), & \text{if } T_i \text{ is even}. \end{cases} \tag{A.42}$$

For simplicity of notation, we prove only

$$\lim p^{-1} E \mathrm{Tr}\Big(\prod_{l=1}^{T} \mathcal{R}_l\Big) = \begin{cases} 0, & \text{if } T \text{ is odd} \\ \varphi(\prod_{l=1}^{T} \alpha_l), & \text{if } T \text{ is even}. \end{cases} \tag{A.43}$$

Similar argument works to prove the more general (A.42).

Let $A(i,j)$ be the (i,j)-th element of the matrix A. For convenience we write $\mathrm{Tr}(\prod_{l=1}^{T} \mathcal{R}_l)$ in the form

$$\mathrm{Tr}\Big(\prod_{l=1}^{T} \mathcal{R}_l\Big) = \sum_{\substack{u_{l,3k_l+1}=u_{l+1,1} \\ u_{T+1,1}=u_{11}}} \prod_{l=1}^{T} \mathcal{R}_l(u_{l,1}, u_{l,3k_l+1}), \quad \text{where} \tag{A.44}$$

$$\mathcal{R}_l(u_{l,1}, u_{l,3k_l+1}) = \sqrt{\frac{n}{p}}\big(\mathbb{P}_l(u_{l,1}, u_{l,3K_l+1}) - \mathbb{G}_l(u_{l,1}, u_{l,3K_l+1})\big), \tag{A.45}$$

$$\mathbb{P}_l(u_{l,1}, u_{l,3k_l+1}) = \frac{1}{n^{k_l}} \sum_{\substack{u_{l,j}, v_{l,j} \\ u_{l,j} \neq u_{l,1}, u_{l,3K_l+1}}} \prod_{\substack{i: u_{l,3i}=u_{l,3i+1} \\ 1 \leq i \leq k_l-1}} A_{l,2i-1}(u_{l,3i-2}, u_{l,3i-1})$$

$$\times \varepsilon_{u_{l,3i-1}, v_{l,2i-1}} A_{l,2i}(v_{l,2i-1}, v_{l,2i}) \varepsilon_{u_{l,3i}, v_{l,2i}} A_{l,2k_l+1}(u_{l,3k_l}, u_{l,3k_l+1}). \quad (A.46)$$

For each $1 \leq l \leq T$, we define

$$\mathcal{I}_l = \{(u_{l,3i+\delta}, v_{l,2i+\delta}) : \delta = -1, 0, \ 1 \leq i \leq k_l\}. \quad (A.47)$$

Note that $\mathcal{I}_l$ is the set of all indices attached with ε's in the expansion of $\mathcal{R}_l$ given in (A.44)-(A.46). An index $(u_{l,3k+\delta}, v_{l,2k+\delta})$ is said to be matched if there is at least one $(k', \delta', l') \neq (k, \delta, l)$ with $(u_{l,3k+\delta}, v_{l,2k+\delta}) = (u_{l',3k'+\delta'}, v_{l',2k'+\delta'})$. Now note that $E\left(\text{Tr}(\prod_{l=1}^T \mathcal{R}_l)\right)$ involves all indices in $\cup_{l=1}^T \mathcal{I}_l$. As $\{\varepsilon_{i,j}\}$ are independent and have mean 0, all indices in $\cup_{l=1}^T \mathcal{I}_l$ need to be matched to guarantee a non-zero contribution. For each $1 \leq l \leq T$, consider the following sets of matched indices.

$$B_l = \text{ all matches where for each } (k, \delta), \exists (k', \delta') \neq (k, \delta) \text{ with}$$
$$(u_{l,3k+\delta}, v_{l,2k+\delta}) = (u_{l,3k'+\delta'}, v_{l,2k'+\delta'}) \text{ and for } l \neq l',$$
$$\nexists (k', \delta', l') \text{ so that } (u_{l,3k+\delta}, v_{l,2k+\delta}) = (u_{l',3k'+\delta'}, v_{l',2k'+\delta'}). \quad (A.48)$$

Consider the disjoint decomposition $\cup_{l=1}^{T+1} C_l$ of all possible matches of indices in $\cup_{l=1}^T \mathcal{I}_l$, where

$$C_1 = B_1, \ C_l = (\cap_{j=1}^{l-1} B_j^c) \cap B_l \ \forall 2 \leq l \leq T, \ C_{T+1} = \cap_{l=1}^T B_l^c. \quad (A.49)$$

Let for any set A, E_A be the usual expectation restricting on the set A. Then we have the following lemma.

Lemma A.3.2. *Suppose Assumptions (A1), (A2), and (A3a) hold. Then*
(i) $E(y_n^{-1/2} \mathbb{P}_l(u_2, u_3)) = y_n^{-1/2} \mathbb{G}_l(u_2, u_3) + O(y_n^{1/2})$.
(ii) $L_1 = L_2$ *where*

$$L_1 = \lim \frac{1}{p} E_{C_l} \text{Tr}\left(y_n^{-l/2} \prod_{i=1}^l \mathbb{P}_i \prod_{i=l+1}^T \mathcal{R}_i\right)$$

$$L_2 = \lim \frac{1}{p} E_{\cap_{i=1}^{l-1} B_i^c} \text{Tr}\left(y_n^{-l/2} \prod_{i=1}^{l-1} \mathbb{P}_i \mathbb{G}_l \prod_{i=l+1}^T \mathcal{R}_i\right).$$

(iii) $\lim \frac{1}{p} E \text{Tr}(\prod_{l=1}^T \mathcal{R}_l) = \lim \frac{1}{p} E_{C_{T+1}} \text{Tr}(y_n^{-T/2} \prod_{l=1}^T \mathbb{P}_l)$.

Proof. *(i)* Recall that

$$\mathbb{P}_l = A_{l,1} \frac{Z}{\sqrt{n}} A_{l,2} \frac{Z^*}{\sqrt{n}} A_{l,3} \frac{Z}{\sqrt{n}} \cdots \frac{Z^*}{\sqrt{n}} A_{l,2k_l+1}.$$

Consider the partition $\sigma^* = \{\{1, 2\}, \{3, 4\}, \ldots, \{2k_l - 1, 2k_l\}\}$. Note that

$$E_{\sigma^*}(y_n^{-1/2}\mathbb{P}_l(u_2, u_3)) = y_n^{-1/2}\mathbb{G}_l(u_2, u_3). \tag{A.50}$$

Recall $\mathcal{P}_{2k}$ in (A.33). Let $\mathcal{P}_{2k}^c = $ set of all partitions of $\{1, 2 \ldots, 2k\} - \mathcal{P}_{2k}$. Note that,

$$\begin{aligned}
E(y_n^{-1/2}\mathbb{P}_l(u_2, u_3)) &= E_{\sigma^*}(y_n^{-1/2}\mathbb{P}_l(u_2, u_3)) + \sum_{\sigma \in \mathcal{P}_{2k_l}^c} E_\sigma(y_n^{-1/2}\mathbb{P}_l(u_2, u_3)) \\
&\quad + \sum_{\sigma \in NC(2k_l) \cap \mathcal{P}_{2k_l} - \{\sigma^*\}} E_\sigma(y_n^{-1/2}\mathbb{P}_l(u_2, u_3)) \\
&\quad + \sum_{\sigma \in \mathcal{P}_{2k_l} - NC(2k_l)} E_\sigma(y_n^{-1/2}\mathbb{P}_l(u_2, u_3)) \\
&= T_1 + T_2 + T_3 + T_4, \ (say). \tag{A.51}
\end{aligned}$$

As each partition in $\mathcal{P}_{2k_l}^c$ has at least one singleton block, $T_2 = 0$. Also a partition in $NC(2k_l) \cap \mathcal{P}_{2k_l} - \{\sigma^*\}$ contains either a block of length 2 and starts with an even index or a block of length longer than 2. Hence, by Lemma A.3.1 (b), $T_3 = O(y_n^{1/2})$. Moreover, crossing partitions in $\mathcal{P}_{2k_l} - NC(2k_l)$ have more restrictions on indices than that of partitions in $\mathcal{P}_{2k_l} \cap NC(2k_l) - \{\sigma^*\}$. Therefore contribution of $\mathcal{P}_{2k_l} - NC(2k_l)$ in $E(y_n^{-1/2}\mathbb{P}_l(u_2, u_3))$ is smaller than the contribution of the latter. Therefore, $T_4 = O(y_n^{1/2})$. Hence, by (A.50) and (A.51), the proof of Lemma A.3.3(i) is complete.

(ii) To prove this part, we need more analysis for the set C_l. Define the sets

$$\begin{aligned}
\mathcal{S}_l &= \text{all matches of indices in } \mathcal{I}_l, \\
\mathcal{S}_{-l} &= \text{all matches of indices in } \cup_{j \neq l} \mathcal{I}_j \text{ so that for each } j < l, \text{ there are} \\
&\quad \text{matched indices between } \mathcal{I}_j \text{ and } \mathcal{I}_k, \ k \neq j, l.
\end{aligned}$$

Note that

$$C_l = (\cap_{j=1}^{l-1} B_j^c) \cap B_l = \{(\sigma_1 \cup \sigma_2) : \sigma_1 \in \mathcal{S}_l, \ \sigma_2 \in \mathcal{S}_{-l}\}.$$

Let us denote

$$W_{P,l} := (np^{-1})^{l/2} \prod_{i=1}^{l} \mathbb{P}_i, \quad W_{R,l} := \prod_{i=l+1}^{T} \mathcal{R}_i.$$

Then for all $2 \leq l \leq T$, we have

$$\frac{1}{p} E_{C_l} \operatorname{Tr}(W_{P,l} W_{R,l}) \tag{A.52}$$

$$= \sum_{\sigma \in C_l} \frac{1}{p} E_\sigma \operatorname{Tr}(W_{P,l-1} \sqrt{np^{-1}} \mathbb{P}_l W_{R,l}) \left(\frac{n}{p}\right)^{l/2} \sum_{\sigma \in C_l} E_\sigma(\Pi_{j=1}^{l} \mathbb{P}_j \Pi_{j=l+1}^{T} \mathcal{R}_j)$$

$$= \sum_{\sigma \in C_l} \frac{1}{p} \sum_u E_\sigma \left(W_{P,l-1}(u_1, u_2) \sqrt{np^{-1}} \mathbb{P}_l(u_2, u_3) W_{R,l}(u_3, u_1) \right)$$

$$= \sum_{\sigma_1 \in \mathcal{S}_l,\ \sigma_2 \in \mathcal{S}_{-l}} p^{-1} \sum_u E_{\sigma_1} (\sqrt{np^{-1}} \mathbb{P}_l(u_2, u_3))$$

$$\times E_{\sigma_2} (W_{P,l-1}(u_1, u_2) W_{R,l}(u_3, u_1)) \text{ [as } C_l \subset B_l \text{ and under}$$
$$B_l, \{\varepsilon_{u,v} \text{ for } (u, v) \in \mathcal{I}_l \text{ and } \cup_{j \neq l} \mathcal{I}_j \text{ are independent]}$$

$$= \frac{1}{p} \sum_u \left(\sum_{\sigma_1 \in \mathcal{S}_l} E_{\sigma_1} (\sqrt{np^{-1}} \mathbb{P}_l(u_2, u_3)) \right)$$

$$\times \left(\sum_{\sigma_2 \in \mathcal{S}_{-l}} E_{\sigma_2} (W_{P,l-1}(u_1, u_2) W_{R,l}(u_3, u_1)) \right)$$

$$= \frac{1}{p} \sum_u E(\sqrt{\frac{n}{p}} \mathbb{P}_l(u_2, u_3)) \sum_{\sigma_2 \in \mathcal{S}_{-l}} E_{\sigma_2} W_{P,l-1}(u_1, u_2) W_{R,l}(u_3, u_1)$$

$$\times (E_{\cap_{i=1}^{l-1} B_i^c} (W_{P,l-1}(u_1, u_2) W_{R,l}(u_3, u_1))), \quad \text{[by } (a)]$$

$$= \frac{1}{p} E_{\cap_{i=1}^{l-1} B_i^c} \text{Tr} \left(W_{P,l-1} \sqrt{np^{-1}} \mathbb{G}_l W_{R,l} \right) + O((p/n)^{1/2}).$$

Hence, the proof of (ii) is complete.

$$(iii) \qquad \lim \frac{1}{p} E \text{Tr} \left[\Pi_{i=1}^T \mathcal{R}_i \right]$$

$$= \lim \frac{1}{p} E_{B_1} \text{Tr}(\sqrt{np^{-1}} \mathbb{P}_1 W_{R,1}) + \lim p^{-1} E_{B_1^c} \text{Tr}(\sqrt{np^{-1}} \mathbb{P}_1 W_{R,1})$$

$$- \lim \frac{1}{p} E \text{Tr} \left(\sqrt{np^{-1}} \mathbb{G}_1 W_{R,1} \right)$$

$$= \lim \frac{1}{p} E_{B_1^c} \text{Tr}(\sqrt{np^{-1}} \mathbb{P}_1 W_{R,1}), \quad \text{(by } (ii) \text{ for } l = 1)$$

$$= \lim \frac{1}{p} E_{B_1^c \cap B_2} \text{Tr}(W_{P,2} W_{R,2}) + \lim p^{-1} E_{B_1^c \cap B_2^c} \text{Tr}(W_{P,2} W_{R,2})$$

$$- \lim \frac{1}{p} E_{B_1^c} \text{Tr}(W_{P,1} \sqrt{np^{-1}} \mathbb{G}_2 W_{R,2})$$

$$= \lim \frac{1}{p} E_{B_1^c \cap B_2^c} \text{Tr}(W_{P,2} W_{R,2}), \quad \text{(by } (ii), \text{ for } l = 2)$$

$$\vdots$$

$$= \lim \frac{1}{p} E_{B_1^c \cap B_2^c \cap \cdots \cap B_T^c} \text{Tr}(\Pi_{i=1}^T \mathbb{P}_i), \text{ by repeated use of } (ii) \text{ for } l \geq 3$$

$$= \lim (np^{-1})^{T/2} p^{-1} E_{C_{T+1}} \text{Tr}(\Pi_{i=1}^T \mathbb{P}_i).$$

Therefore, (iii) is established.

Thus, proof of Lemma A.3.2 is complete.   □

Now we get back to the proof of the Theorem. By Lemma A.3.2(iii), we have

$$\lim \frac{1}{p} E\left(\text{Tr}\left(\prod_{l=1}^{T} \mathcal{R}_l\right)\right) = \lim \left(\frac{n}{p}\right)^{T/2} p^{-1} E_{C_{T+1}} \text{Tr}(\Pi_{l=1}^{T} \mathbb{P}_l). \quad (A.53)$$

Next we shall analyze the set C_{T+1} and identify the set of matches which contribute in the limit.

Two index sets $\mathcal{I}_i$ and $\mathcal{I}_{i'}$ are said to be connected if there is (k, δ) and (k', δ') with $(u_{i,3k+\delta}, v_{i,2k+\delta}) = (u_{i',3k'+\delta'}, v_{i',2k'+\delta'})$, where $(u_{i,3k+\delta}, v_{i,2k+\delta}) \in \mathcal{I}_i$ and $(u_{i',3k'+\delta'}, v_{i',2k'+\delta'}) \in \mathcal{I}_{i'}$. Also a collection of index sets $\{\mathcal{I}_{i_1}, \mathcal{I}_{i_2}, \ldots, \mathcal{I}_{i_s}\}$, $s \geq 2$, is said to form a connected group if for each $1 \leq k \leq s - 1$, $\mathcal{I}_{i_k}$ and $\mathcal{I}_{i_{k+1}}$ is connected. Note that, in a typical match in C_{T+1}, for each i, $\mathcal{I}_i$ is connected with some other $\mathcal{I}_{i'}$, $i' \neq i$. Therefore, each match in C_{T+1} corresponds to some disjoint connected groups each of length at least 2. Consider the following disjoint decomposition of C_{T+1}.

$$C_{T+1} = \bigcup_{\substack{2 \leq g_1, g_2, \ldots, g_R \leq T \\ \sum_{j=1}^{R} g_j = T, \ R \geq 1}} G(g_1, g_2, \ldots, g_R), \text{ where} \quad (A.54)$$

$$G(g_1, g_2, \ldots, g_R) = \text{set of all such matches in } C_{T+1} \text{ which form exactly}$$
$$R \text{ connected groups of length } g_1, \ldots, g_R. \quad (A.55)$$

Note that $R \leq T/2$ and equality holds if T is even and $g_i = 2, \forall i$. Then we have the following lemma.

Lemma A.3.3. *Suppose Assumptions (A1), (A2) and (A3a) hold. Then*

$$\lim \frac{1}{p} E_{G(g_1, \ldots, g_R)} \text{Tr}(n^{T/2} p^{-T/2} \prod_{i=1}^{T} \mathbb{P}_i) = O(y_n^{T/2-R}).$$

Proof. Let $D(g_1, \ldots, g_R)$ be the set of all non-crossing pair matches in C_{T+1} which form exactly R connected groups of lengths $g_1, \ldots, g_R$. Note that $D(g_1, \ldots, g_R) \subset G(g_1, \ldots, g_R)$. We shall first show that

$$\lim \frac{1}{p} E_{D(g_1, \ldots, g_R)} \text{Tr}(n^{T/2} p^{-T/2} \prod_{i=1}^{T} \mathbb{P}_i) = O(y_n^{T/2-R}). \quad (A.56)$$

Under $D(g_1, \ldots, g_R)$, to connect two index sets $\mathcal{I}_l$ and $\mathcal{I}_{l'}$, there must be a match of the type $(u_{l,3i}, v_{l,2i}) = (u_{l',3i'-1}, v_{l',2i'-1})$ for some i and i'. Note that they respectively correspond to the i-th Z^* in $\mathbb{P}_l$ and i'-th Z in $\mathbb{P}_{l'}$. Therefore, under $D(g_1, \ldots, g_R)$, to connect $\mathcal{I}_l$ and $\mathcal{I}_{l'}$, there must be a block which starts with an even index. Now to form a connected group of length g (say), we need to connect g many index sets $\mathcal{I}_{l_1}, \mathcal{I}_{l_2}, \ldots, \mathcal{I}_{l_g}$ (say) and hence there must be $(g - 1)$ matches of the form $(u_{l_k,3i'_k}, v_{l_k,2i'_k}) = (u_{l_{k+1},3i_{k+1}-1}, v_{l_{k+1},2i_{k+1}-1})$ for some $i_k \leq i'_k$ and for all $1 \leq k \leq g - 1$. Therefore, under $D(g_1, \ldots, g_R)$, to form a connected group of length g, there must be $(g - 1)$ blocks which start with an even index. Hence, by Lemma A.3.1, as we have R connected groups of lengths $g_1, \ldots, g_R$,

$$\lim \frac{1}{p} E_{D(g_1,\ldots,g_R)} \text{Tr}(n^{T/2} p^{-T/2} \prod_{i=1}^{T} \mathbb{P}_i)$$
$$= O(y_n^{-T/2+\sum(g_i-1)}) = O(y_n^{-T/2+T-R}) = O(y_n^{T/2-R}). \qquad (A.57)$$

Let
$$F(g_1, \ldots, g_R) = G(g_1, \ldots, g_R) - D(g_1, \ldots, g_R).$$

Then, by Lemma A.3.1 and (A.57),

$$\lim \frac{1}{p} E_{F(g_1,\ldots,g_R)} \text{Tr}(n^{T/2} p^{-T/2} \prod_{i=1}^{T} \mathbb{P}_i) = o(y_n^{T/2-R}). \qquad (A.58)$$

Hence, by (A.57) and (A.58), proof of Lemma A.3.3 is complete. □

Getting back to the proof of the theorem, by (A.53) and Lemma A.3.3, we have

$$\lim p^{-1} E\text{Tr}\left[\Pi_{i=1}^{T} \mathcal{R}_i\right] = \lim p^{-1} E_{C_{T+1}} \text{Tr}(n^T p^{-T} \Pi_{i=1}^{T} \mathbb{P}_i)$$
$$= \sum_{\substack{2 \le g_1,\ldots,g_R \le T \\ \sum_{j=1}^{R} g_j = T, \; R \ge 1}} \lim p^{-1} E_{G(g_1,\ldots,g_R)} \text{Tr}(n^T p^{-T} \Pi_{i=1}^{T} \mathbb{P}_i)$$
$$= \begin{cases} 0, & \text{if } T \text{ is odd} \\ \lim p^{-1} E_{G(2,2,\ldots,2)} \text{Tr}(n^{T/2} p^{-T/2} \Pi_{i=1}^{T} \mathbb{P}_i), & \text{if } T \text{ is even.} \end{cases} \qquad (A.59)$$

Therefore, (A.43) is proved for odd T.

It remains to show that (A.43) and (A.59) are equivalent when T is even. Let $T = 2m$ and $D(2,2,\ldots,2)$ be the set of all non-crossing pair matches in $G(2,2,\ldots,2)$. Then from the proof of Lemma A.3.3, it is obvious that

$$\lim \frac{1}{p} E_{G(2,\ldots,2)} \text{Tr}(n^m p^{-m} \Pi_{i=1}^{2m} \mathbb{P}_i) = \lim \frac{1}{p} E_{D(2,\ldots,2)} \text{Tr}(n^m p^{-m} \Pi_{i=1}^{2m} \mathbb{P}_i). \qquad (A.60)$$

Note that $D(2,\ldots,2)$ is the set of all non-crossing pair matches each of which has $T/2$ many connected groups of length 2. Moreover, observe that at least one block must start with an even index to get a connected group of length 2. Hence, each match in $D(2,2,\ldots,2)$ has at least $T/2$ blocks which start with an even index. Now consider $\mathcal{C} \subset D(2,\ldots,2)$ of matches which have exactly $T/2$ many blocks starts with even index.

$$\mathcal{C} = \{\sigma_{\tau,(i_1,\ldots,i_{2m})} : \tau \in NC_2(2m)\}, \qquad (A.61)$$

where for each $\tau = \{(l_1, l_2), (l_3, l_4), \ldots, (l_{2m-1}, l_{2m})\} \in NC_2(2m)$, $l_{2k-1} < l_{2k}$

for all $k \leq m$, we have

$$\sigma_{\tau,(i_1,\ldots,i_{2m})} = \{(u_{l_{2k-1},3i_{2k-1}}, v_{l_{2k-1},2i_{2k-1}}) = (u_{l_{2k},3i_{2k}-1}, v_{l_{2k},2i_{2k}-1}),$$
$$(u_{l,3i-1}, v_{l,2i-1}) = (u_{l,3i}, v_{l,2i}), \forall i \neq i_l, l \leq 2m\}. \quad (A.62)$$

Note that $D(2,\ldots,2) - \mathcal{C}$ has more than $T/2$ blocks that start with an even index. Therefore, by Lemma A.3.1,

$$\lim \frac{1}{p} E_{D(2,\ldots,2)-\mathcal{C}} \mathrm{Tr}(n^m p^{-m} \Pi_{i=1}^{2m} \mathbb{P}_i) = 0. \quad (A.63)$$

Hence, by (A.59), (A.60), and (A.63), we have

$$\lim p^{-1} E \mathrm{Tr}\left[\Pi_{i=1}^{2m} \mathcal{R}_i\right] = \lim p^{-1} E_{\mathcal{C}} \mathrm{Tr}(n^m p^{-m} \Pi_{i=1}^{2m} \mathbb{P}_i). \quad (A.64)$$

Hence, it remains to show that the right sides of (A.43) and (A.64) match for $T = 2m$. Now it is easy to show that

$$\sum_{\tau \in NC_2(2m)} \lim p^{-1} E_{\sigma_{\tau,(i_1,i_2,\ldots,i_{2m})}} \mathrm{Tr}(n^m p^{-m} \Pi_{i=1}^{2m} \mathbb{P}_i) \quad (A.65)$$

$$= \sum_{\tau \in NC_2(2m)} \left[\varphi_{0K(\tau)} \left(\tilde{c}_{k,i_k} c_{(k+1) \bmod 2m, \, i_{(k+1) \bmod 2m}} : 1 \leq k \leq 2m \right) \right.$$
$$\left. \times \kappa_\tau (w_{k,i_k} : 1 \leq k \leq 2m) \right]$$

$$= \varphi_0 \left(\prod_{k=1}^{2m} c_{k,i_k} w_{k,i_k} \tilde{c}_{k,i_k} \right).$$

Now

$$\frac{1}{p} E_{\mathcal{C}} \mathrm{Tr}(n^m p^{-m} \Pi_{i=1}^{2m} \mathbb{P}_i)$$

$$= \sum_{i_1,i_2,\ldots,i_{2m}} \sum_{\tau \in NC_2(2m)} \lim p^{-1} E_{\sigma_{\tau,(i_1,i_2,\ldots,i_{2m})}} \mathrm{Tr}(n^m p^{-m} \Pi_{i=1}^{2m} \mathbb{P}_i)$$

$$= \sum_{i_1,i_2,\ldots,i_{2m}} \varphi_0 \left(\prod_{k=1}^{2m} c_{k,i_k} w_{k,i_k} \tilde{c}_{k,i_k} \right) = \varphi_0 \left(\prod_{k=1}^{2m} \sum_{i_k} c_{k,i_k} w_{k,i_k} \tilde{c}_{k,i_k} \right)$$

$$= \varphi_0 \left(\prod_{l-1}^{2m} \alpha_l \right). \quad (A.66)$$

Hence, by (A.66), $\lim p^{-1} E \mathrm{Tr}\left[\Pi_{i=1}^{2m} \mathcal{R}_i\right] = \varphi(\prod_{l=1}^{2m} \alpha_l)$. Therefore, proof of Theorem 7.2.1(a) is complete.

(b) Proof of (b) is immediate from Theorem 7.2.1(a) by observing the fact that proof of (a) will go through if instead of $\{A_{l,2i-1}\} \subset \{B_{2i-1}, B_{2i-1}^*\}$, we assume $\{A_{l,2i-1}\} \subset \mathrm{Span}\{B_{2i-1}, B_{2i-1}^*\}$.

Hence, the proof of Theorem 7.2.1 is complete. $\qquad\square$

A.4   Proof of Lemma 8.2.1

Here we shall only prove (a). We omit the proof of (b) since it requires similar arguments.

Note that X_t can be written as

$$X_t = (\psi_0 \ \psi_1 \ \psi_2 \ldots \psi_q)(\varepsilon_t^* \ \varepsilon_{t-1}^* \ \varepsilon_{t-2}^* \ldots \varepsilon_{t-q}^*)^* \quad \forall t, n \geq 1.$$

This can be used to compute the sample autocovariance matrices for $\{X_t\}$ in terms of the "sample" autocovariance matrices for $\{\varepsilon_t\}$. Forming inner products and averaging yields,

$$
\begin{aligned}
n\hat{\Gamma}_k &= \sum_{j,j'=0}^{q} \psi_j \ \hat{\Gamma}_{j'-j+k}(\varepsilon) \ \psi_{j'}^* - \sum_{\substack{j,j'=0 \\ j-j' \neq k}}^{q} \sum_{t=n-j+1}^{n} \psi_j \varepsilon_{t,p} \varepsilon_{t-(j'+k-j)}^* \psi_{j'}^* \\
&\quad + \sum_{\substack{j,j'=0 \\ j-j' \neq k}}^{q} \sum_{t=k-j+1}^{j'+k-j} \psi_j \varepsilon_t \varepsilon_{t-(j'+k-j)}^* \psi_{j'}^* + \sum_{j=0}^{q} \sum_{t=n-j+1}^{n} \psi_j \varepsilon_t \varepsilon_t^* \psi_{j-k.p}^* \\
&\quad + \sum_{j=0}^{q} \sum_{t=k-j+1}^{0} \psi_j^{(n)} \varepsilon_t \varepsilon_t^* \psi_{j-k}^* \\
&= n\Delta_k + R_{1n} + R_{2n} + R_{3n} + R_{4n}, \text{ (say).} \tag{A.67}
\end{aligned}
$$

Note that, for any k matrices $A_1, A_2, \ldots, A_k$ of order p and integers $\{r_i, p_i\}$, $\sum_{j=1}^{k} r_j^{-1} = 2$, $\sum_{j=1}^{k} p_j^{-1} = 1$, we have

$$\frac{1}{p} E\mathrm{Tr}(A_1 A_2 \cdots A_k) \leq E\Big[\prod_{j=1}^{k} \big(\frac{1}{p}\mathrm{Tr}(A_j^* A_j)^{r_j}\big)^{\frac{1}{2r_j}}\Big]$$

$$\leq \prod_{j=1}^{k} \big[E\big(\frac{1}{p}\mathrm{Tr}(A_j^* A_j)^{r_j}\big)^{\frac{p_j}{2r_j}}\big]^{1/p_j}. \tag{A.68}$$

Moreover, for any polynomial Π, $(\Pi(\hat{\Gamma}_i, \hat{\Gamma}_i^* : i \geq 0) - \Pi(\Delta_i, \Delta_i^* : i \geq 0))$ involves monomials with at least one of R_{1n}, R_{2n}, R_{3n} or R_{4n}. Hence, to show (a), by (A.68), it suffices to show, for all $r, s \geq 1$ and $i = 1, 2, 3, 4$,

$$(i) \ \lim E\left(p^{-1}\mathrm{Tr}\left(\Delta_u^* \Delta_u\right)^r\right)^s < \infty, \text{ and} \tag{A.69}$$

$$(ii) \ E\left(p^{-1}\mathrm{Tr}\left(n^{-2} R_{in}^* R_{in}\right)^r\right)^s \to 0. \tag{A.70}$$

Now

$$E\big(p^{-1}\mathrm{Tr}(\Delta_u^* \Delta_u)^r\big)^s \leq E\left(p^{-1}\mathrm{Tr}\left(\Delta_u^* \Delta_u\right)^K\right)^{rs/K}, \text{ where } K > rs$$

$$\leq \left(p^{-1}E\mathrm{Tr}\left(\Delta_u^* \Delta_u\right)^K\right)^{rs/K}.$$

In the course of the proof of Theorem 6.2.1, we have already shown that $\lim p^{-1} E \text{Tr} \left(\Delta_u^* \Delta_u \right)^K < \infty$. This establishes (i).

Now we shall prove (A.70). Note that for every $r \geq 1$,

$$\left| E \left(\text{Tr} \left(n^{-2} R_{in}^* R_{in} \right)^r \right)^s \right| \leq E \left(\text{Tr} \left(n^{-2} R_{in}^* R_{in} \right) \right)^{rs}. \tag{A.71}$$

Hence, it is enough to show that there is $C_r > 0$, such that

$$\left| E \left(\text{Tr} \left(n^{-2} R_{in}^* R_{in} \right) \right)^r \right| < C_r, \ \forall n \geq 1. \tag{A.72}$$

Let us first prove (A.72) for $i = 1$. Similar idea works for $i > 1$.

Recall the definition of R_{1n} in (A.67). Note that for $i = 1$, it suffices to show, for $r \geq 1$, $s_k > 0$ and $\{A_k\} \in \text{Span}\{\psi_j, \psi_j^* : j \geq 0\}$, there is $C_r > 0$ such that for all $n \geq 1$,

$$\left| E \prod_{k=1}^r \left(\frac{1}{n^2} \text{Tr}(A_{2k-1} \varepsilon_{t_{2k-1}} \varepsilon_{t_{2k-1}-s_{2k-1}}^* A_{2k} \varepsilon_{t_{2k}-s_{2k}} \varepsilon_{t_{2k}}^*) \right) \right| < C_r. \tag{A.73}$$

To prove (A.73), we use induction on r. For $r = 1$, by Assumption ($A3$), for all $n \geq 1$ and for some $C_1 > 0$,

$$\left| E \left(n^{-2} \text{Tr}(A_1 \varepsilon_{t_1} \varepsilon_{t_1-s_1}^* A_2 \varepsilon_{t_2-s_2} \varepsilon_{t_2}^*) \right) \right| \ \leq \ \left| \prod_{i=1}^2 \frac{1}{n} \text{Tr}(A_1) \right| < C_1.$$

Suppose (A.73) is true for all $r \leq m$. Now for $r = m + 1$, consider

$$E \prod_{k=1}^{m+1} \left(n^{-2} \text{Tr}(A_{2k-1} \varepsilon_{t_{2k-1}} \varepsilon_{t_{2k-1}-s_{2k-1}}^* A_{2k} \varepsilon_{t_{2k}-s_{2k}} \varepsilon_{t_{2k}}^*) \right). \tag{A.74}$$

As $\{\varepsilon_{i,j}\}$ are independent and of mean 0, $\{\varepsilon_t\}$ has to be matched for a non-zero contribution. Now the following two cases may happen.

Case 1. Matches are such that no index in $\{t_{2k_u-1}, t_{2k_u-1}-s_{2k_u-1}, t_{2k_u}, t_{2k_u} - s_{2k_u} : 1 \leq u \leq U < m + 1\}$ matches with any index in $\{t_{2k-1}, t_{2k-1} - s_{2k-1}, t_{2k}, t_{2k} - s_{2k} : k \neq k_u, \forall 1 \leq u \leq U < m + 1\}$. Then (A.74) would become

$$\left| \left(E \prod_{u=1}^U \left(n^{-2} \text{Tr}(A_{2k_u-1} \varepsilon_{t_{2k_u}} \varepsilon_{t_{2k_u}-1-s_{2k_u}-1}^* A_{2k_u} \varepsilon_{t_{2k_u}-s_{2k_u}} \varepsilon_{t_{2k_u}}^*) \right) \right) \right.$$

$$\left. \times \left(E \prod_{k \neq k_u} \left(n^{-2} \text{Tr}(A_{2k-1} \varepsilon_{t_{2k-1}} \varepsilon_{t_{2k-1}-s_{2k-1}}^* A_{2k} \varepsilon_{t_{2k}-s_{2k}} \varepsilon_{t_{2k}}^*) \right) \right) \right|$$

$$\leq \ C, \text{ using (A.73) for } r = U \leq m \text{ and } r = m + 1 - U \leq m.$$

Case 2. A typical match which is not covered in Case 1, is of the form

$$t_{2k} = t_{2k+1}, \ s_{2k} = s_{2k+1}, \ \forall 1 \leq k \leq m. \tag{A.75}$$

Then (A.74) reduces to

$$n^{-2(m+1)} E\mathrm{Tr}\Big[\Big(\prod_{k=1}^{m+1} A_{t_{2k}}\varepsilon_{t_{2k}-s_{2k}}\varepsilon^*_{t_{2k+1}-s_{2k+1}}\Big) \tag{A.76}$$

$$\times\Big(\prod_{k=0}^{m} A_{2(m-k)+1}\varepsilon_{t_{2(m-k)+1}}\varepsilon^*_{t_{2(t-m)}}\Big)\Big],$$

where $t_{2m+3} - s_{2m+3} = t_{2m+2}$ and (A.75) holds. Now, using the idea that was used in the proof of (M1) condition for $\{\Delta_i, \Delta_i^*\}$ in Theorem 6.2.1, it is easy to see that (A.76) is bounded for all $n \geq 1$. Hence, (A.73) is established for $r = m + 1$. Therefore, proof of (A.73) and hence Lemma 8.2.1(b) is complete. □

A.5   Proof of Corollary 8.2.1(c)

We need a few inequalities. Recall the Lévy metric from (8.22) and $||\cdot||_2$ from (1.8). For any matrix M, let F^M denotes its ESD. The following inequalities are taken respectively from Theorems A.43 and A.45 and, Corollaries A.41 and A.42 of Bai and Silverstein [2009].

Lemma A.5.1. *Let A, B, C and D be $p \times p$ matrices where A and B are symmetric. Then*

(a) $L(F^A, F^B) \leq \frac{1}{p} rank(A - B)$,

(b) $L(F^A, F^B) \leq ||A - B||_2^2$,

(c) $L^3(F^{C+C^*}, F^{D+D^*}) \leq \frac{1}{p} Tr((C - D)(C - D)^*)$,

(d) $L^4(F^{CC^*}, F^{DD^*}) \leq \frac{2}{p^2} Tr(CC^* + DD^*) Tr((C - D)(C - D)^*)$.

Let $X_t \sim MA(q)$, $q \geq 1$ process and suppose Assumptions $(B1)$, (B) hold, $\{\varepsilon_{i,j}\} \in \mathcal{L}_{2+\delta} \cap U(\delta)$ for some $\delta \in (0, 2]$ and $p/n \to y \in (0, \infty)$.
   Let

$$\tilde{\varepsilon}_{t,i} = \varepsilon_{t,i} I(|\varepsilon_{t,i}| < \eta_n n^{\frac{1}{2+\delta}}), \quad \hat{\varepsilon}_{t,i} = \tilde{\varepsilon}_{t,i} - E(\tilde{\varepsilon}_{t,i}), \ \forall t, i \text{ and some } \eta_n \downarrow 0,$$

$$\sigma_{t,i}^2 = E|\hat{\varepsilon}_{t,i}|^2, \ \Delta = n^{-\frac{\delta}{4+2\delta}}, X_{t,i} = 2Ber(0.5) - 1, \text{ i.i.d. for all } t, i,$$

$$\bar{\varepsilon}_{t,i} = \begin{cases} X_{t,i}, & \text{if } \sigma_{t,i}^2 < 1 - \Delta, \\ \frac{\hat{\varepsilon}_{t,i}}{\sigma_{t,i}}, & \text{otherwise,} \end{cases}$$

$$\hat{\Gamma}_i(\varepsilon), \tilde{\Gamma}_i(\varepsilon), \hat{\tilde{\Gamma}}_i(\varepsilon), \bar{\Gamma}_i(\varepsilon) \ = \ i\text{-th order sample autocovariance matrix of}$$
$$\{\varepsilon_{t,i}\}, \{\tilde{\varepsilon}_{t,i}\}, \{\hat{\varepsilon}_{t,i}\}, \{\bar{\varepsilon}_{t,i}\}(\text{respectively}),$$

$$\hat{T}_i = \sum_{j,j'=0}^{q} \psi_j \hat{\Gamma}_{j-j'+i}(\varepsilon)\psi_{j'}^*, \quad \tilde{T}_i = \sum_{j,j'=0}^{q} \psi_j \tilde{\Gamma}_{j-j'+i}(\varepsilon)\psi_{j'}^*,$$

$$\hat{\tilde{T}}_i = \sum_{j,j'=0}^{q} \psi_j \hat{\tilde{\Gamma}}_{j-j'+i}(\varepsilon)\psi_{j'}^*, \quad \bar{T}_i = \sum_{j,j'=0}^{q} \psi_j \bar{\Gamma}_{j-j'+i}(\varepsilon)\psi_{j'}^*.$$

Note that the existence of the LSD of $\{\bar{T}_i + \bar{T}_i^*\}_{i \geq 0}$ follows by an application of Theorem 8.2.1. We will now show that the LSD of $\{\hat{\Gamma}_i + \hat{\Gamma}_i^*\}_{i \geq 0}$ is same as that of $\{\bar{T}_i + \bar{T}_i^*\}_{i \geq 0}$.

Let F^A denote the ESD of the matrix A and L denote the Lévy metric on the space of probability distribution functions (see (8.22)). It is then enough to show that

$$L(F^{\hat{\Gamma}_i + \hat{\Gamma}_i^*}, F^{\bar{T}_i + \bar{T}_i^*}) \to 0 \ \forall i \geq 0 \ \text{ almost surely.}$$

Note that

$$
\begin{aligned}
L(F^{\hat{\Gamma}_i + \hat{\Gamma}_i^*}, F^{\bar{T}_i + \bar{T}_i^*}) &\leq L(F^{\hat{\Gamma}_i + \hat{\Gamma}_i^*}, F^{\hat{T}_i + \hat{T}_i^*}) + L(F^{\hat{T}_i + \hat{T}_i^*}, F^{\tilde{T}_i + \tilde{T}_i^*}) \\
&\quad + L(F^{\tilde{T}_i + \tilde{T}_i^*}, F^{\hat{\bar{T}}_i + \hat{\bar{T}}_i^*}) + L(F^{\hat{\bar{T}}_i + \hat{\bar{T}}_i^*}, F^{\bar{T}_i + \bar{T}_i^*}) \\
&= B_1 + B_2 + B_3 + B_4, \text{ (say).}
\end{aligned}
$$

We will show that, for each $1 \leq i \leq 4$, $B_i \to 0$ almost surely.

Proof of $B_1 \to 0$. By Lemma A.5.1(a), we have

$$
\begin{aligned}
B_1 &\leq 2p^{-1} \left(\operatorname{rank}(R_{1n}) + \operatorname{rank}(R_{2n}) + \operatorname{rank}(R_{3n}) + \operatorname{rank}(R_{4n}) \right) \\
&\leq \frac{8q}{p} \to 0 \text{ almost surely}
\end{aligned}
$$

where R_{1n}, R_{2n}, R_{3n} and R_{4n} are as in (A.67).

Proof of $B_2 \to 0$. As mentioned earlier, we will use the truncation arguments of Jin et al. [2014]. By Lemma A.5.1(a), we have for some $C > 0$

$$
\begin{aligned}
B_2 &\leq \frac{1}{p}\operatorname{rank}(\hat{T}_i + \hat{T}_i^* - \tilde{T}_i - \tilde{T}_i^*) \leq \frac{2}{p}\operatorname{rank}(\hat{T}_i - \tilde{T}_i) \\
&\leq \frac{1}{p}\operatorname{rank}\Big[\sum_{j,j'=0}^{q} \psi_j \left(\hat{\Gamma}_{j-j'+i}(\varepsilon) - \tilde{\Gamma}_{j-j'+i}(\varepsilon) \right) \psi_{j'}^* \Big] \\
&\leq \frac{C}{p}\operatorname{rank}(\hat{\Gamma}_i(\varepsilon) - \tilde{\Gamma}_i(\varepsilon)) \to 0 \text{ a.s. (Jin et al. [2014], p. 1210)}
\end{aligned}
$$

Proof of $B_3 \to 0$. Recall $\| \ \|_2$ in (1.8). By Lemma A.5.1(b), we have for some $C > 0$,

$$
\begin{aligned}
B_3 &\leq \|\tilde{T}_i + \tilde{T}_i^* - \hat{\bar{T}}_i - \hat{\bar{T}}_i^*\|_2^2 \\
&\leq C\|\tilde{\Gamma}_i(\varepsilon) - \hat{\bar{\Gamma}}_i(\varepsilon)\|_2^2 \to 0 \text{ a.s. (page 1211, Jin et al. [2014])}
\end{aligned}
$$

Proof of $B_4 \to 0$. By Lemma A.5.1(c), we have

$$B_4^3 \leq \frac{1}{p}\mathrm{Tr}\left[(\hat{\bar{T}}_i + \hat{\bar{T}}_i^* - \bar{T}_i - \bar{T}_i^*)(\hat{\bar{T}}_i + \hat{\bar{T}}_i^* - \bar{T}_i - \bar{T}_i^*)^*\right]$$

$$\leq \frac{4}{p}\mathrm{Tr}\left((\hat{\bar{T}}_i - \bar{T}_i)(\hat{\bar{T}}_i - \bar{T}_i)^*\right)$$

$$= \frac{4}{p}\sum_{j,j',k,k'=0}^{q}\mathrm{Tr}\left[\psi_j\left(\hat{\bar{\Gamma}}_{j-j'+i}(\varepsilon) - \bar{\Gamma}_{j-j'+i}(\varepsilon)\right)\psi_{j'}^*\right.$$

$$\left.\psi_{k'}\left(\hat{\bar{\Gamma}}_{k-k'+i}(\varepsilon) - \bar{\Gamma}_{k-k'+i}(\varepsilon)\right)^*\psi_k^*\right]$$

$$\leq 4\left(\sum_{j,j'=0}^{q}\left[p^{-1}\mathrm{Tr}(\psi_j(\hat{\bar{\Gamma}}_{j-j'+i}(\varepsilon) - \bar{\Gamma}_{j-j'+i}(\varepsilon))\psi_{j'}^*\right.\right.$$

$$\left.\left.\psi_{j'}\left(\hat{\bar{\Gamma}}_{j-j'+i}(\varepsilon) - \bar{\Gamma}_{j-j'+i}(\varepsilon)\right)^*\psi_j^*)\right]^{1/2}\right)^2.$$

Therefore, it is enough to show that

$$p^{-1}\mathrm{Tr}(A(\hat{\bar{\Gamma}}_i(\varepsilon) - \bar{\Gamma}_i(\varepsilon))BB^*(\hat{\bar{\Gamma}}_i(\varepsilon) - \bar{\Gamma}_i(\varepsilon))^*A^*) \to 0, \text{ a.s.,} \qquad \text{(A.77)}$$

for any $A, B \in \mathrm{Span}\{\psi_j, \psi_j^* : j \geq 0\}$.

The proof of (A.77) given below goes along the same lines as the proof of $p^{-1}\mathrm{Tr}((\hat{\bar{\Gamma}}_i(\varepsilon) - \bar{\Gamma}_i(\varepsilon))(\hat{\bar{\Gamma}}_i(\varepsilon) - \bar{\Gamma}_i(\varepsilon))^*) \to 0$ given in Jin et al. [2014]. Here we have the extra factors of A, B etc. Let

$$\hat{\alpha}_k = (2n)^{-1/2}(\hat{\varepsilon}_{k,1}, \hat{\varepsilon}_{k,2}, \ldots, \hat{\varepsilon}_{k,p})^T, \quad \bar{\alpha}_k = (2n)^{-1/2}(\bar{\varepsilon}_{k,1}, \bar{\varepsilon}_{k,2}, \ldots, \bar{\varepsilon}_{k,p})^T,$$

$$\hat{U} = (\hat{\alpha}_1, \hat{\alpha}_2, \ldots, \hat{\alpha}_{n-i}), \qquad\qquad \bar{U} = (\bar{\alpha}_1, \bar{\alpha}_2, \ldots, \bar{\alpha}_{n-i}),$$

$$\hat{V} = (\hat{\alpha}_{1+i}, \hat{\alpha}_{2+i}, \ldots, \hat{\alpha}_n), \qquad\qquad \bar{V} = (\bar{\alpha}_{1+i}, \bar{\alpha}_{2+i}, \ldots, \bar{\alpha}_n).$$

Then

$$\frac{1}{p}\mathrm{Tr}(A(\hat{\bar{\Gamma}}_i - \bar{\Gamma}_i)BB^*(\hat{\bar{\Gamma}}_i - \bar{\Gamma}_i)^*A^*)$$

$$= \frac{1}{p}\mathrm{Tr}(A(\hat{U}\hat{V}^* - \bar{U}\bar{V}^*)BB^*(\hat{U}\hat{V}^* - \bar{U}\bar{V}^*)^*A^*)$$

$$= \frac{1}{p}\mathrm{Tr}(A((\hat{U} - \bar{U})\hat{V}^* + \bar{U}(\hat{V} - \bar{V})^*)BB^*((\hat{U} - \bar{U})\hat{V}^* + \bar{U}(\hat{V} - \bar{V})^*)^*A^*)$$

$$\leq 2\frac{1}{p}\mathrm{Tr}(A(\hat{U} - \bar{U})\hat{V}^*BB^*\hat{V}(\hat{U} - \bar{U})^*A^*)$$

$$+2\frac{1}{p}\mathrm{Tr}(A\bar{U}(\hat{V} - \bar{V})^*BB^*(\hat{V} - \bar{V})\bar{U}^*A^*).$$

Now, we have for some $C > 0$, with $A = ((a_{ij}))$ and $B = ((b_{ij}))$,

$$p^{-1}\mathrm{Tr}(A(\hat{U} - \bar{U})\hat{V}^* BB^* \hat{V}(\hat{U} - \bar{U})^* A^*)$$

$$\leq \frac{C}{n^3} \sum_{u,v} |\sum_{l,k,j} a_{ul}(\hat{\varepsilon}_{k,l} - \bar{\varepsilon}_{k,l})\hat{\varepsilon}^*_{(k+i),j} b_{jv}|^2$$

$$= \frac{C}{n^3} \sum_{u,v} \sum_{l_1,k_1,j_1} \sum_{l_2,k_2,j_2} \left(a_{ul_1}(\hat{\varepsilon}_{k_1,l_1} - \bar{\varepsilon}_{k_1,l_1})\hat{\varepsilon}^*_{(k_1+i),j_1} b_{j_1v} b^*_{j_2v} \right.$$

$$\left. \times \hat{\varepsilon}_{(k_2+i),j_2}(\hat{\varepsilon}_{k_2,l_2} - \bar{\varepsilon}_{k_2,l_2})^* a^*_{ul_2} \right)$$

$$= J_1 + J_2 + J_3 + J_4 + J_5,$$

where,

$$\sum_1 = \sum_{u_1,v_1} \sum_{\substack{l_1,l_2,j_1,j_2 \\ k_1>k_2,\ k_1\neq k_2+i}} , \quad \sum_2 = \sum_{u_1,v_1} \sum_{\substack{l_1,j_1,l_2, \\ j_2,k_2}} ,$$

$$\sum_3 = \sum_{u_1,v_1} \sum_{\substack{l_1,l_2,j_1,j_2 \\ k_2>k_1,\ k_2\neq k_1+i}} , \quad \sum_4 = \sum_{u_1,v_1} \sum_{\substack{l_1,j_1,l_2, \\ j_2,k_1}} ,$$

$$\sum_5 = \sum_{\substack{u_1,v_1,l_1,l_2 \\ j_1,j_2,k}} \tag{A.78}$$

and J_1, J_2, J_3, J_4, J_5 are given by, respectively,

$$\frac{C}{n^3} \sum_1 a_{ul_1}(\hat{\varepsilon}_{k_1,l_1} - \bar{\varepsilon}_{k_1,l_1})\hat{\varepsilon}^*_{(k_1+i),j_1} b_{j_1v} b^*_{j_2v} \hat{\varepsilon}_{(k_2+i),j_2}(\hat{\varepsilon}_{k_2,l_2} - \bar{\varepsilon}_{k_2,l_2})^* a^*_{ul_2},$$

$$\frac{C}{n^3} \sum_2 \left(a_{ul_1}(\hat{\varepsilon}_{(k_2+i),l_1} - \bar{\varepsilon}_{(k_2+i),l_1})\hat{\varepsilon}^*_{(k_2+2i),j_1} b_{j_1v} b^*_{j_2v} \hat{\varepsilon}_{(k_2+i),j_2} \right.$$

$$\left. \times (\hat{\varepsilon}_{k_2,l_2} - \bar{\varepsilon}_{k_2,l_2})^* a^*_{ul_2} \right),$$

$$\frac{C}{n^3} \sum_3 \left(a_{ul_1}(\hat{\varepsilon}_{k_1,l_1} - \bar{\varepsilon}_{k_1,l_1})\hat{\varepsilon}^*_{(k_1+i),j_1} b_{j_1v} b^*_{j_2v} \hat{\varepsilon}_{(k_2+i),j_2} \right.$$

$$\left. \times (\hat{\varepsilon}_{k_2,l_2} - \bar{\varepsilon}_{k_2,l_2})^* a^*_{ul_2} \right),$$

$$\frac{C}{n^3} \sum_4 \left(a_{ul_1}(\hat{\varepsilon}_{k_1,l_1} - \bar{\varepsilon}_{k_1,l_1})\hat{\varepsilon}^*_{(k_1+i),j_1} b_{j_1v} b^*_{j_2v} \hat{\varepsilon}_{(k_1+2i),j_2} \right.$$

$$\left. \times (\hat{\varepsilon}_{(k_1+i),l_2} - \bar{\varepsilon}_{(k_1+i),l_2})^* a^*_{ul_2} \right),$$

$$\frac{C}{n^3} \sum_5 \left(a_{ul_1}(\hat{\varepsilon}_{k,l_1} - \bar{\varepsilon}_{k,l_1})\hat{\varepsilon}^*_{(k+i),j_1} b_{j_1v} b^*_{j_2v} \hat{\varepsilon}_{(k+i),j_2}(\hat{\varepsilon}_{k,l_2} - \bar{\varepsilon}_{k,l_2})^* a^*_{ul_2} \right). \tag{A.79}$$

Note that $E(J_i) = 0$ for all i. Let

$$\sum_6 = \sum_{u_2,v_2} \sum_{\substack{l_3,l_4,j_3,j_4 \\ k_3>k_4,\ k_3\neq k_4+i}} , \quad \sum_7 = \sum_{u_2,v_2} \sum_{\substack{l_3,j_3,l_4, \\ j_4,k_4}} .$$

Then for some $C_1, C_2, C_3 > 0$,

$$\text{Var}(J_1) = E(J_1)^2 \tag{A.80}$$

$$\leq \frac{C_1}{n^6} \sum_1 \sum_6 E\bigg[a_{u_1 l_1}(\hat{\varepsilon}_{k_1,l_1} - \bar{\varepsilon}_{k_1,l_1})\hat{\varepsilon}^*_{(k_1+i),j_1} b_{j_1 v_1} b^*_{j_2 v_1}\hat{\varepsilon}_{(k_2+i),j_2}(\hat{\varepsilon}_{k_2,l_2} - \bar{\varepsilon}_{k_2,l_2})^*$$

$$\times a^*_{u_1 l_2} a_{u_2 l_3}(\hat{\varepsilon}_{k_3,l_3} - \bar{\varepsilon}_{k_3,l_3})\hat{\varepsilon}^*_{(k_3+i),j_3} b_{j_3 v_2} b^*_{j_4 v_2}\hat{\varepsilon}_{(k_4+i),j_4}(\hat{\varepsilon}_{k_4,l_4} - \bar{\varepsilon}_{k_4,l_4})^* a^*_{u_2 l_4} \bigg]$$

$$\leq \frac{C_2}{n^4} \sum_{\substack{u_1,u_2 \\ v_1,v_2}} \sum_{\substack{l_1,l_2 \\ j_1,j_2}} \big(a_{u_1 l_1} b_{j_1 v_1} b^*_{j_2 v_1} a^*_{u_1 l_2} a_{u_2 l_1} b_{j_1 v_2} b^*_{j_2 v_2} a^*_{u_2 l_2} \big)$$

$$\leq \frac{C_3}{n^2}(n^{-1}\text{Tr}(A^2 A^{*2}))(n^{-1}\text{Tr}(B^2 B^{*2})) = O(n^{-2}).$$

Also for some $C_1, C_2 > 0$,

$$\text{Var}(J_2) = E(J_2)^2$$

$$\leq \frac{C_1}{n^6} \sum_2 \sum_7 E\bigg[a_{u_1 l_1}(\hat{\varepsilon}_{(k_2+i),l_1} - \bar{\varepsilon}_{(k_2+i),l_1})\hat{\varepsilon}^*_{(k_2+2i),j_1} b_{j_1 v_1} b^*_{j_2 v_1}$$

$$\times \hat{\varepsilon}_{(k_2+i),j_2}(\hat{\varepsilon}_{k_2,l_2} - \bar{\varepsilon}_{k_2,l_2})^* a^*_{u_1 l_2} a_{u_2 l_3}(\hat{\varepsilon}_{(k_4+i),l_3} - \bar{\varepsilon}_{(k_4+i),l_3})$$

$$\times \hat{\varepsilon}^*_{(k_4+2i),j_3} b_{j_3 v_2} b^*_{j_4 v_2}\hat{\varepsilon}_{(k_4+i),j_4}(\hat{\varepsilon}_{k_4,l_4} - \bar{\varepsilon}_{k_4,l_4})^* a^*_{u_2 l_4} \bigg]$$

$$\leq \frac{C_2}{n^4} \sum_{\substack{u_1,v_1 \\ u_2,v_2}} \sum_{\substack{l_1,j_1,l_2, \\ j_2}} \sum_{\substack{l_3,j_3,l_4, \\ j_4}} \big(a_{u_1 l_1} b_{j_1 v_1} b^*_{j_2 v_1} a^*_{u_1 l_2} a_{u_2 l_3} b_{j_1 v_2} b^*_{j_4 v_2} a^*_{u_2 l_2} \big)$$

$$= O(n^{-2}).$$

Similarly one can show that

$$\text{Var}(J_3) = O(n^{-2}), \quad \text{Var}(J_4) = O(n^{-2}).$$

Let $\tilde{\tilde{\varepsilon}}_{ti} = \varepsilon_{ti} I(|\varepsilon_{ti}| > \eta_n n^{\frac{1}{2+\delta}})$, $\forall t, i$. Therefore, as $E(\varepsilon_{ti}) = 0$, note that $E(\tilde{\varepsilon}_{ti}) = -E(\tilde{\tilde{\varepsilon}}_{ti})$, $\forall t, i$. Also note that

$$1 = \text{Var}(\varepsilon_{t,i}) = \text{Var}\left(\tilde{\varepsilon}_{t,i} - E(\tilde{\varepsilon}_{t,i}) + \tilde{\tilde{\varepsilon}}_{t,i} - E(\tilde{\tilde{\varepsilon}}_{t,i})\right) \tag{A.81}$$

$$= \sigma^2_{t,i} + \text{Var}(\tilde{\tilde{\varepsilon}}_{t,i}) + 2(E(\tilde{\tilde{\varepsilon}}_{t,i}))^2.$$

Therefore, as $\{\varepsilon_{i,j}\} \in U(\delta)$, for some $C > 0$

$$1 - \sigma^2_{t,i} \leq 2E(\tilde{\tilde{\varepsilon}}^2_{t,i}) \leq 2CP(|\varepsilon_{t,i}| > \eta_n n^{\frac{1}{2+\delta}})^{\frac{\delta}{2+\delta}} \leq 2C\eta_n^{-\delta} n^{-\frac{\delta}{2+\delta}}. \tag{A.82}$$

Let $E = \{(t,i): \sigma^2_{t,i} < 1 - \Delta\}$. Then if $(t,i) \notin E$, we have for some $C > 0$,

$$(1 - \sigma^{-1}_{t,i})^2 \leq C\eta_n^{-2\delta} n^{-\frac{2\delta}{2+\delta}}. \tag{A.83}$$

Moreover note that if $(t, i) \in E$, then $\frac{1-\sigma_{t,i}^2}{\Delta} > 1$. Then by (A.82) and (A.83), we have for some $C_1, C_2 > 0$,

$$
\begin{aligned}
E(J_5) &= \frac{C}{n^3} \sum_{u,v,l_1,j_1,k} a_{ul_1} E|\hat{\varepsilon}_{k,l_1} - \bar{\varepsilon}_{k,l_1}|^2 E|\hat{\varepsilon}_{(k+i),j_1}|^2 b_{j_1 v} b_{j_1 v}^* a_{ul_1}^* \\
&\leq \frac{C_1}{n^3} \sum_{\substack{u,v,j_1 \\ (k,l_1) \in E}} a_{ul_1} b_{j_1 v} b_{j_1 v}^* a_{ul_1}^* \left(\frac{1 - \sigma_{kl_1}^2}{\Delta} \right) \\
&\quad + \frac{C_2}{n^3} \sum_{\substack{u,v,j_1 \\ (k,l_1) \notin E}} a_{ul_1} b_{j_1 v} b_{j_1 v}^* a_{ul_1}^* \left(1 - \sigma_{kl}^{-1} \right)^2 \\
&= O(n^{\frac{-\delta}{4+2\delta}}) + O(n^{\frac{-2\delta}{2+\delta}}).
\end{aligned}
$$

Therefore,

$$
E(p^{-1}\mathrm{Tr}(A(\hat{U} - \bar{U})\hat{V}^* BB^* \hat{V}(\hat{U} - \bar{U})^* A^*)) \to 0.
$$

Similarly one can show that for some $\epsilon > 0$, $\mathrm{Var}(J_5) = O(n^{-1-\epsilon})$ and as a consequence we have

$$
\mathrm{Var}(p^{-1}\mathrm{Tr}(A(\hat{U} - \bar{U})\hat{V}^* BB^* \hat{V}(\hat{U} - \bar{U})^* A^*)) = O(n^{-1-\epsilon}).
$$

Hence,

$$
p^{-1}\mathrm{Tr}(A(\hat{U} - \bar{U})\hat{V}^* BB^* \hat{V}(\hat{U} - \bar{U})^* A^*) \to 0, a.s.. \tag{A.84}
$$

Similarly,

$$
p^{-1}\mathrm{Tr}(A\bar{U}(\hat{V} - \bar{V})^* BB^* (\hat{V} - \bar{V})\bar{U}^* A^*) \to 0, a.s.. \tag{A.85}
$$

Hence, by (A.84) and (A.85), (A.77) is established and $B_4 \to 0$ almost surely. Therefore, proof of Corollary 8.2.1 (c) is complete. $\square$

A.6 Proof of Corollary 8.2.4(c)

Additionally if we assume $\sup_{t,i} E|\varepsilon_{ti}|^4 < M < \infty$, then we need to show that the LSD of $\{\hat{\Gamma}_i \hat{\Gamma}_i^*\}_{i \geq n}$ exists. Proof of this is along the same lines as the proof of Corollary 8.2.1(c). Hence, we omit the detailed calculations and briefly outline the steps. The convergence below are all in the almost sure sense. Let L denote the Lévy metric (see (8.22)).

1. $\begin{aligned}[t] L(F^{\hat{\Gamma}_i \hat{\Gamma}_i^*}, F^{\hat{T}_i \hat{T}_i^*}) &\leq p^{-1}\mathrm{rank}(\hat{\Gamma}_i \hat{\Gamma}_i^* - \hat{T}_i \hat{T}_i^*) \\ &\leq p^{-1}\mathrm{rank}((\hat{\Gamma}_i - \hat{T}_i)\hat{\Gamma}_i^* + \hat{T}_i(\hat{\Gamma}_i - \hat{T}_i)^*) \\ &\leq 2p^{-1}\mathrm{rank}(\hat{\Gamma}_i - \hat{T}_i) \to 0 \text{ (proof similar to } B_1 \to 0). \end{aligned}$

2. $L(F^{\hat{T}_i \hat{T}_i^*}, F^{\tilde{T}_i \tilde{T}_i^*}) \leq 2p^{-1}\mathrm{rank}(\hat{T}_i - \tilde{T}_i) \to 0 \text{ (proof similar to } B_2 \to 0).$

By Lemma A.5.1(d),

3.

$$L^4(F^{\tilde{T}_i\tilde{T}_i^*}, F^{\hat{\tilde{T}}_i\hat{\tilde{T}}_i^*}) \leq 2p^{-2}\mathrm{Tr}(\tilde{T}_i\tilde{T}_i^* + \hat{\tilde{T}}_i\hat{\tilde{T}}_i^*)\mathrm{Tr}((\tilde{T}_i - \hat{\tilde{T}}_i)(\tilde{T}_i - \hat{\tilde{T}}_i)^*)$$
$$\leq 2p^{-1}\mathrm{Tr}(\tilde{T}_i\tilde{T}_i^* + \hat{\tilde{T}}_i\hat{\tilde{T}}_i^*)\|\tilde{T}_i - \hat{\tilde{T}}_i\|_2^2,$$
$$\to 0 \text{ (proof similar to } B_3 \to 0).$$

4.

$$L^4(F^{\bar{T}_i\bar{T}_i^*}, F^{\hat{\bar{T}}_i\hat{\bar{T}}_i^*}) \leq 2p^{-2}\mathrm{Tr}(\bar{T}_i\bar{T}_i^* + \hat{\bar{T}}_i\hat{\bar{T}}_i^*)\mathrm{Tr}((\bar{T}_i - \hat{\bar{T}}_i)(\bar{T}_i - \hat{\bar{T}}_i)^*)$$
$$\to 0 \text{ (proof similar to } B_4 \to 0).$$

This completes the proof of Corollary 8.2.4(c). $\qquad\square$

A.7 Proof of Corollary 8.3.1(c)

Here we show that Corollary 8.3.1(a) remains true even if we drop (B3) and use the more relaxed Assumption $\{\varepsilon_{i,j}\} \in \mathcal{L}_{2+\delta} \cap U(\delta)$ for some $\delta \in (0, 2]$.

Let $X_t \sim \mathrm{MA}(q)$ and suppose Assumptions (B1), (B4) hold, $\{\varepsilon_{i,j}\} \in \mathcal{L}_4 \cap U(\delta)$ for some $\delta > 0$ and $p/n \to 0$. Let

$$\tilde{\varepsilon}_{t,i} = \varepsilon_{t,i}I(|\varepsilon_{t,i}| < \eta n^{\frac{1}{2+\delta}}), \quad \hat{\varepsilon}_{t,i} = \tilde{\varepsilon}_{t,i} - E(\tilde{\varepsilon}_{t,i}), \quad \forall t, i \text{ and some } \eta > 0,$$

$$\sigma_{t,i}^2 = E|\hat{\varepsilon}_{t,i}|^2, \quad \Delta = n^{-\frac{\delta}{4+2\delta}}, \quad B_{t,i} = 2Ber(0.5) - 1, \text{ i.i.d. for all } t, i,$$

$$\bar{\varepsilon}_{t,i} = \begin{cases} B_{t,i}, & \text{if } \sigma_{t,i}^2 < 1 - \Delta, \\ \frac{\hat{\varepsilon}_{t,i}}{\sigma_{t,i}}, & \text{otherwise,} \end{cases}$$

$$C_p = \Gamma_i + \Gamma_i^*,$$

$$\hat{\Gamma}_i(\varepsilon), \tilde{\Gamma}_i(\varepsilon), \hat{\tilde{\Gamma}}_i(\varepsilon), \bar{\Gamma}_i(\varepsilon) = i\text{-th order sample autocovariance matrix of}$$
$$\{\varepsilon_{t,i}\}, \{\tilde{\varepsilon}_{t,i}\}, \{\hat{\varepsilon}_{t,i}\}, \{\bar{\varepsilon}_{t,i}\}(\text{respectively}),$$

$$\hat{T}_i = \sum_{j,j'=0}^{q} \psi_j\hat{\Gamma}_{j-j'+i}(\varepsilon)\psi_{j'}^*, \quad \tilde{T}_i = \sum_{j,j'=0}^{q} \psi_j\tilde{\Gamma}_{j-j'+i}(\varepsilon)\psi_{j'}^*,$$

$$\hat{\tilde{T}}_i = \sum_{j,j'=0}^{q} \psi_j\hat{\tilde{\Gamma}}_{j-j'+i}(\varepsilon)\psi_{j'}^*, \quad \bar{T}_i = \sum_{j,j'=0}^{q} \psi_j\bar{\Gamma}_{j-j'+i}(\varepsilon)\psi_{j'}^*.$$

Since $\{\bar{\varepsilon}_{t,i}\}$ satisfy the stronger assumption $(B3)$, the existence of the LSD of $\sqrt{n}p^{-1}(\bar{T}_i + \bar{T}_i^* - C_p)$ is guaranteed by Corollary 8.3.1(a).

We will actually show that the LSD of $\sqrt{n}p^{-1}(\hat{\Gamma}_i + \hat{\Gamma}_i^* - C_p)$ is same as that of $\sqrt{n}p^{-1}(\bar{T}_i + \bar{T}_i^* - C_p)$. Let L be the Lévy metric between two distribution functions. For any matrix A, let F^A denote the cumulative distribution function of the ESD of A. Then note that

$$L(F\sqrt{np^{-1}}(\hat{\Gamma}_i + \hat{\Gamma}_i^* - C_p), F\sqrt{np^{-1}}(\bar{T}_i + \bar{T}_i^* - C_p))$$

$$\leq L(F\sqrt{np^{-1}}(\hat{\Gamma}_i + \hat{\Gamma}_i^* - C_p), F\sqrt{np^{-1}}(\hat{T}_i + \hat{T}_i^* - C_p))$$

$$+ L(F\sqrt{np^{-1}}(\hat{T}_i + \hat{T}_i^* - C_p), F\sqrt{np^{-1}}(\tilde{T}_i + \tilde{T}_i^* - C_p))$$

$$+ L(F\sqrt{np^{-1}}(\tilde{T}_i + \tilde{T}_i^* - C_p), F\sqrt{np^{-1}}(\hat{\tilde{T}}_i + \hat{\tilde{T}}_i^* - C_p))$$

$$+ L(F\sqrt{np^{-1}}(\hat{\tilde{T}}_i + \hat{\tilde{T}}_i^* - C_p), F\sqrt{np^{-1}}(\bar{T}_i + \bar{T}_i^* - C_p)).$$

$$= T_1 + T_2 + T_3 + T_4, \text{ (say)}. \tag{A.86}$$

It is enough to show that $T_i \to 0$ almost surely for all $i = 1, 2, 3, 4$.

To prove $T_1 \to 0$ almost surely, note that

$$n\hat{\Gamma}_{i.p}$$

$$= \sum_{j,j'=0}^{q} \psi_{j.p}^{(n)} \hat{\Gamma}_{j'-j+i}(\varepsilon) \psi_{j'.p}^{(n)*} - \sum_{\substack{j,j'=0 \\ j-j'\neq i}}^{q} \sum_{t=n-j+1}^{n} \psi_{j.p}^{(n)} \varepsilon_{t.p}\varepsilon_{t-(j'+i-j)}^{*} \psi_{j'.p}^{(n)*}$$

$$+ \sum_{\substack{j,j'=0 \\ j-j'\neq i}}^{q} \sum_{t=i-j+1}^{j'+i-j} \psi_{j.p}^{(n)} \varepsilon_{t.p}\varepsilon_{t-(j'+i-j)}^{*} \psi_{j'.p}^{(n)*} + \sum_{j=0}^{q} \sum_{t=n-j+1}^{n} \psi_{j.p}^{(n)} \varepsilon_{t.p}\varepsilon_{t.p}^{*} \psi_{j-i.p}^{(n)*}$$

$$+ \sum_{j=0}^{q} \sum_{t=i-j+1}^{0} \psi_{j.p}^{(n)} \varepsilon_{t.p}\varepsilon_{t.p}^{*} \psi_{j-i.p}^{(n)*}$$

$$= \hat{T}_i + R_{1p} + R_{2p} + R_{3p} + R_{4p}, \text{ (say)}. \tag{A.87}$$

By Lemma A.5.1(a), we have for some $C > 0$, with R_{1n}, R_{2n}, R_{3n} and R_{4n} as in (A.87),

$$T_1 \leq p^{-1} (\text{rank}(R_{1p}) + \text{rank}(R_{2p}) + \text{rank}(R_{3p}) + \text{rank}(R_{4p}))$$

$$\leq \frac{4Cq}{p} \to 0 \text{ a.s..} \tag{A.88}$$

By Lemma A.5.1(a), we have for some $C, C_1 > 0$

$$T_2 \leq \frac{1}{p}\text{rank}(\hat{T}_i + \hat{T}_i^* - \tilde{T}_i - \tilde{T}_i^*) \leq \frac{2}{p}\text{rank}(\hat{T}_i - \tilde{T}_i)$$

$$\leq \frac{1}{p}\text{rank}\left(\sum_{j,j'=0}^{q} \psi_j \left(\hat{\Gamma}_{j-j'+i}(\varepsilon) - \tilde{\Gamma}_{j-j'+i}(\varepsilon) \right) \psi_{j'}^{*} \right)$$

$$\leq \frac{C}{p}\text{rank}(\hat{\Gamma}_i(\varepsilon) - \tilde{\Gamma}_i(\varepsilon))$$

$$\leq \frac{C_1}{p} \sum_{j=1}^{p} \sum_{t=1}^{n+i} I(|\varepsilon_{t,j}| \geq \eta p^{1/(2+\delta)}). \tag{A.89}$$

Also, we have

$$E\Big(\frac{1}{p}\sum_{j=1}^{p}\sum_{t=1}^{n+i}I(|\varepsilon_{t,j}| \geq \eta p^{1/(2+\delta)})\Big)$$

$$\leq \frac{1}{\eta^{2+\delta}p^2}\sum_{j=1}^{p}\sum_{t=1}^{n+i}E\big(|\varepsilon_{t,j}|^{(2+\delta)}I(|\varepsilon_{t,j}| \geq \eta p^{1/(2+\delta)})\big) = o(1) \quad (A.90)$$

and

$$\mathrm{Var}\Big(\frac{1}{p}\sum_{j=1}^{p}\sum_{t=1}^{n+i}I(|\varepsilon_{t,j}| \geq \eta p^{1/(2+\delta)})\Big)$$

$$\leq \frac{1}{\eta^{2+\delta}p^3}\sum_{j=1}^{p}\sum_{t=1}^{n+i}E\big(|\varepsilon_{t,j}|^{(2+\delta)}I(|\varepsilon_{t,j}| \geq \eta p^{1/(2+\delta)})\big) = o(p^{-1}). \quad (A.91)$$

Applying Bernstein's inequality and (A.90), (A.91), for all $\epsilon > 0$ and large p, we have for some $C, C_1 > 0$,

$$P\Big(\frac{1}{p}\sum_{j=1}^{p}\sum_{t=1}^{n+i}I(|\varepsilon_{t,j}| > \eta p^{1/(2+\delta)}) \geq \epsilon\Big) \leq Ce^{-C_1 p}.$$

Therefore, by Borel–Cantelli lemma, we have

$$T_2 \to 0 \text{ a.s.}. \quad (A.92)$$

Let $\hat{\gamma}_k = n^{-1/2}(\hat{\varepsilon}_{k,1}, \hat{\varepsilon}_{k,2}, \ldots, \hat{\varepsilon}_{k,p})'$ and $\tilde{\gamma}_k = n^{-1/2}(\tilde{\varepsilon}_{k,1}, \tilde{\varepsilon}_{k,2}, \ldots, \tilde{\varepsilon}_{k,p})'$. By Lemma A.5.1(b), we have for some $C, C_1 > 0$,

$$\begin{aligned}
T_3 &\leq \sqrt{np^{-1}}||\tilde{T}_i + \tilde{T}_i^* - \hat{\tilde{T}}_i - \hat{\tilde{T}}_i^*||_2 \leq C\sqrt{np^{-1}}||\tilde{\Gamma}_i(\varepsilon) - \hat{\Gamma}_i(\varepsilon)||_2 \\
&\leq C_1\sqrt{np^{-1}}||\sum_{k=1}^{n}(\hat{\gamma}_k E\tilde{\gamma}_{k+i}^* + \hat{\gamma}_{k+i}E\tilde{\gamma}_k^*)||_2 \\
&\quad + C_1\sqrt{np^{-1}}||\sum_{k=1}^{n}(E\hat{\gamma}_k E\tilde{\gamma}_{k+i}^* + E\hat{\gamma}_{k+i}E\tilde{\gamma}_k^*)||_2.
\end{aligned} \quad (A.93)$$

For the second part, we have for some $C > 0$,

$$\sqrt{np^{-1}}||\sum_{k=1}^{n}(E\hat{\gamma}_k E\tilde{\gamma}_{k+i}^* + E\hat{\gamma}_{k+i}E\tilde{\gamma}_k^*)||_2$$

$$\leq \sqrt{(np)^{-1}}\sum_{k=1}^{n}\sum_{j=1}^{p}|E(\varepsilon_{k,j}I(|\varepsilon_{k,j}| > \eta p^{1/(2+\delta)}))$$

$$\times E(\varepsilon_{k+i,j}I(|\varepsilon_{k+i,j}| > \eta p^{1/(2+\delta)}))|$$

$$\leq C\frac{p}{\sqrt{np}}p^{-2}\sum_{k=1}^{n+i}\sum_{j=1}^{p}E(|\varepsilon_{k,j}|^{2+\delta}I(|\varepsilon_{k,j}| > \eta p^{1/(2+\delta)})) = o(1). \quad (A.94)$$

Proof of Corollary 8.3.1(c)

For the first part, note that

$$np^{-1}||\sum_{k=1}^{n}(\hat{\gamma}_k E\tilde{\gamma}^*_{k+i} + \hat{\gamma}_{k+i}E\tilde{\gamma}^*_k)||_2^2$$

$$\leq 2np^{-1}(||\sum_{k=1}^{n}\hat{\gamma}_k E\tilde{\gamma}^*_{k+i}||_2^2 + ||\sum_{k=1}^{n}\hat{\gamma}_{k+i}E\tilde{\gamma}^*_k||_2^2). \qquad (A.95)$$

Now, for some $C > 0$, we have

$$np^{-1}||\sum_{k=1}^{n}\hat{\gamma}_k E\tilde{\gamma}^*_{k+i}||_2^2 \leq C(np)^{-1}\sum_{j=1}^{p}\sum_{l=1}^{p}\left(\sum_{k=1}^{n}\hat{\varepsilon}_{k,j}E\tilde{\varepsilon}_{k+i,l}\right)^2$$

$$= C(np)^{-1}\sum_{j=1}^{p}\sum_{l=1}^{p}\sum_{k_1=1}^{n}\sum_{k_2=1}^{n}(\hat{\varepsilon}_{k_1,j}E\tilde{\varepsilon}_{k_1+i,l}\hat{\varepsilon}_{k_2,j}E\tilde{\varepsilon}_{k_2+i,l})$$

$$= C(np)^{-1}\sum_{j=1}^{p}\sum_{l=1}^{p}\left(\sum_{k_1=1}^{n}\hat{\varepsilon}^2_{k_1,j}(E\tilde{\varepsilon}_{k_1+i,l})^2 + \sum_{k_1\neq k_2}\hat{\varepsilon}_{k_1,j}E\tilde{\varepsilon}_{k_1+i,l}\hat{\varepsilon}_{k_2,j}E\tilde{\varepsilon}_{k_2+i,l}\right)$$

$$= J_{11} + J_{12}, \text{ (say)}.$$

As $E(\hat{\varepsilon}^4_{t,i}), E(\tilde{\varepsilon}^4_{t,i}) < \infty$, there exists constant C_1, C_2 and C_3 such that

$$EJ_{11} = \frac{C}{np}\sum_{j=1}^{p}\sum_{l=1}^{p}\sum_{k_1=1}^{n}\hat{\varepsilon}^2_{k_1,j}(E\tilde{\varepsilon}_{k_1+i,l})^2$$

$$\leq \frac{C}{np}\sum_{j=1}^{p}\sum_{l=1}^{p}\sum_{k_1=1}^{n}(E(|\varepsilon_{k_1,l}|I(|\varepsilon_{k_1,l}| > \eta p^{1/(2+\delta)})))^2$$

$$\leq \frac{C}{np}(\eta p^{\frac{1}{2+\delta}})^{-2(1+\delta)}\sum_{j,l=1}^{p}\sum_{k_1=1}^{n}E(|\varepsilon_{k_1,l}|^{2+\delta}I(|\varepsilon_{k_1,l}| > \eta p^{1/(2+\delta)}))^2$$

$$= O(p^{-\delta/(2+\delta)}). \qquad (A.96)$$

and

$$\text{Var}J_{11} = \frac{C^2}{(np)^2}\sum_{j=1}^{p}\sum_{k_1=1}^{n}E(\hat{\varepsilon}^2_{k_1,j} - E\hat{\varepsilon}^2_{k_1,j})^2\left(\sum_{l=1}^{p}(E\tilde{\varepsilon}_{k_1+i,l})^2\right)^2$$

$$\leq \frac{C_2}{(np)^2}\sum_{j=1}^{p}\sum_{k_1=1}^{n}E(\tilde{\varepsilon}^4_{k_1,j})\left(p\eta^{-2(1+\delta)}p^{-2(1+\delta)/(2+\delta)}\right)^2$$

$$= O(p^{-1-4\delta/(2+\delta)}). \qquad (A.97)$$

Therefore, by (A.96), (A.97) and Borel–Cantelli Lemma, $J_{11} \to 0$ a.s.. Further, we have $E(J_{12}) = 0$ and

$$
\begin{aligned}
\mathrm{Var} J_{12} &= \frac{C}{(np)^2} \sum_{j=1}^{p} \sum_{k_1 \neq k_2} E\hat{\varepsilon}_{k_1,j}^2 E\hat{\varepsilon}_{k_2,j}^2 \Big(\sum_{l=1}^{p} E\tilde{\varepsilon}_{k_1+i,l} E\tilde{\varepsilon}_{k_2+i,l} \Big)^2 \\
&\leq C_3 (np)^{-2} \sum_{j=1}^{p} \sum_{k_1 \neq k_2} \Big(p\eta^{-2(1+\delta)} p^{-2(1+\delta)/(2+\delta)} \Big)^2 \\
&= O(p^{-1-2\delta/(2+\delta)}).
\end{aligned}
\tag{A.98}
$$

By Borel–Cantelli Lemma these relations imply $J_{12} \to 0$, a.s.. Hence, we have

$$
\| \sum_{k=1}^{n} \hat{\gamma}_k E \tilde{\gamma}_{k+i}^* \|_2^2 \to 0, \quad \text{a.s.}
$$

Similarly,

$$
\| \sum_{k=1}^{n} \hat{\gamma}_{k+i} E \tilde{\gamma}_k^* \|_2^2 \to 0, \quad \text{a.s.}
$$

Thus, by (A.93)–(A.95)

$$
T_3 \to 0, \quad \text{a.s..}
\tag{A.99}
$$

We now finally prove $T_4 \to 0$ almost surely. By Lemma A.5.1(c), we have

$$
\begin{aligned}
T_4^3 &\leq \frac{n}{p^2} \mathrm{Tr}\big[(\hat{\hat{T}}_i + \hat{\hat{T}}_i^* - \bar{T}_i - \bar{T}_i^*)(\hat{\hat{T}}_i + \hat{\hat{T}}_i^* - \bar{T}_i - \bar{T}_i^*)^* \big] \\
&\leq \frac{4n}{p^2} \mathrm{Tr}\big[(\hat{\hat{T}}_i - \bar{T}_i)(\hat{\hat{T}}_i - \bar{T}_i)^* \big]) \\
&= \frac{4n}{p^2} \sum_{j,j',k,k'=0}^{q} \mathrm{Tr}\Big(\psi_j \big(\hat{\hat{\Gamma}}_{j-j'+i}(\varepsilon) - \bar{\Gamma}_{j-j'+i}(\varepsilon) \big) \psi_{j'}^* \\
&\qquad\qquad\qquad \psi_{k'} \big(\hat{\hat{\Gamma}}_{k-k'+i}(\varepsilon) - \bar{\Gamma}_{k-k'+i}(\varepsilon) \big)^* \psi_k^* \Big).
\end{aligned}
$$

Therefore, it is enough to show that

$$
np^{-2} \mathrm{Tr}\big(A (\hat{\hat{\Gamma}}_i(\varepsilon) - \bar{\Gamma}_i(\varepsilon)) BB^* (\hat{\hat{\Gamma}}_i(\varepsilon) - \bar{\Gamma}_i(\varepsilon))^* A^* \big) \to 0, \quad \text{a.s.}
\tag{A.100}
$$

for any $A, B \in \mathrm{Span}\{\psi_j, \psi_j^* : j \geq 0\}$. The proof of (A.100) given below goes along the same lines as the proof of $p^{-1} \mathrm{Tr}((\hat{\hat{\Gamma}}_i(\varepsilon) - \bar{\Gamma}_i(\varepsilon))(\hat{\hat{\Gamma}}_i(\varepsilon) - \bar{\Gamma}_i(\varepsilon))^*) \to 0$ given in page $1210 - 1217$ of Jin et al. [2014]. Here we have the extra factors of A, B etc. Let

$$
\begin{aligned}
\hat{\alpha}_k &= (n)^{-1/2}(\hat{\varepsilon}_{k,1}, \hat{\varepsilon}_{k,2}, \ldots, \hat{\varepsilon}_{k,p})^T, & \bar{\alpha}_k &= (n)^{-1/2}(\bar{\varepsilon}_{k,1}, \bar{\varepsilon}_{k,2}, \ldots, \bar{\varepsilon}_{k,p})^T, \\
\hat{U} &= (\hat{\alpha}_1, \hat{\alpha}_2, \ldots, \hat{\alpha}_{n-i}), & \bar{U} &= (\bar{\alpha}_1, \bar{\alpha}_2, \ldots, \bar{\alpha}_{n-i}), \\
\hat{V} &= (\hat{\alpha}_{1+i}, \hat{\alpha}_{2+i}, \ldots, \hat{\alpha}_n), & \bar{V} &= (\bar{\alpha}_{1+i}, \bar{\alpha}_{2+i}, \ldots, \bar{\alpha}_n).
\end{aligned}
$$

Then,

$$\frac{n}{p^2}\mathrm{Tr}(A(\hat{\Gamma}_i(\varepsilon) - \bar{\Gamma}_i(\varepsilon))BB^*(\hat{\Gamma}_i(\varepsilon) - \bar{\Gamma}_i(\varepsilon))^*A^*)$$

$$= \frac{n}{p^2}\mathrm{Tr}(A(\hat{U}\hat{V}^* - \bar{U}\bar{V}^*)BB^*(\hat{U}\hat{V}^* - \bar{U}\bar{V}^*)^*A^*)$$

$$= \frac{n}{p^2}\mathrm{Tr}(A((\hat{U} - \bar{U})\hat{V}^* + \bar{U}(\hat{V} - \bar{V})^*)BB^*((\hat{U} - \bar{U})\hat{V}^* + \bar{U}(\hat{V} - \bar{V})^*)^*A^*)$$

$$\leq 2\frac{n}{p^2}\mathrm{Tr}(A(\hat{U} - \bar{U})\hat{V}^*BB^*\hat{V}(\hat{U} - \bar{U})^*A^*)$$

$$+ 2\frac{n}{p^2}\mathrm{Tr}(A\bar{U}(\hat{V} - \bar{V})^*BB^*(\hat{V} - \bar{V})\bar{U}^*A^*).$$

Recall $J_1 - J_5$ from (A.79). Now, we have for some $C > 0$, with $A = ((a_{ij}))$ and $B = ((b_{ij}))$,

$$p^{-1}\mathrm{Tr}(A(\hat{U} - \bar{U})\hat{V}^*BB^*\hat{V}(\hat{U} - \bar{U})^*A^*)$$

$$\leq \frac{Cn}{p^2 n^2}\sum_{u,v}|\sum_{l,k,j}a_{ul}(\hat{\varepsilon}_{k,l} - \bar{\varepsilon}_{k,l})\hat{\varepsilon}^*_{(k+i),j}b_{jv}|^2$$

$$= \frac{Cn}{p^2 n^2}\sum_{u,v}\sum_{l_1,k_1,j_1}\sum_{l_2,k_2,j_2}(a_{ul_1}(\hat{\varepsilon}_{k_1,l_1} - \bar{\varepsilon}_{k_1,l_1})\hat{\varepsilon}^*_{(k_1+i),j_1}b_{j_1v}$$

$$\times b^*_{j_2v}\hat{\varepsilon}_{(k_2+i),j_2}(\hat{\varepsilon}_{k_2,l_2} - \bar{\varepsilon}_{k_2,l_2})^*a^*_{ul_2})$$

$$= n^2 p^{-2}(J_1 + J_2 + J_3 + J_4 + J_5). \tag{A.101}$$

Note that $E(J_1) = E(J_2) = E(J_3) = E(J_4) = 0$.

Moreover by (A.80), (A.81) and for some $C_1, C_2, C_3 > 0$,

$$\mathrm{Var}(n^2 p^{-2} J_1) = \frac{n^4}{p^4}E(J_1)^2$$

$$\leq \frac{C_3}{p^2}(p^{-1}\mathrm{Tr}(A^2 A^{*2}))(p^{-1}\mathrm{Tr}(B^2 B^{*2})) = O(p^{-2})$$

and

$$\mathrm{Var}(n^2 p^{-2} J_2) = \frac{n^4}{p^4}E(J_2)^2$$

$$\leq \frac{C_2}{p^4}\sum_{\text{all indices}}(a_{u_1l_1}b_{j_1v_1}b^*_{j_2v_1}a^*_{u_1l_2}a_{u_2l_3}b_{j_1v_2}b^*_{j_4v_2}a^*_{u_2l_2})$$

$$= O(p^{-2}).$$

Similarly one can show that

$$\mathrm{Var}(J_3) = O(p^{-2}), \text{ and } \mathrm{Var}(J_4) = O(p^{-2}).$$

Let

$$\tilde{\bar{\varepsilon}}_{t,i} = \varepsilon_{t,i}I(|\varepsilon_{t,i}| > \eta_n n^{\frac{1}{2+\delta}}), \ \forall t, i.$$

Since $E(\varepsilon_{t,i}) = 0$, we have

$$E(\tilde{\varepsilon}_{t,i}) = -E(\tilde{\tilde{\varepsilon}}_{t,i}), \quad \forall t, i.$$

Also note that

$$
\begin{aligned}
1 = \mathrm{Var}(\varepsilon_{t,i}) &= \mathrm{Var}\left(\tilde{\varepsilon}_{t,i} - E(\tilde{\varepsilon}_{t,i}) + \tilde{\tilde{\varepsilon}}_{t,i} - E(\tilde{\tilde{\varepsilon}}_{t,i})\right) \\
&= \sigma_{t,i}^2 + \mathrm{Var}(\tilde{\tilde{\varepsilon}}_{t,i}) + 2(E(\tilde{\tilde{\varepsilon}}_{t,i}))^2.
\end{aligned}
$$

Therefore, using (A6), for some $C > 0$

$$1 - \sigma_{t,i}^2 \leq 2E(\tilde{\tilde{\varepsilon}}_{t,i}^2) \leq 2C(P(|\varepsilon_{t,i}| > \eta_n n^{\frac{1}{2+\delta}}))^{\frac{\delta}{2+\delta}} \leq 2C\eta_n^{-\delta} p^{-\frac{\delta}{2+\delta}} \quad \text{(A.102)}$$

Let $E = \{(t,i) : \sigma_{t,i}^2 < 1 - \Delta\}$. Then if $(t,i) \notin E$, we have for some $C > 0$

$$(1 - \sigma_{t,i}^{-1})^2 \leq C\eta_n^{-2\delta} p^{-\frac{2\delta}{2+\delta}}. \quad \text{(A.103)}$$

Moreover note that if $(t,i) \in E$, then $\frac{1-\sigma_{t,i}^2}{\Delta} > 1$. Then by (A.102) and (A.103), we have for some $C_1, C_2 > 0$,

$$
\begin{aligned}
E(J_5) &= \frac{Cn}{p^2 n^2} \sum_{u,v,l_1,j_1,k} a_{ul_1} E|\hat{\varepsilon}_{k,l_1} - \bar{\varepsilon}_{k,l_1}|^2 E|\hat{\varepsilon}_{(k+i),j_1}|^2 b_{j_1 v} b_{j_1 v}^* a_{ul_1}^* \\
&\leq \frac{C_1 n}{p^2 n^2} \sum_{\substack{u,v,j_1 \\ (k,l_1) \in E}} a_{ul_1} b_{j_1 v} b_{j_1 v}^* a_{ul_1}^* \left(\frac{1 - \sigma_{k,l_1}^2}{\Delta}\right) \\
&\quad + \frac{C_2}{n^3} \sum_{\substack{u,v,j_1 \\ (k,l_1) \notin E}} a_{ul_1} b_{j_1 v} b_{j_1 v}^* a_{ul_1}^* \left(1 - \sigma_{k,l}^{-1}\right)^2 \\
&= O(p^{\frac{-\delta}{4+2\delta}}) + O(p^{\frac{-2\delta}{2+\delta}}).
\end{aligned}
$$

Similarly one can show that for some $\epsilon > 0$, $\mathrm{Var}(J_5) = O(p^{-1-\epsilon})$ Therefore, using (A.101) and the estimate for $E(J_i)$ and $V(J_i)$,

$$E\left(\frac{n}{p^2}\mathrm{Tr}(A(\hat{U} - \bar{U})\hat{V}^* BB^* \hat{V}(\hat{U} - \bar{U})^* A^*)\right) \to 0, \text{ and} \quad \text{(A.104)}$$

$$\mathrm{Var}\left(\frac{n}{p^2}\mathrm{Tr}(A(\hat{U} - \bar{U})\hat{V}^* BB^* \hat{V}(\hat{U} - \bar{U})^* A^*)\right) = O(p^{-1-\epsilon}). \quad \text{(A.105)}$$

Hence, by (A.104), (A.105) and Borel–Cantelli Lemma,

$$np^{-2}\mathrm{Tr}(A(\hat{U} - \bar{U})\hat{V}^* BB^* \hat{V}(\hat{U} - \bar{U})^* A^*) \to 0, \text{ a.s.} \quad \text{(A.106)}$$

Similarly,

$$np^{-2}\mathrm{Tr}(A\bar{U}(\hat{V} - \bar{V})^* BB^* (\hat{V} - \bar{V})\bar{U}^* A^*) \to 0, \text{ a.s.} \quad \text{(A.107)}$$

Hence, by (A.106) and (A.107), (A.100) is proved. Also by (A.88), (A.89), (A.93), and (A.100),

$$T_4 \to 0, \text{ a.s.} \quad \text{(A.108)}$$

Since we have shown $T_i \to 0$ almost surely for all $i = 1, 2, 3, 4$ in (A.88), (A.92), (A.99), and (A.108), the proof of Corollary 8.3.1(c) is now complete. $\qquad\square$

A.8   Proof of Lemma 8.2.2

Following lemma is useful in this proof.

Lemma A.8.1. *Suppose (B4) holds. Let* $\epsilon_i = 1, * \ \forall i \geq 1$ *and* $C = \sum_{j=0}^{\infty} \sup_p \|\psi_j\|_2^2$. *Then for all* $K \geq 1$,

$$\sum_{\substack{1 \leq j_i, j_i' \leq \infty \\ 1 \leq i \leq K}} |\bar{\varphi}(\prod_{i=1}^{K} \bar{\eta}_{j_i} s \underline{c}_{j_i - j_1' + u_i}^{\epsilon_i} s \bar{\eta}_{j_i'}^*)| \leq (2C)^{2K}. \tag{A.109}$$

Proof. Let us use the temporary notation

$$H = \bar{\varphi} \left(\prod_{i=1}^{K} \bar{\eta}_{j_i} s \underline{c}_{j_i - j_i' + u_i}^{\epsilon_i} s \bar{\eta}_{j_i'}^* \right).$$

By Lemma 5.4.2,

$$H = \sum_{\pi \in NC_2(2K)} \bar{\varphi}_{K(\pi)}[\bar{\eta}_{j_1}, \underline{c}_{j_1 - j_1' + u_1}^{\epsilon_1}, \bar{\eta}_{j_1'}^*, \ldots, \bar{\eta}_{j_k}, \underline{c}_{j_k - j_k' + u_k}^{\epsilon_k}, \bar{\eta}_{j_k'}^*].$$

Therefore, by Lemma 5.1.1 (b) and as $\#NC_2(2K) \leq 2^{2K}$, we have some $h_i, r_i \geq 1$ such that

$$|H| \leq 2^{2k} \prod_{i=1}^{K} (\bar{\varphi}(\bar{\eta}_{j_i}^* \bar{\eta}_{j_i})^{h_i})^{1/h_i} \prod_{i=1}^{K} (\bar{\varphi}(\underline{c}_{j_i - j_i' + u_i}^* \underline{c}_{j_i - j_i' + u_i})^{r_i})^{1/r_i}. \tag{A.110}$$

Now, by (8.7),

$$\bar{\varphi}(\underline{c}_u^* \underline{c}_u)^r \leq 1, \quad \forall r, u \geq 1. \tag{A.111}$$

Also, for all $j \geq 1$, we have

$$\bar{\varphi}(\bar{\eta}_j^* \bar{\eta}_j) = \lim p^{-1} \mathrm{Tr}(\bar{\psi}_j^* \bar{\psi}_j) \leq \sup_p \|\psi_j\|_2^2. \tag{A.112}$$

Hence, by (A.110), (A.111), and (A.112), we have

$$|H| \leq 2^{2K} \prod_{i=1}^{K} \sup_p \|\psi_{j_i}\|_2^2. \tag{A.113}$$

Hence, under Assumption (B4), Lemma A.8.1 holds by summing both sides of (A.113) over j_i, j_i' for all $1 \leq i \leq K$.   □

Now we continue the proof of Lemma 8.2.2. Note that without loss of generality, we can take

$$\Pi(\bar{\gamma}_{uq}, \bar{\gamma}_{uq}^* : u \geq 0) = \sum_{j=1}^{r} m_{l_j}, \quad \text{where} \ \ m_{l_j} = \prod_{i=1}^{l_j} \bar{\gamma}_{u_{j,i}q}^{\epsilon_{j,i}}, \ \epsilon_{j,i} = 1, *. \tag{A.114}$$

Now, by (8.8)

$$\bar{\varphi}(\Pi(\bar{\gamma}_{uq}, \bar{\gamma}_{uq}^* : u \geq 0)) = \sum_{j=1}^{r} \sum_{\substack{1 \leq k_{i,j}, k'_{i,j} \leq q \\ 1 \leq i \leq l_j}} \bar{\varphi}(\prod_{i=1}^{l_j} \bar{\eta}_{k_{j,i}}^{\epsilon_{j,i}} s\underline{c}_{k_{j,i}-k'_{j,i}+u_{j,i}}^{\epsilon_{j,i}} s\bar{\eta}_{k'_{i,j}}^{*(1-\epsilon_{j,i})}).$$

Hence, by Lemma A.8.1, under Assumption (B4), we can let $q \to \infty$ above to conclued that $\lim_{q\to\infty} \bar{\varphi}(\Pi(\bar{\gamma}_{uq}, \bar{\gamma}_{uq}^* : u \geq 0))$ is given by

$$\sum_{j=1}^{r} \sum_{\substack{1 \leq k_{i,j}, k'_{i,j} \leq \infty \\ 1 \leq i \leq l_j}} \bar{\varphi}(\prod_{i=1}^{l_j} \bar{\eta}_{k_{j,i}}^{\epsilon_{j,i}} s\underline{c}_{k_{j,i}-k'_{j,i}+u_{j,i}}^{\epsilon_{j,i}} s\bar{\eta}_{k'_{i,j}}^{*(1-\epsilon_{j,i})}),$$

which is finite. This completes the proof of Lemma 8.2.2.   □

A.9   Proof of Lemma 8.2.3

First note that by (8.9), the right side of (8.20) is a moment sequence. By using (8.8) and Lemma A.8.1, it is easy to see that for all $u_1, u_2, \ldots, u_k \geq 0$, $\epsilon_1, \epsilon_2, \ldots, \epsilon_k = 1, *$ and $K \geq 1$, we have

$$|\bar{\varphi}\left(\prod_{i=1}^{K} \bar{\gamma}_{u_i q}^{\epsilon_i}\right)| \leq (2C)^K,$$

where C is as in Lemma A.8.1.

Therefore, expressing Π in the form (A.114), we have

$$|\bar{\varphi}(\Pi(\bar{\gamma}_{uq}, \bar{\gamma}_{uq}^* : u \geq 0))^K| = \sum_{1 \leq j_1 \ldots j_K \leq r} |\bar{\varphi}(\prod_{u=1}^{K} m_{l_{j_u}})|$$

$$\leq \sum_{1 \leq j_1 \ldots j_K \leq r} (2C)^{\sum_{u=1}^{K} l_{j_u}} \leq (C')^K,$$

where $C' > 0$ does not depend on q. Hence, by Lemma 4.1.3(b), proof of Lemma 8.2.3(a) is complete.

Now note that by (8.18) and (8.20), the right side of (8.21) is a moment sequence. As $C' > 0$ does not depend on q, by (8.18), we have

$$|\bar{\varphi}_\infty(\Pi(\bar{\gamma}_{u\infty}, \bar{\gamma}_{u\infty}^* : u \geq 0))^K| \leq (C')^K.$$

Hence, by Lemma 4.1.3(a), proof of Lemma 8.2.3(b) is complete.

Lemma 8.2.3(c) is trivial by (8.18).   □

A.10   Lemmas for Theorem 8.2.2

Lemma A.10.1. *For any non-commutative variables* $\{a_i, b_i : 1 \leq i \leq k\}$ *we have for all* $k \geq 2$,

$$\prod_{i=1}^{k} a_i - \prod_{i=1}^{k} b_i = \sum_{j=1}^{k} \Big(\prod_{i=1}^{j-1} a_i\Big)(a_j - b_j)\Big(\prod_{i=j+1}^{k} b_i\Big), \quad a_0 = b_{k+1} = 1. \quad (A.115)$$

Proof. Note that (A.115) is true for $k = 2$, as $a_1 a_2 - b_1 b_2 = (a_1 - b_1) b_2 + a_1 (a_2 - b_2)$. Suppose (A.115) is true for $k = m$. Then for $k = m+1$, note that

$$\Big(\prod_{i=1}^{m} a_i\Big) a_{m+1} - \Big(\prod_{i=1}^{m} b_i\Big) b_{m+1}$$

$$= \Big(\prod_{i=1}^{m} a_i - \prod_{i=1}^{m} b_i\Big) b_{m+1} + \Big(\prod_{i=1}^{m} a_i\Big)(a_{m+1} - b_{m+1}),$$

$$= \sum_{j=1}^{m} \Big(\prod_{i=1}^{j-1} a_i\Big)(a_j - b_j)\Big(\prod_{i=j+1}^{m} b_i\Big) b_{m+1} + \Big(\prod_{i=1}^{m} a_i\Big)(a_{m+1} - b_{m+1}),$$

$$\text{using (A.115) for } k = m$$

$$= \sum_{j=1}^{m+1} \Big(\prod_{i=1}^{j-1} a_i\Big)(a_j - b_j)\Big(\prod_{i=j+1}^{m+1} b_i\Big).$$

Hence, the proof is completed by induction on k. $\square$

Recall $F_{p,q}$ and $F_{p,\infty}$ respectively from (8.23) and (8.24).

Lemma A.10.2. $\lim_{q \to \infty} \lim_{p \to \infty} L(F_{p,q}, F_{p,\infty}) = 0$ *almost surely.*

Proof. For convenience, in this proof, let us denote the sample autocovariance matrices for the MA(q) and the MA(∞) processes respectively by $\hat{\Gamma}_{uq}$ and $\hat{\Gamma}_{u\infty}$. Let,

$$g_q = \Pi(\hat{\Gamma}_{uq}, \hat{\Gamma}_{uq}^* : u \geq 0), \quad g_\infty = \Pi(\hat{\Gamma}_{u\infty}, \hat{\Gamma}_{u\infty}^* : u \geq 0).$$

To prove Lemma A.10.2, by Lemma A.5.1(c), it is enough to show

$$\lim_{q \to \infty} \lim_{p \to \infty} \frac{1}{p} \mathrm{Tr}(g_q - g_\infty)(g_q - g_\infty)^* \to 0, \quad \text{almost surely.} \quad (A.116)$$

Let us first prove (A.116) in the simplest case when $g_q = \hat{\Gamma}_{0q}$ and $g_\infty = \hat{\Gamma}_{0\infty}$. Recall the matrices $\{\Delta_u\}$ in (6.3). For convenience, in this proof, let us denote these matrices respectively for $q < \infty$ and $q = \infty$ by $\{\Delta_{uq}\}$ and $\{\Delta_{u\infty}\}$. Note that there is a $C > 0$ such that

$$\frac{1}{p}\mathrm{Tr}(\hat{\Gamma}_{0q} - \hat{\Gamma}_{0\infty})^2 \leq C\Big(\frac{1}{p}\mathrm{Tr}(\hat{\Gamma}_{0q} - \Delta_{0q})^2 + \frac{1}{p}\mathrm{Tr}(\Delta_{0q} - \Delta_{0\infty})^2$$

$$+ \frac{1}{p}\mathrm{Tr}(\Delta_{0\infty} - \hat{\Gamma}_{0\infty})^2\Big). \quad (A.117)$$

Using techniques used in the proof of Lemma 8.2.1, it can be proved that as $p \to \infty$ (for (A.119), we additionally need Assumption (B4)),

$$\frac{1}{p}\mathrm{Tr}(\hat{\Gamma}_{0q} - \Delta_{0q})^2 \overset{\text{a.s.}}{\to} 0 \ \forall q \geq 0, \text{ and} \tag{A.118}$$

$$\frac{1}{p}\mathrm{Tr}(\Delta_{0\infty} - \hat{\Gamma}_{0\infty})^2 \overset{\text{a.s.}}{\to} 0. \tag{A.119}$$

We omit the details. Therefore, by (A.117)–(A.119), proof of (A.116) when $g_q = \hat{\Gamma}_{0q}$ and $g_\infty = \hat{\Gamma}_{0\infty}$ will be completed if we can show

$$\lim_{q \to \infty} \lim_{p \to \infty} \frac{1}{p}\mathrm{Tr}(\Delta_{0q} - \Delta_{0\infty})^2 = 0 \text{ almost surely.} \tag{A.120}$$

To show (A.120), now note that

$$\frac{1}{p}\mathrm{Tr}\Big[\sum_{j=0}^{\infty}\sum_{j'=0}^{\infty} \psi_j Z P_{j'-j} Z^* \psi_{j'}^* - \sum_{j=0}^{q}\sum_{j'=0}^{q} \psi_j Z P_{j'-j} Z^* \psi_{j'}^*\Big]^2$$

$$= \frac{1}{p}\mathrm{Tr}\Big[\sum_{j=q+1}^{\infty}\sum_{j'=q+1}^{\infty} \psi_j Z P_{j'-j} Z^* \psi_{j'}^* + \sum_{j=q+1}^{\infty}\sum_{j'=0}^{q} \psi_j Z P_{j'-j} Z^* \psi_{j'}^*$$

$$+ \sum_{j=0}^{q}\sum_{j'=q+1}^{\infty} \psi_j Z P_{j'-j} Z^* \psi_{j'}^*\Big]^2$$

$$= \frac{1}{p}\mathrm{Tr}\Big[\sum_{j,j'=q+1}^{\infty}\sum_{k,k'=q+1}^{\infty} \psi_j Z P_{j'-j} Z^* \psi_{j'}^* \psi_k Z P_{k'-k} Z^* Z_{k'}^*$$

$$+ \sum_{j,k=q+1}^{\infty}\sum_{j',k'=0}^{q} \psi_j Z P_{j'-j} Z^* \psi_{j'}^* \psi_k Z P_{k'-k} Z^* \psi_{k'}^*$$

$$+ \sum_{j,k=0}^{q}\sum_{j',k'=q+1}^{\infty} \psi_j Z P_{j'-j} Z \psi_{j'}^* \psi_k Z P_{k'-k} Z^* \psi_{k'}^*$$

$$+ \sum_{j,k'=q+1}^{\infty}\sum_{j',k=0}^{q} \psi_j Z P_{j'-j} Z^* \psi_{j'}^* \psi_k Z P_{k'-k} Z^* \psi_{k'}^*$$

$$+ \sum_{j,k'=0}^{q}\sum_{j',k=q+1}^{\infty} \psi_j Z P_{j'-j} Z^* \psi_{j'}^* \psi_k Z P_{k'-k} Z^* \psi_{k'}^*$$

$$+ \sum_{j,j',k=q+1}^{\infty}\sum_{k'=0}^{q} \psi_j Z P_{j'-j} Z^* \psi_{j'}^* \psi_k Z P_{k'-k} Z^* \psi_{k'}^*$$

$$+ \sum_{j,k,k'=q+1}^{\infty}\sum_{j'=0}^{q} \psi_j Z P_{j'-j} Z^* \psi_{j'}^* \psi_k Z P_{k'-k} Z^* \psi_{k'}^*$$

$$+ \sum_{j,j',k'=q+1}^{\infty}\sum_{k=0}^{q} \psi_j Z P_{j'-j} Z^* \psi_{j'}^* \psi_k Z P_{k'-k} Z^* \psi_{k'}^*$$

$$+ \sum_{j=0}^{q} \sum_{j',k,k'=q+1}^{\infty} \psi_j Z P_{j'-j} Z^* \psi_{j'}^* \psi_k Z P_{k'-k} Z^* \psi_{k'}^*]$$

$$= \sum_{i=1}^{9} T_i, \text{ say.} \qquad (A.121)$$

Using the technique used in the proof of Lemma 6.3.1, under (B4), it can be shown that as $p \to \infty$,

$$E(T_i - ET_i)^4 = O(p^{-4}), \ \forall \ 1 \leq i \leq 9, \ q \geq 1. \qquad (A.122)$$

Moreover, under (B4), one can easily show that

$$\lim_{q \to \infty} \lim_{p \to \infty} E(T_i) \to 0. \qquad (A.123)$$

For example, note that

$$\lim_{p \to \infty} E(T_1) = \sum_{j,j',k,k'=q+1}^{\infty} \lim_{p \to \infty} \frac{1}{p} \text{Tr}(\psi_j \psi_{j'}^* \psi_k \psi_{k'}^*)$$

$$+ (\sum_{j,j'=q+1}^{\infty} \lim_{p \to \infty} \frac{1}{p} \text{Tr}(\psi_j \psi_{j'}^*))^2$$

$$\leq 2(\sum_{j=q+1}^{\infty} \sup_p ||\psi_j||_2)^4 \to 0, \ (\text{as } q \to \infty) \qquad (A.124)$$

by (B4). Similar arguments work for $2 \leq i \leq 9$. Therefore, by Borel–Cantelli Lemma, (A.122) and (A.123), $T_i \to 0$ almost surely. By (A.121), proof of (A.116) when $g_q = \hat{\Gamma}_{0q}$ and $g_\infty = \hat{\Gamma}_{0\infty}$, is complete. Using similar arguments, it is easy to prove for all $u \geq 0$ and $k \geq 1$,

$$\lim_{q \to \infty} \lim_{p \to \infty} \frac{1}{p} \text{Tr}((\hat{\Gamma}_{uq} - \hat{\Gamma}_{u\infty})(\hat{\Gamma}_{uq} - \hat{\Gamma}_{u\infty})^*)^k = 0 \text{ almost surely.} \qquad (A.125)$$

Now we prove (A.116) when g_q and g_∞ are monomials. Without loss of generality, suppose for some $k \geq 1$,

$$g_q = \prod_{i=0}^{k} \hat{\Gamma}_{u_i q}, \ y_\infty = \prod_{i=0}^{k} \hat{\Gamma}_{u_i \infty}.$$

Then by Lemma A.10.1, we have

$$(g_q - g_\infty)(g_q - g_\infty)^* = \sum_{j,j'=1}^{k} [(\prod_{i=1}^{j-1} \hat{\Gamma}_{u_i q})(\hat{\Gamma}_{u_j q} - \hat{\Gamma}_{u_j \infty})(\prod_{i=j+1}^{k} \hat{\Gamma}_{u_i \infty})$$

$$\times (\prod_{i=1}^{j'-1} \hat{\Gamma}_{u_i q})(\hat{\Gamma}_{u'_j q} - \hat{\Gamma}_{u'_j \infty})(\prod_{i=j'+1}^{k} \hat{\Gamma}_{u_i \infty})].$$

Keeping in view Lemma 5.1.1(b) and the relation (A.125), to establish (A.116), it is enough to prove that for all $u \geq 0$, $k \geq 1$,

$$\lim_{q \to \infty} \lim_{p \to \infty} \frac{1}{p} \mathrm{Tr}(\hat{\Gamma}_{uq} \hat{\Gamma}_{uq}^*)^k < \infty \text{ almost surely and} \tag{A.126}$$

$$\lim_{q \to \infty} \lim_{p \to \infty} \frac{1}{p} \mathrm{Tr}(\hat{\Gamma}_{u\infty} \hat{\Gamma}_{u\infty}^*)^k < \infty \text{ almost surely.} \tag{A.127}$$

Using the same arguments as in the proof of (8.15), it can be proved that

$$E\left(\frac{1}{p} \mathrm{Tr}(\hat{\Gamma}_{uq} \hat{\Gamma}_{uq}^*)^k - \frac{1}{p} E \mathrm{Tr}(\hat{\Gamma}_{uq} \hat{\Gamma}_{uq}^*)^k\right)^4 = O(p^{-4}) \text{ and} \tag{A.128}$$

$$E\left(\frac{1}{p} \mathrm{Tr}(\hat{\Gamma}_{u\infty} \hat{\Gamma}_{u\infty}^*)^k - \frac{1}{p} E \mathrm{Tr}(\hat{\Gamma}_{u\infty} \hat{\Gamma}_{u\infty}^*)^k\right)^4 = O(p^{-4}). \tag{A.129}$$

To show (A.129), we additionally need Assumption (B4). We omit the details. Hence, by (A.128), (A.129), and Borel–Cantelli Lemma, for $u \geq 0$ and $k \geq 1$,

$$\lim_{p \to \infty} \left(\frac{1}{p} \mathrm{Tr}(\hat{\Gamma}_{uq} \hat{\Gamma}_{uq}^*)^k - \frac{1}{p} E \mathrm{Tr}(\hat{\Gamma}_{uq} \hat{\Gamma}_{uq}^*)^k\right) = 0, \text{ a.s. and} \tag{A.130}$$

$$\lim_{p \to \infty} \left(\frac{1}{p} \mathrm{Tr}(\hat{\Gamma}_{u\infty} \hat{\Gamma}_{u\infty}^*)^k - \frac{1}{p} E \mathrm{Tr}(\hat{\Gamma}_{u\infty} \hat{\Gamma}_{u\infty}^*)^k\right) = 0, \text{ a.s.} . \tag{A.131}$$

Again by (8.9) and (8.18), we have for all $u \geq 0$ and $k \geq 1$,

$$\lim_{q \to \infty} \lim_{p \to \infty} \frac{1}{p} E \mathrm{Tr}(\hat{\Gamma}_{uq} \hat{\Gamma}_{uq}^*)^k < \infty \text{ and} \tag{A.132}$$

$$\lim_{q \to \infty} \lim_{p \to \infty} \frac{1}{p} E \mathrm{Tr}(\hat{\Gamma}_{u\infty} \hat{\Gamma}_{u\infty}^*)^k < \infty. \tag{A.133}$$

Therefore, by (A.130)–(A.133), we conclude that (A.126) and (A.127) hold.

Hence, (A.116) holds for monomials. Similar arguments work for polynomials. For example, if $g_q = \pi_{1q} + \pi_{2q}$ and $g_q = \pi_{1\infty} + \pi_{2\infty}$ where π_{1q}, π_{2q}, $\pi_{1\infty}$ and $\pi_{2\infty}$ are monomials, then

$$|p^{-1} \mathrm{Tr}(g_q - g_\infty)(g_q - g_\infty)^*|$$
$$\leq p^{-1} \mathrm{Tr}(\pi_{1q} - \pi_{1\infty})(\pi_{1q} - \pi_{1\infty})^* + p^{-1} \mathrm{Tr}(\pi_{2q} - \pi_{2\infty})(\pi_{2q} - \pi_{2\infty})^*$$
$$+ 2\sqrt{p^{-1} \mathrm{Tr}(\pi_{1q} - \pi_{1\infty})(\pi_{1q} - \pi_{1\infty})^* p^{-1} \mathrm{Tr}(\pi_{2q} - \pi_{2\infty})(\pi_{2q} - \pi_{2\infty})^*}$$
$$\to 0, \text{ as (A.116) holds for monomials}$$

Therefore, (A.116) is proved for any polynomial and the proof of Lemma A.10.2 is complete. $\qquad \square$

Bibliography

G. I. Allen and R. Tibshirani. Transposable regularized covariance models with an application to missing data imputation. *Ann. Appl. Stat.*, 4(2): 764–790, 2010.

G. Anderson, A. Guionnet, and O. Zeitouni. *An Introduction to Random Matrices.* Cambridge University Press, Cambridge, UK, 2009.

T. W. Anderson. *An Introduction to Multivariate Statistical Analysis (3rd edition).* Wiley Series in Probability and Statistics, 2003.

L. Arnold. On the asymptotic distribution of the eigenvalues of random matrices. *J. Math. Anal. Appl.*, 20(2):262–268, 1967.

L. Arnold. On Wigner's semicircle law for the eigenvalues of random matrices. *Z. Wahr. und Verw. Gebiete*, 19(3):191–198, 1971.

R. B. Ash. *Probability and Measure Theory*, 2nd Edition. A Harcourt Science and technology company, 2000.

P. Assouad. Deux remarques sur l'estimation. *C. R. Acad. Sci. Paris Sér. I Math.*, 296:1021–1024, 1983.

Z. D. Bai and J. W. Silverstein. *Spectral Analysis of Large Dimensional Random Matrices.* Springer, 2009.

Z. D. Bai and Y. Q. Yin. Convergence to the semicircle law. *Ann. Probab.*, 16(2):863–875, 1988.

Z. D. Bai and L. X. Zhang. The limiting spectral distribution of the product of the Wigner matrix and a nonnegative definite matrix. *J. Multivariate Anal.*, 101(9):1927–1949, 2010.

T. Banica, S. T. Belinschi, M. Capitaine, and B. Collins. Free Bessel laws. *Canad. J. Math.*, 63(1):3–37, 2011.

Z. Bao. Strong convergence of ESD for the generalized sample covariance matrices when $p/n \to 0$. *Statist. Probab. Lett.*, 82(5):894–901, 2012.

A. Basak, A. Bose, and S. Sen. Limiting spectral distribution of sample autocovariance matrices. *Bernoulli*, 20(3):1234–1259, 2014.

S. Basu and G. Michailidis. Regularized estimation in sparse high-dimensional time series models. *Ann. Statist.*, 43(4):1535–1567, 2015.

F. Benaych-Georges. Rectangular random matrices, related convolution. *Probab. Theory Related Fields*, 144(3-4):471–515, 2009.

F. Benaych-Georges. On a surprising relation between the Marčenko–Pastur law, rectangular and square free convolutions. *Ann. de l'institut Henri Poincaré (B)*, 46(3):644–652, 2010.

F. Benaych-Georges and R. R. Nadakuditi. The singular values and vectors of low rank perturbations of large rectangular random matrices. *J. Mult. Anal.*, 111:120–135, 2012.

R. Bhatia. *Notes on Functional Analysis*. Hindustan Book Agency, India, 2009.

M. Bhattacharjee and A. Bose. Consistency of large dimensional sample covariance matrix under weak dependence. *Statistical Methodology*, 20: 11–26, 2014a.

M. Bhattacharjee and A. Bose. Estimation of autocovariance matrices for infinite dimensional vector linear process. *J. Time Series Anal.*, 35(3): 262–281, 2014b.

M. Bhattacharjee and A. Bose. Large sample behaviour of high dimensional autocovariance matrices. *Ann. Statist.*, 44(2):598–628, 2016a.

M. Bhattacharjee and A. Bose. Joint convergence of sample variance-covariance matrices when $p/n \to 0$ with application. *manuscript*, 2016b.

M. Bhattacharjee and A. Bose. Polynomial generalizations of sample variance-covariance matrices when $pn^{-1} \to 0$. *Random Matrices: Theory and Applications*, 5(4):1650014, 2016c.

M. Bhattacharjee and A. Bose. Matrix polynomial generalizations of the sample variance-covariance matrix when $pn^{-1} \to y \in (0, \infty)$. *Indian Journal of Pure and Applied Mathematics*, 48(4):575–607, 2017.

P. J. Bickel and E. Levina. Regularized estimation of large covariance matrices. *Ann. Statist.*, 36(1):199–227, 2008a.

P. J. Bickel and E. Levina. Covariance regularization by thresholding. *Ann. Statist.*, 36(6):2577–2604, 2008b.

P. Billingsley. *Probability and Measure,* 3rd Edition. Wiley series in Probability and Mathematical Statistics, 1995.

A. Bose. *Patterned Random Matrices*. Chapman and Hall, 2018.

A. Bose and S. Gangopadhyay. Convergence of a class of Hankel-type matrices. *Acta Physica Polonica B*, 46(9), 2015.

P. J. Brockwell and R. A. Davis. *Time Series: Theory and Methods*. Springer, 2009.

P. Bühlmann and S. van de Geer. *Statistics for High-Dimensional Data: Methods, Theory and Applications*. Springer, 2011.

T. T. Cai and H. H. Zhou. Optimal rates of convergence for sparse covariance matrix estimation. *Annals of Statistics*, 40(5):2389–2420, 2012.

T. T. Cai, C. H. Zhang, and H. H. Zhou. Optimal rates of convergence for covariance matrix estimation. *Annals of Statist.*, 38(4):2118–2144, 2010.

T. T. Cai, Z. Ren, and H. H. Zhou. Optimal rates of convergence for estimating Toeplitz covariance matrices. *Probability Theory and Related Fields*, 156(1-2):101–143, 2013.

X. Chen, M. Xu, and W. B. Wu. Covariance and precision matrix estimation for high-dimensional time series. *Ann. Statist.*, 41(6):2994–3021, 2013.

R. Couillet and M. Debbah. *Random Matrix Methods for Wireless Communications*. Cambridge University Press, Cambridge, UK, 2011.

P. H. Edelman. Chain enumeration and non-crossing partitions. *Discrete Mathematics*, 31(2):171–180, 1980.

B. Efron. Are a set of microarrays independent of each other? *Ann. Appl. Stat.*, 3(3):922–942, 2009.

P. Franaszczuk, K. Blinowska, and M. Kowalczyk. The application of parametric multichannel spectral estimates in the study of electrical brain activity. *Biological Cybernetics*, 51(4):239–247, 1985.

K. J. Friston, P. Jezzard, and R. Turner. Analysis of functional MRI time-series. *Human Brain Mapping*, 1(2):153–171, 1994.

D. R. Fuhrmann. Application of Toeplitz covariance estimation to adaptive beamforming and detection. *IEEE Transactions on Signal Processing*, 39:2194–2198, 1991.

J. S. Geronimo and T. P. Hill. Necessary and sufficient condition that the limit of Stieltjes transforms is a Stieltjes transform. *Annals of Probability*, 121(1):54–60, 2003.

G. H. Golub and C. F. van Loan. *Matrix Computations, 3rd ed.* The Johns Hopkins University Press, Baltimore, 1996.

U. Grenander and G. Szegö. *Toeplitz Forms and Their Applications*. University of California Press, 1958.

E. Hannan. *Multiple Time Series*. John Wiley & Sons, Inc., New York, 1970.

B. Jin, C. Wang, Z. D. Bai, K. K. Nair, and M. Harding. Limiting spectral distribution of a symmetrized auto-cross covariance matrix. *Ann. Appl. Probab.*, 24(3):1199–1225, 2014.

H. Liu, A. Aue, and D. Paul. On the Marčenko-Pastur law for linear time series. *Ann. Statist.*, 43(2):675–712, 2015.

V. Marčenko and L. Pastur. Distribution of eigenvalues for some sets of random matrices. *Mathematics of the USSR-Sbornik*, 1:457–483, 1967.

T. L. McMurry and D. N. Politis. Banded and tapered estimates for auto-covariance matrices and the linear process bootstrap. *Journal of Time Series Analysis*, 1:471–482, 2010.

A. Nica and R. Speicher. *Lectures on the Combinatorics of Free Probability*. Cambridge University Press, Cambridge, UK, 2006.

O. Pfaffel and E. Schlemm. Eigenvalue distribution of large sample covariance matrices of linear processes. *Probab. Math. Statist.*, 31(2):313–329, 2011.

M. Pourahmadi. *High-Dimensional Covariance Estimation: With High-Dimensional Data.* Wiley, New York, 2013.

D. Quah. Internet cluster emergence. *European Economic Review*, 44(4): 1032–1044, 2000.

A. J. Rothman, P. J. Bickel, E. Levina, and J. Zhu. Sparse permutation invariant covariance estimation. *Electronic Journal of Statistics*, 2:494–515, 2008.

L. Saulis and V. A. Statulevičius. *Limit Theorems for Large Deviations.* Kluwer Academic Publishers, Dordrecht, 1991.

R. J. Serfling. *Approximation Theorems of Mathematical Statistics.* John Wiley & Sons, 2002.

H. Visser and J. Molenaar. Trend estimation and regression analysis in climatological time series: an application of structural time series models and the Kalman filter. *Journal of Climate*, 8(5):969–979, 1995.

K. W. Wachter. The strong limits of random matrix spectra for sample matrices of independent elements. *Ann. Probab.*, 6(1):1–18, 1978.

L. Wang and D. Paul. Limiting spectral distribution of renormalized separable sample covariance matrices when $p/n \to 0$. *J. Multivariate Anal.*, 126:25–52, 2014.

L. Wang, A. Aue, and D. Paul. Spectral analysis of linear time series in moderately high dimensions. *Bernoulli*, 23(4A):2181–2209, 2017.

E. P. Wigner. On the distribution of the roots of certain symmetric matrices. *Ann. Math.*, 67(2):325–328, 1958.

W. B. Wu. Nonlinear system theory: Another look at dependence. *Proc. Natl. Acad. Sci. USA*, 102(40):14150–14154, 2005.

W. B. Wu and M. Pourahmadi. Nonparametric estimation of large covariance matrices of longitudinal data. *Biometrica*, 90:831–844, 2003.

W. B. Wu and Y. N. Wu. High-dimensional linear models with dependent observations. *Preprint*, 2014.

Y. Q. Yin. Limiting spectral distribution for a class of random matrices. *J. Multivariate Anal.*, 20(1):50–68, 1986.

B. Yu. Assouad, Fano and Le Cam. In *Festschrift for Lucien Le Cam* (D. Pollard, E. Torgersen and G. Yang, eds.). 423-435:Springer, Berlin, 1997.

A. Zygmund. *Trigonometric Series. Vol. I, II.* Cambridge Mathematical Library (3rd ed.), Cambridge University Press, 2002.

Index